CHEMISTRY RESEARCH AND APPLICATIONS

A CLOSER LOOK AT HORMONES

CHEMISTRY RESEARCH AND APPLICATIONS

Additional books and e-books in this series can be found on Nova's website under the Series tab.

CHEMISTRY RESEARCH AND APPLICATIONS

A CLOSER LOOK AT HORMONES

ALEXANDER V. KNYAZEV
EDITOR

nova science publishers
New York

Library of Congress Cataloging-in-Publication Data

ISBN: 978-1-53617-254-6

Published by Nova Science Publishers, Inc. † New York

CONTENTS

PREFACE

Hormones are tiny chemicals messengers delivering the message to organs and tissues. At present, a lot of hormones have been studied and synthesized, and interest in these compounds is constantly growing.

This book is devoted to highlight the recent progress on the extraction, analysis, physico-chemical properties and the use of hormones.

The first chapter presents the main terpenoid and steroid compounds that are present in medicinal mushrooms. High antitumor activity of extracts and fractions obtained from fungi and individual terpenoid and steroid compounds was shown. An interesting point is the discussion of the main mechanisms providing high antitumor activity of terpenoid and steroid compounds.

The second chapter discusses the content of phytohormones in seaweed. The main attention is paid to red algae, in which phytohormones such as auxin, ethylene, abscisic acid, cytokinins and gibberellins were found.

The third chapter provides an overview of the most used and latest methodologies for identifying and quantifying steroid hormones in environmental matrices. Effective extraction and concentration techniques, as well as selective and sensitive methods for determining steroid hormones, are discussed.

The fourth chapter is also devoted to steroid hormones. A systematic analysis of structural and thermodynamic data is carried out. It was shown that steroid hormones are characterized by rich isomorphism, but there are too few accurate physico-chemical data to quantitatively describe the areas of existence of a particular modification.

The fifth chapter is devoted to the activities of "vitamin" D on the female reproductive system. The chapter discloses the role of steroid hormones (in particular, sekosteroid) in the menstrual cycle and their interplay with progesterone during the early stages of pregnancy and the physiologic development to delivery.

The sixth chapter presents a comparative genomic analysis that revealed unexpected and complex patterns of change during the evolution of primates. It is shown that phylogenetic analysis makes it possible to reconstruct the evolutionary history of the human genome.

The seventh chapter presents the results of using hormones as a tool for algal biosynthesis and, particularly, for fatty acids profile manipulation in the biofuel and pharmaceutical industries. It has been shown that the nanoconcentration of phyto- and animal hormones commonly found in agricultural wastewater significantly stimulates the productivity of microalgae.

The final chapter discusses the brassinosteroids that regulate seed germination, primary photochemistry, photosynthesis, different biochemical enzymes, elemental composition, carbohydrate metabolism and redox status of plants.

This collection of contributions was only possible thanks to the tough work and effort of all authors, who have gently shared their knowledge with us and devoted their valuable time to this project.

Alexander Knyazev, Professor, DSc
Lobachevsky University, Nizhni Novgorod, Russia
E-mail: knyazevav@gmail.com

In: A Closer Look at Hormones
Editor: Alexander V. Knyazev
ISBN: 978-1-53617-254-6

Chapter 1

CHEMISTRY AND BIOLOGICAL ACTIVITY OF STEROID AND TERPENOID COMPOUNDS OF MEDICINAL MUSHROOMS

M. A. Sysoeva, S. A. Kovalenko, E. V. Sysoeva [*]
and V. R. Khabibrakhmanova
Department of Food Biotechnology
Kazan National Research Technological University
Kazan, Russia

ABSTRACT

This Chapter presents the main terpenoid and steroid compounds of *Ganoderma lucidum, Lentinus edodes, Poria cocos, Fomitopsis pinicola, Fomitopsis betulina* and *Inonotus obliguus* medicinal mushrooms. High antitumor activity of extracts and fractions obtained from fungi and individual terpenoid and steroid compounds are shown. The main

[*] Corresponding Author's Email: inonotus@yandex.ru.

possible mechanisms providing high antitumor activity of terpenoid and steroid compounds are discussed.

Keywords: terpenoid, steroid, antitumor, activity, medicinal mushrooms

INTRODUCTION

Biologically active substances that are isolated from higher fungi have a significant impact on human health, in recent years; such mushrooms have become known as "medical" or "medicinal". Some of them are recognized as official medicine and are included in the state Pharmacopoeia. For instance, the Russian Federation Pharmacopoeia includes *Inonotus obliquus*, commonly known as chaga; the European Pharmacopoeia contains article about Poria mushroom (*Wolfiporia extensa*), also termed as *Poria cocos*; the U.S. Pharmacopeia comprises information about Reishi Mushroom (*Ganoderma lucidum*) as well named Lingzhi [1-3]. These mushrooms are also accepted by the official medicine of China, Japan and other countries of Southeast Asia. Since ancient times, more than 100 mushrooms species have been used in traditional medicine of East Asian and Western European countries. They are an extensive and still largely unused source of biologically active compounds: triterpenoids, polysaccharides, phenolic substances.

Mushrooms extracts and infusions are used for treatment and prevention of gastrointestinal and cold-related diseases, bronchial asthma, chronic fatigue syndrome, different cancer types; and exert immunomodulatory effect, improve hormonal balance, exhibit antioxidant properties [4-12].

The study of the biological activity of higher fungi showed that many of their therapeutic properties provide terpenoid and steroid compounds. Among them, bioactive tetracyclic triterpenoids of lanostane structure and ergosterol derivatives predominate. The main obstacle to the development of drugs based on steroids and terpenoids is that they often accumulate in fungi in very low concentrations and are usually found in a complex mixture with other compounds. Many of the found compounds of these classes have a pronounced activity, for example, against tumor cells without harming normal cells [13].That is why, the search for alternative sources of biologically active compounds of these classes, non-toxic and more effective, today remains an urgent task in the research.

CLASSIFICATION AND BIOSYNTHESIS OF TRITERPENES AND STEROIDS

Steroids and triterpenoids are biogenetic relatives. They belong to the group of isoprenoids, i.e., compounds whose carbon skeleton is composed of isoprene (isopentane) fragments of C_5H_8.The number of isoprene units determines their classification to mono-, di-, sesqui-, tri- and tetraterpenoids. Triterpenoids are widely distributed in nature: they are found in plants, animals, marine organisms, microorganisms, geological sediments. Some triterpenoids can cause the degradation of the carbon skeleton, from C30 to C18. Such triterpenoids form a specific group of steroids.

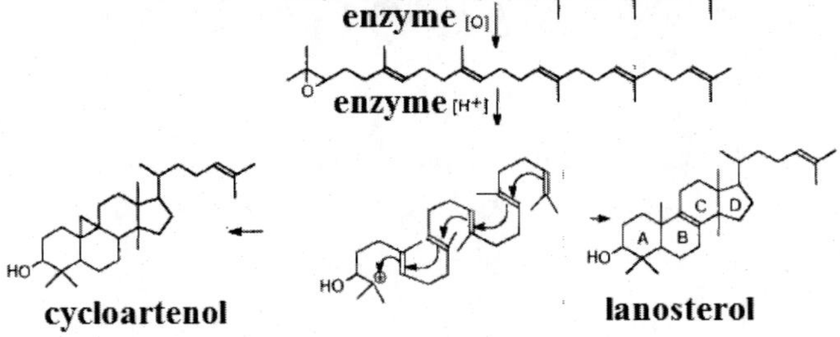

Figure 1. Triterpenoid biosynthesis.

Triterpenoids biosynthesis begins with the condensation of two molecules of farnesyl pyrophosphate (C15) with the formation of squalene (Figure 1). Then the process of epoxidation of terminal olefin bond of squalene occurs, and thereafter synchronous multicenter cyclization of squalene oxide progresses. Multiple oxidation steps that are catalyzed by cytochrome P450 monooxygenases, modify the triterpene skeleton and generate various triterpenoid metabolites. The result of this process are polycyclic condensed systems with cyclohexane fragments, mainly represented by the lanostane skeleton. Lanostane type triterpenoids are characteristic of the trans junction of rings A/B, B/C and C/D, β-oriented methyl at C10, C13, α-oriented methyl at C14, β-oriented sidechain at C17 and R configuration for C20 [14-18].

In most cases, the lanostane compounds structure becomes complicated owing to secondary biosynthetic reactions of oxidation, alkylation and the formation of oxygen-containing heterocycles. A significant number of carboxy-derivatives, hydroxy- and oxo-substituted and seco-structures of lanosterol makes up a wide variety of these compounds.

The subgroup of lanostane triterpenoids includes derivatives of cycloartane, which is a precursor of the common fungal sterols, β-sitosterol and campesterol.

The main sterol of the animal world, cholesterol C27 and sterol, which performs the same functions in fungi is ergosterol, which forms from lanosterol by consecutive loss of the methyl groups from positions C14 and C4. Fungal steroids are biogenetic precursors of D vitamins group, as well as lanostane compounds, largely determine the therapeutic properties of higher fungi [14, 19, 20].

The most interesting from the point of medical use mushrooms are fungi-destroyers of wood. *Ganoderma lucidum, Poria cocos, Fomitopsis pinicola, Fomitopsis betulina, Lentinula edodes* are found mainly on stumps and dead wood, but Inonotus obliquus grows only on living trees. That influences on the metabolism of these mushrooms and determines the significant differences in their biologically active compounds accumulation. The synthesis of terpenoids and steroids in fungi usually takes place at low rH values, which leads to the production of these compounds in forms of alcohols, hydrocarbons and rarely acids. However, a distinctive feature of the Reishi mushroom is the production of highly oxidized triterpenoids, which are most often found in pairs of C3 α/β stereoisomers and C3/C15 positional isomers. Numerous studies of various scientists have convincingly demonstrated that the biological activity of higher fungi largely depends on the presence of tetracyclic lanostane triterpenoids in them.

Content of Terpenoids and Steroids in Medicinal Mushrooms and Their Physiological Properties

Ganoderma lucidum

The most common and popular worldwide fungy is *Ganoderma lucidum (Leyss. ex Fr.) Karst. (Ganodermataceae)* (Figure 2). It is called «Lingzhi» in China, «Reishi» in Japan.

Figure 2. *Ganoderma lucidum*.

In China and other Asian countries for more than 2000 years ago, it was named as "mushroom of immortality", the "elixir of life", "divine mushroom," "magic fungus" and so on. More than 280 biologically active secondary metabolites were isolated from the fruiting bodies, mycelium and spores of *G. lucidum*, among them more than 130 are triterpenoids and steroids (Figure 3, Table 1).

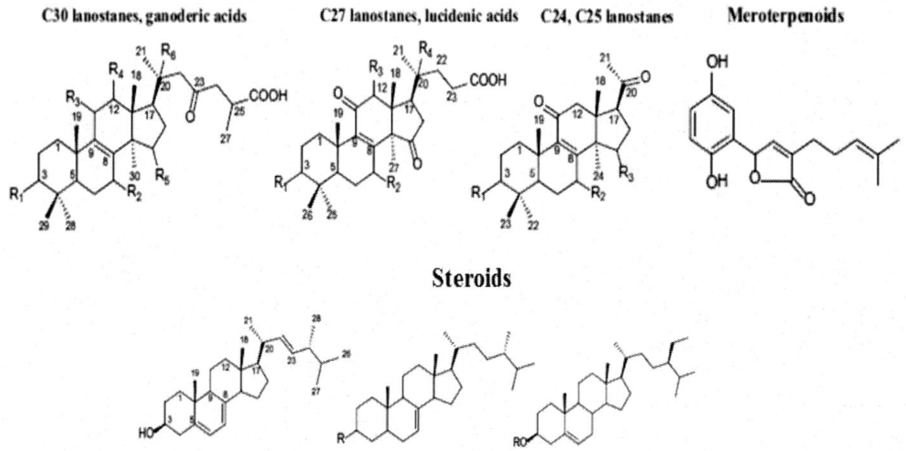

Figure 3. The structure of the main terpenoid compounds of *Ganoderma lucidum*.

Extracts from *G. lucidum* containing triterpenoid compounds are able to stop the development of various types of cancer cells by stopping the cell cycle in phases G1 and G2. For example, the *G. lucidum* alcoholic extracts have demonstrated abilities to inhibit proliferation, induce apoptosis of a breast cancer and melanoma cells [43] and to stop the cell cycle in lung cancer [44]. Aqueous extracts of *Ganoderma* can inhibit a breast and prostate cancer cells growth [45], as well as to prevent a neuronal apoptosis, and to induce a neuronal differentiation [46]. *G. lucidum* extracts with high triterpenoids content can inhibit the hepatoma cells growth without affecting on a liver normal cells development. The inhibitory effect of these extracts on cancer cells was caused by a decreasing of protein kinase C (PKC) activity, activation of JNK kinase (c-Jun N-terminal kinase) and p38 MAPK, as a result the cell cycle growth stops in the G2 phase [47].

The research data of the "LING ZHI", commercial preparation based on the *Ganoderma* extracts have been summarized in the review [48]. A number of *in vivo* studies have revealed a direct cytostatic and antiangiogenic effect of this medication. In two randomized and one non-randomized clinical studies, it was shown that the drug improved the quality of life in cancer patients by 65% and significantly increased their cellular immunity reactions.

Table 1. Triterpenoids and steroids of *Ganoderma lucidum* [17, 21-42]

Chemical group	Compounds
Terpenoids	
acids	ganoderic acids A-Z, B8, C1, C2, C6, AM1, Df, DM, Ma, Mb, Mc, Md, Me, Mg, Mh, Mi, Mj, Mk, Mf, TR, Ja, Jb, P1, P2, T-N, T-O, T-Q, Sz, LM2, α, β, γ, δ, ε, Σ η, ζ, θ; 12-hydroxyganoderic acid C2; 8β,9α-dihydroganoderic acid C,J; 15-hydroxy ganoderic acid S; 20-hydroxylganoderic acid G; 20-hydroxy-ganoderic acid G; 12-deacetylganoderic acid H; 8β,9α-dihydroganoderic acid J; 3-O-acetyl-ganoderic acids B, H, K; 12-acetoxy-ganoderic acid D; 7-O-ethyl ganoderic acid O; 7-oxo-ganoderic acid Z; Ganodernoid C; 12β-hydroxy-3,7,11,15,23-pentaoxo-5α-lanosta-8-en-26-oic acid; 12,15-bis(acetyloxy)-3-hydroxy-7,11,23-trioxo-lanost-8-en-26-oic acid; 3α,22β-diacetoxy-7α-hydroxyl-5α-lanost-8,24E-dien-26-oic acid; 3β,15α-diacetoxy lanosta-8,24-dien-26-oic acid; 11α-hydroxy-3,7-dioxo-5α-lanosta-8,24(E)-dien-26-oic acid; Lanosta-7,9(11),24-trien-3β-acetoxy-15α,22β-dihydroxy-26-oic acid; Lanosta-7,9(11),24-trien-3β,15α,22β-triacetoxy-26-oic acid; 3α,15α,22α-trihydroxylanosta-7,9(11),24-trien-26-oic acid; 3β,15α,22β-trihydroxylanosta-7,9(11),24-trien-26-oic acid; 3α,15α-diacetoxy-22α-hydroxylanosta-7,9(11),24-trien-26-oic acid; 3β,15α-diacetoxy-22α-hydroxylanosta-7,9(11),24-trien-26-oic acid; 22β-acetoxy-3α,15α-dihydroxylanosta-7,9(11),24-trien-26-oic acid; Lanosta-7,9(11),24-trien-3α-acetoxy-15α-hydroxy-23-oxo-26-oic acid; Lanosta-7,9(11),24-trien-15α-acetoxy-3α-hydroxy-23-oxo-26-oic acid; Lanosta-7,9(11),24-trien-3α,l5α-diacetoxy-23-oxo-26-oic acid; Ganolucidic acids: A, B, C, D, E; Ganoderenic acids: A, B, C, D, H, F, K; Elfvingic acid A; Ganosporeric acid A; Lucideric acid A Lucidenic acids: A-P, D1, D2, E1, E2, N, P, Q; 20(21)-dehydrolucidenic acid A; 20-hydroxylucidenic acids D2, F, E2, N, P; ganolucidinic acids; 3β-hydroxy-4,4,14-trimethyl-7,11,15-trioxochol-8-en-24-oic acid; 4,4,14α-trimethyl-5α-chol-7,9(11)-dien-3-oxo-24-oic acid; 4,4,14α-trimethyl-3,7-dioxo-5α-chol-8-en-24-oic acid

Table 1. (Continued)

Chemical group	Compounds
Terpenoids	
alcohols	Ganoderol A, B; Ganodermatriol; Ganodermanontriol; Ganoderiol A-I; Lucidumol A, B; Ganodermanondiol; Ganodermenonol; Lucidadiol; Ganodermadiol; Lanosta-7,9(11)-dien-3β-ol; Ganoderitriol M; Epoxyganoderiol A,B,C
aldehydes	(24*E*)-3β-hydroxy-5α-lanosta-7,9(11),24-trien-26-al (lucialdehyde A); (24E)-3,7-dioxo-5α-lanosta-8,24-dien-26-al (lucialdehyde B); (24E)-3β-hydroxy-7-oxo-5α-lanosta-8,24-dien-26-al (lucialdehyde C); Lucialdehyde D-E; 3β-hydroxy-11-oxolanost-8,24-dien-26-al; Ganoderal A-B; Ganoderic aldehyde A, TR; Lucidal
ketones	Ganoderone A, C; 3β,26-dihydroxylanosta-8,24-dien-11-one; 3*S*,24*S*,25,26-tetrahydroxy-5α-lanost-8-en-11-one; 26-hydroxy-5α-lanosta-7,9(11),24-triene-3,22-dione; 26,27-dihydroxy-5α-lanosta-7,9(11),24-triene-3,22-dione, Lucidone A-C
esters	Methyl lucidenate A-F, N, P, Q, D1; Ethyl lucidenate A; Butyl lucidenate A-P; Methyl 20(21)-dehydrolucidenate A; Methyl ganoderate A-J, P; Methyl-O-acetyl ganoderate C; 3β,7β-Dihydroxy-12β-acetoxy-11,15,23-trioxo-5α-lanosta-8-en-26-oic acid methyl ester; Ethyl ganoderate J; Ethyl 3-O-acetyl ganoderate B; 12β-acetoxy-3,7,11,15,23-pentaoxo-5α-lanosta-8-en-26-oic acid ethyl ester, Butyl ganoderate A-H; 12β-acetoxy-3β,7β-dihydroxy-11,15,23-trioxolanost-8-en-26-oic acid butyl ester; 3β,7β-dihydroxy-11,15,23-trioxo-lanost-8,16-dien-26-oic acid methyl ester; Methyl ganoderate A acetonide; Ganodermaside A-D; Ganoleuconin M, N, P

Chemical group	Compounds
Terpenoids	
Lactones	Ganolactone; Ganosporelactones A-B; 7-oxo-ganoderlactone D; 7-oxo-12-O-acetyl ganoderlactone D; 21-hydroxyganoderlactone D; 12α-hydroxyganolactone; Ganoderlactone A-F
meroterpenoid	Lingzhiol
steroids	
ergosterol and its derivates	Ergosta-7,22-dien-3β-ol; 22,23-dihydroergosterol; Ergosterol peroxide (5α,8α-epidioxy-22E-ergosta-6,22-dien-3β-ol); 5α,8α-epidioxyergosta-6,9(11),22-trien-3β-ol (9,11-dehydroergosterol peroxide); 22E,24R-ergosta-7,22-diene-3β,5α,6β-triol; 22E,4R-ergosta-7,22-diene-3β,5α,6β,9α,14α-pentol; 22E,24R-ergosta-7,22-diene-3β,5α,6β,9α-tetraol; 3β,5α-dihydroxy-(22E,24R)-ergosta-7,22-diene-6-one; 3β,5α,9α-trihydroxy-(22E,24R)-ergosta-7,22-diene-6-one; (22E)-4α,5α -epoxy-6β-methoxyergosta-7, 22-diene-3β-ol; (3β,5α,6β,22E)-ergosta-7,9(11),22-triene-3,5,6-triol; (3β,5α,6β,9α,22E)-ergosta-7,22-diene-3,5,6,9-tetrol; (3β,5α,6β,22E)-ergosta-7,22-diene-3,5,6-triol; (3β,5α,6α,22E)-ergosta-7,22-diene-3,5,6-triol; (3β,5α,6α,9α,22E)-ergosta-7,22-diene-3,5,6,9-tetrol; (3β,5α,6β,14α,22E)-ergosta-7,9(11),22-triene-3,5,6,14-tetrol; (3β,5α,6β,9α,14α,22E)-ergosta-7,22-diene-3,5,6,9,14-pentol; (3β,5α,6α,7α,22E)-5,6- epoxy-ergosta-8,22-diene-3,7-diol; (3β,5α,6α,7α,22E)-5,6-epoxy-ergosta-8(14),22-diene-3,7-diol; (3β,5α,8α,22E)-5,8-epidioxy-ergosta-6,22-diene-3-ol; (3β,22E)-ergosta-5,7,22-triene-3-ol; (3β,5α,8α,22E)-5,8-epidioxy-ergosta-6,9(11),22-triene-3-ol; (3β,5α,6α,14α,22E)-3,14-dihydroxy-5,6-epoxy-ergosta-8,22-diene-7-one; Ergosta-7,22-diene-3β-yl pentadecanoate; Ergosta-7,22-dien-3β-yl palmitate
other sterols	Fungisterol (5α-ergost-7-en-3b-ol); β-sitosterol; Kaempferol 3-xyloside; 17β-estradiol 3-sulfate; Ganodosterone

In [49] have been shown that the fraction consisting of only triterpenes inhibits the proliferation of SW 480 colorectal cancer cells. This fraction had a greater antiproliferative effect on cancer cells and, in addition, suppressed DNA synthesis and caused the formation of highly mitogenic radicals. *In vivo* experiments it was shown that triterpenoid fractions of *G. lucidum* are effective in the treatment of benign prostatic hyperplasia [50]. Triterpenoid fractions also inhibited the growth of primary solid tumors in the spleen and prevented liver metastases in mice [51].

Triterpenoids, in particular T-Z ganoderic acids, have demonstrated *in vitro* cytotoxic activity on hepatoma cells [52]. Lanostanoid, 3β-hydroxyl-26-oxo-5α-lanosta-8,24-diene-11-Oh, and steroid, ergosta-7,22-diene-3β,3α,9α-triol isolated from fruit bodies demonstrated powerful inhibitory effects on human hepatoma cells PLC/PRF/5 *in vitro* [53]. Triterpenoids isolated from *G. lucidum* inhibited the growth of K562 leukemia cells [54].

Lanostanoids of *G. lucidum* have an inhibitory action on some enzymes. It was determined that the ganoderic acids of A and C are inhibitors of farnesyltransferase (FTP) and suppress oncogenic Ras proteins [55]. Lucidenic acid O, lucidenic lactone and ergosterol peroxide inhibit DNA polymerase, which is one of the antitumor effects [56, 57].

Table 2 presents an antitumor activity of triterpenoid and steroid compounds isolated from *G. Lucidum*.

Table 2. Antitumor activity of triterpenoids and steroids of *Ganoderma lucidum* [27, 29, 31, 32, 40, 44, 58-60]

Compound	Cell line	Inhibition concentration (IC50)
Ganoderic acid A	HepG2 human hepatoblastoma cells	CC_{50} = 38.7μM
Ganoderic acid C2	HepG2 human hepatoblastoma cells	CC_{50} > 0.100 mM
Ganoderic acid D	HepG2 human hepatoblastoma cells	CC_{50} = 22.9 μM
	RAW264.7 mouse macrophages	CC_{50} > 0.400 mM
	Osteoclast formation (RANKL/tumour necrosis factor-α-induced), inhibition	0.400 mM
Ganoderic acid DM	RAW264.7 mouse macrophages	CC_{50} = 65 μM

Compound	Cell line	Inhibition concentration (IC_{50})
Ganoderic acid X	Mitogenesis,inhibition (Apoptosis induction)	20 mg/L
Ganoderic acid Y	Mitogenesis,inhibition (Apoptosis induction)	0.180mM
Lucialdehyde B	LLC Lewis lung carcinoma	14.3 mg/ml
	T-47D human breast cancer	15.0 mg/ml
	Meth-A sarcoma	4.0 mg/ml
Lucialdehyde C	LLC Lewis lung carcinoma	10.7 mg/ml
	T-47D human breast cancer	4.7 mg/ml
	S-180 sarcoma	7.1 mg/ml
	Meth-A sarcoma	3.8 mg/ml
Ganodermanonol	T-47D human breast cancer	4.8 mg/ml
	S-180 sarcoma	10.0 mg/ml
	Meth-A sarcoma	2.8 mg/ml
Ganodermanondiol	LLC Lewis lung carcinoma	14.0 mg/ml
	T-47D human breast cancer	4.7 mg/ml
	S-180 sarcoma	11.0 mg/ml
	Meth-A sarcoma	9.2 mg/ml
(3β,5α,6β,22E)-ergosta-7,9(11),22-triene-3,5,6-triol	HepG2 human hepatoblastoma cells	62.5μM
	MDA-MB-231 human breast cancer	56.3 μM
(3β,5α,6α,7α,22E)-5,6-epoxy-ergosta-8, 22-diene-3, 7-diol	HepG2 human hepatoblastoma cells	22.1 μM
	MDA-MB-231 human breast cancer	20.3 μM
(3β,5α,6α,7α 22E)-5,6-epoxy-ergosta-8(14),22-diene-3,7-diol	HepG2 human hepatoblastoma cells	50.6 μM
	MDA-MB-231 human breast cancer	46.7 μM
(3β,5α,8α, 22E)-5,8-epidioxy-ergosta-6, 22-dien-3-ol	HepG2 human hepatoblastoma cells	34.8 μM
	MDA-MB-231 human breast cancer	44.6 μM
(3β,22E)-ergosta-5,7,22-trien-3-ol	HepG2 human hepatoblastoma cells	75.6 μM
	MDA-MB-231 human breast cancer	69.7 μM
(3β,5α,8α,22E)-5,8-epidioxy-ergosta-6,9(11), 22-trien-3-ol	HepG2 human hepatoblastoma cells	44.5 μM
	MDA-MB-231 human breast cancer	32.1 μM

Consequently, the researchers discuss several main possible mechanisms by which *G. lucidum* triterpenes exhibit antitumor effects - cell cycle blockade in G0/G1, S or G2/M phases, apoptosis induction, and metastatic potential reduction. It was found that *G. Lucidum* triterpenoids may exhibit other types of activity. Ganoderic acid α has ability to act on HIV-1 [61]. ganoderenic acid A has a strong hepatoprotective effect on liver damages caused by tetrachloromethane [62]. Ganoderic acid A can inhibit histamine release and improve digestive function in humans [26], regulate liver function, reduce blood fat level and blood pressure [63].

Lentinula edodes

Lentinula edodes (Figure 4) called by its Japanese name of shiitake is one of the most cultivated and widely consumed edible mushrooms in the world. *Lentinula edodes* belongs to saprotrophic basidiomycetes and contain wide range of steroid compounds. Moreover, it accumulates ergosterol in a large amount up to 83-89%. Ergosterol in fungal cells presents as free form, as form of peroxides, esters with higher fatty acids, glycosides, or as complexes with polysaccharides (Table 3).

Figure 4. *Lentinula edodes*.

Table 3. Triterpenoids and steroids of *Lentinula edodes* [63-72]

Chemical group	Compounds
Terpenoids	
pentacyclic	Friedelin
steroids	
ergosterol and its derivates	Ergosterol; Ergosterol peroxide; 5α,6α,8α,9α-diepoxy-(22E,24R)-ergost-22-ene-3β,7α-diol; (22E,24R)-23-methylergosta-7,22-diene-3β,5α,6β-triol; (22E,24R)-23-methylergosta-7,22-diene-3β,5α,6β,9α-tetrol; (24S)-ergost-7-ene-3β,5α,6β-triol, ergosta-7,22-dien-3β-ol; (22E)-ergosta-5,7,9(11),22-tetraen-3β-ol; (22E)-ergosta-7,9(11),22-trien-3β-ol; (22E)-ergosta-6,8,22-trien-β-ol; (22E)-norergosta-5,7,9,22-tetraen-3β-ol; 3β,5α-dihydroxy-(22E)-ergosta-7,22-dien-6-one; (22E)-ergosta-4,6,8(14),22-tetraen-3-one;
ergosterol and its derivates	(22E)-ergosta-6,22-diene-3β,5α,8α-triol; (22)E-23-methylergosta-5,7,22-triene-3β-ol; 5,6-dihydroxy ergosterol; Ergosta-7,22-dien-3-ol; 5α,6α-epoxy-(22E)-ergosta-8,22-diene-3β,7β,14α-triol
other sterols	Cholestanol; β-sitosterol, (22*E*,24*R*)-methylcholesta-7,22-diene-3β,5α,6β-triol; (22E,24R)-methylcholesta-5,7,22-trien-3β-ol; (22E,24R)-methylcholesta-7,22-dien-3β-ol; (22E,24R)-methylcholesta-6,22-diene-3β,5α,8α-triol; Fungisterol.

Under ultraviolet exposure from ergosterol is generated vitamin D_2 (ergocalciferol). Ergosterol has anti-inflammatory and anti-cancer effects. For example, in mice with hepatitis, the addition of *L. edodes* extract, enriched with vitamin D after UV irradiation, significantly reduced liver damage. Histopathological examination of mice liver tissues showed improvement in their structures. The level of aminotransferases and INF-γ in plasma of mice decreased. Anti-inflammatory activity of vitamin D and mushroom extract causes a synergistic effect [73]. Studies in rats showed an increase in 25-hydroxyvitamin D in serum after a diet based on the fruit bodies of *L. edodes*, which were irradiated with ultraviolet light [74].

(22E,24R)-Methylcholesta-7,22-diene-3β,5α,6β-triol and (22E,24R)-methylcholesta-6,22-diene-3β,5α,8α-triol, isolated from *L. edodes*, exhibit an antimicrobial activity against *Escherichia coli*, *Bacillus cereus*, *Staphyloccocus aureus*, *Erwinia carotovora*, and *Bacillus subtilis* in concentrations from 15.5 to 31.25 μg/mL [75-76].

Ergosterol and its peroxide reduced LPS-induced inflammation and showed an inhibitory effect on the activity of HMG Co-A reductase *in vivo* and *in vitro*, which affects on the reduction of cholesterol in the blood [77,78].

Poria cocos **(Schw.) Wolf (Polyporaceae)**

Poria cocos (Schw.) Wolf (Polyporaceae) is a wood-decay fungus also called Poria, Tuckahoe, Hoelen, China root and Fu-Ling (Figure 5). *Poria cocos* is a unique fungus that grows underground and parasitizes on a moist roots or stumps of a pine trees. Poria is a rare tinder mushroom that resembles a small coconut [79-100].

Today, more than 120 triterpenoids have been isolated from *P. cocos* and classified into 4 main groups: lanostane-8-ene type (type I), lanostane-7,9(11)-diene type (type II), 3,4-seco-lanostane-8-ene type (type III) and 3,4-secolanostane-7,9(11)-diene type (type IV) (Figure 6, Table 4).

Figure 5. *Poria cocos*.

Table 4. Triterpenoids and steroids of *Poria cocos* [79-92]

Chemical group	Compounds
Terpenoids	
acids	3-epidehydrotumulosic acid; Dehydrotumulosic acid; Tumulosic acid; 15α-hydroxydehydrotumulosic acid; 5α,8α-peroxydehydrotumulosic acid; 25-hydroxy-3-epi-dehydrotumulosic acid; 25-hydroxy-3-epi-tumulosic acid; 6*α*-hydroxy-dehydrotumulosic acid; Trametenolic acid; 3-O-acetyl-16a-hydroxytrametenolic acid; 3-epi-dehydrotrametenolic acid; Dehydrotrametenolic acid; 16α,27-dihydroxydehydrotrametenolic acid; 16α-hydroxydehydrotrametenolic acid; 16α-hydroxytrametenolic acid; Polyporenic acid C; 29-hydroxypolyporenic acid C; 25-hydroxyporicoic acid H; 16-deoxyporicoic acid B; Poricoic acids A-H; Poricoic acids AM; Poricoic acid CM; Poricoic acid DM; 25-methoxyporicoic acid A (25-O-methylporicoic acid D); 26-hydroxyporicoic acid DM; 25-hydroxyporicoic acid C; Poricoic acid GM; Poricoic acid HM; 6,7-dehydroporicoic acid H; Dehydropachymic acid; 3-epi-dehydropachymic acid; Pachymic acid; 25-hydroxypachymic acid; Eburicoic acid; Dehydroeburicoic acid; Dehydroeburiconic acid; 16α,25-dihydroxydehydroeburiconic acid; 16α-hydroxyeburiconic acid; 6α,25-dihydroxyeburiconic acid; 16α,29-dihydroxyeburiconic acid; 15α-hydroxyeburiconic acid; Lanosta-8,24-dien-21-oic acid; β,16α-dihydroxylanosta-7,9(11),24-trien-21-oic acid; 3β,16α-dihydroxy-24-oxolanost-7,9(11)-dien-21-oic acid (poriacosone A); 3β,16α-dihydroxy-24-oxolanost-7,9(11)-dien-21-oic acid (poriacosone B); 3β,15α-dihydroxy-24-oxolanosta-8-en-21-oic acid; 3α,16α,25-trihydroxylanosta-8,24-dien-21-oic acid; 3β−acetyloxy-16α,26-dihydroxylanosta-8,24-dien-21-oic acid; 15α-hydroxy-3-oxolanosta-8,24-dien-21-oic acid; 16α-hydroxy-3-oxolanosta-8,24-dien-21-oic acid; 3β,15α-bis(acetyloxy)-24-methylenelanost-8-en-21-oic acid; 3β-acetyloxy-16-hydroxy-24-methylenelanosta-5,7(9),11-tetraene 21-oic acid; 16α-hydroxy-3,4-seco-lanosta-4(28),7,9(11),24(31),25(27)-pentaene-3,21-dioic acid; Coriacoic acid A-D; Daedaleanic acid B; Hispindic acid B; 11β-ethoxydaedaleanic acid; Ceanphytamic acids A-B

Table 4. (Continued)

Chemical group	Compounds
Terpenoids	
alcohols	Poricotriol A; Poricotriol C; Poricotriol G; Pricotriol H; 25-hydroxy-3-epidehydrotumulosol; Dehydroeburicol, such as: Lanosta-8-ene; Lanosta-7,9(11)-diene; 3,4-seco-lanostan-8-ene; 3,4-seco-lanostan-7, 9(11)-dienetriterpene
esters	Dimethyl poricoate C; Dimethyl poricoate G; Dimethyl poricoate H; Methyl 25-hydroxy-3-epi-dehydrotumulosate; Dehydroeburicoic acid monoacetate (6); Eburicoic acid acetate; 16α-hydroxy-3,4-seco-lanosta-4(28),8,24-triene-3,21-dioic acid-3-ethyl ester; 16α-hydroxy-3,4-seco-lanosta-4(28),7(9),11,24-tetraene-3,21-dioic acid- 3-ethyl ester; α-amyrin acetate; Oleanic acid 3-acetate; Dehydroeburicoic acid monoacetate; Eburicoic acid acetate
diterpens	Dehydroabietic acid; 7-oxocallitrisic acid; Pimaric acid
steroids	
ergosterol and its derivates	Ergosta-7,22-dien-3β-ol; Ergosterol peroxide; 9,11-dehydroergosterol peroxide

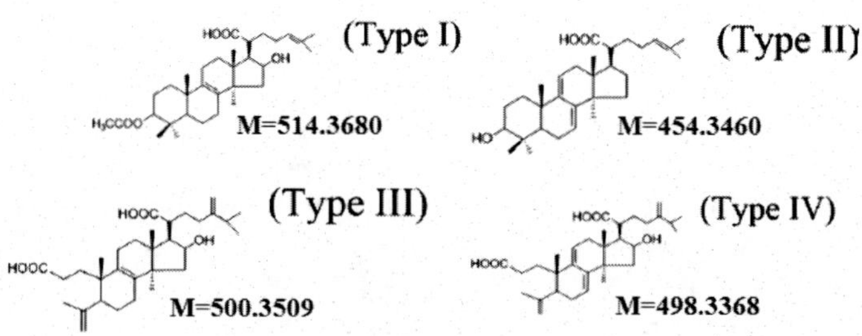

Figure 6. The structure of the main terpenoids of *Poria cocos.*

In traditional Chinese medicine *P. cocos* is mainly used in hyperglycemic disorders treatment; and also in conditions of swelling, urinary difficulty, dizziness, palpitations, low appetite, loose bowels, diarrhea, panic attack and insomnia [79-100]. Lanostanoids, especially pachymic acid, have shown ability to stimulate glucose uptake, GLUT4 gene expression and translocation, as well as to promote the accumulation of triglycerides in adipocytes [93]. Dehydrotrametenolic acid acts as an insulin sensitizer *in vivo* and causes adipocyte differentiation (induces preadipocyte differentiation) [94].

Triterpenoids of *P. cocos* significantly inhibit the action of phorbol 12-myristate 13-acetate (TRA) on the formation of skin tumors in mice [95], demonstrating early activation of Epstein-Barr virus (EBV-EA) antigen [96].

Antitumor activity data of *Poria cocos* substances are presented in Table 5.

Table 5. Antitumor activity of *Poria cocos* compounds [88-101]

Compound	Cell line	Inhibition concentration (IC_{50})
16α-hydroxy-3,4-seco-lanosta-4(28),8,24-triene-3,21-dioic acid-3-ethyl ester	MGC-803 human gastric carcinoma	>100 μM
	HepG2 human hepatocellular carcinoma	>100 μM
	VERO African green monkey kidney	>100 μM
3β-acetyloxy-16α-hydroxy-24- methylenelanosta-5,7(9),11-tetraene-21-oic acid	MGC-803 human gastric carcinoma	42.90 μM
	HepG2 human hepatocellular carcinoma	44.07 μM
	African green monkey kidney VERO	>100 μM
16α-hydroxy-3,4-seco-lanosta-4(28),7,9(11), 24(31),25(27)- pentaene-3,21-dioic acid.	human gastric carcinoma MGC-803	>100 μM
	HepG2 human hepatocellular carcinoma	>100 μM
	African green monkey kidney VERO	>100 μM
poricotriol A	HL-60 leukemia	1.9
	A-549 lung carcinoma	4.2 μM
	CRL1579 melanoma	5.5 μM
	NIH:OVCAR-3 ovarian carcinoma	2.2 μM
	SK-BR-3 breast adenocarcinoma	1.2 μM
	DU145 human prostate cancer	14.5 μM
	AZ521 human gastric carcinoma	1.4 μM
	PANC-1 pancreas cancer	26.3 μM
16R,25-dihydroxy-eburiconic acid	CRL1579 melanoma	28.7 μM
polyporenic acid C	A-549 lung carcinoma	80 mM

Pangamic acid, which belongs to the triterpenoids lanostanovogo type, is one of the main components of *P. cocos* and plays an important role in the therapeutic effect of this fungus. Pachymic acid possesses anti-inflammatory and anti-cancer effects. For example, pachymic acid has an antiproliferative effect on androgen-sensitive prostate carcinoma cells (LNCaP) and androgen-insensitive (DU145) prostate cancer cells, causing their apoptosis [97]. It significantly inhibited (*in vitro* and *in vivo*) gastric cancer cells (SGC-7901) viability and caused G0/G1 cell-cycle block [98].

Triterpenoids mixture and three purified triterpenoids (pachymic acid, dehydropachymic acid и polyporenic acid C) extracted from *P. cocos* can suppress the proliferation of Panc-1, MiaPaca-2, AsPc-1 and BxPc-3 cancer cells in pancreatic cancer. The antiproliferative effects of the extracts on cancer cells in this case relies on G0/G1 cell-cycle block [99].

In addition, the ability to exhibit anti-inflammatory activity has been shown by such *P. cocos* components as coriacoic acid A and B, acetyl eburicoic acid, dehydroeburiconic acid, poricoic acid C [83].

Fomitopsis pinicola

Fomitopsis pinicola is tinder that belongs to the *Polyporaceae* (Figure 7). It also called red-belted polypore, tiaga mushroom, brown crumbly rot and red belt conk. *Fomitopsis pinicola* grows on coniferous and broad-leaved trees and is widely distributed in Asia, Europe and America. By the therapeutic value this fungus traditionally competed with *Ganoderma lucidum* in Far East Asia, and recently throughout the world. Dry fruiting bodies were used in different nations as hemostatic and anti-inflammatory means.

Various extracts from *F. pinicola* demonstrates anti-tumor, anti-bacterial, anti-viral, immuno-modulating, anti-pathogenic, anti-diabetic, anti-inflammatory, adaptogenic action. Among active constituents of this fungus ergosterol, trametenolic acid, polyporenic acid C, lanosterol, inotodiol, pinicolic acid, triterpenes, fomitopinic acid A, adenosine, phenol compounds are usually noted [102-110].

Terpenoids and steroids isolated from *F. pinicola* are listed in Table 6.

Figure 7. *Fomitopsis pinicola.*

For example, *F. pinicola* methanol extracts exhibit antioxidant activity, as well as antitumor activity against SK-Hep 3B, HEP3B, SNU 185, SNU 354, and HeLa cell lines. Chloroform and butanol extracts of this fungus reveal a slight antibacterial effect in concentrations of 4 mg/mL [109]. Polyporenic acid C, 3β- acetyloxylanosta-8,24-dien-21-oic acid, ergosta-7,22-dien-3β-ol, 21-hydroxylanosta-8,24-dien-3-one, pinicolic acid A, trametenolic acid B and pachymic acid of *F. pinicola* demonstrate antibacterial activity against *Bacillus subtilis* [102].

Cell lines of leukemia HL-60, lymphoma U937, Crocker sarcoma S180 and leukemia K562, shows strong sensitivity to chloroform extract *F. Pinicola* which contains up to 75% of triterpenoids. For these lines, IC_{50} were 45.13, 39.07, 104.9, 139.81 μg/ml, respectively, after 48 hours of incubation. The cell lines of SMMC-7721 and Eca-109 esophageal cancer were less sensitive, and IC_{50} was 320.12 and 209.78 μg/ml, respectively, after 48 hours of incubation. The extract also inhibited the growth of S180 sarcoma cells *in vitro* and *in vivo*. Results *in vivo* showed that a daily dose of 200 mg/kg of the extract inhibited the growth of a solid tumor and prolonged the survival days of mice. In addition, it was found that the extract can cause both early and late apoptosis of S180 cells [110].

Table 6. Terpenoids and steroid compounds of *Fomitopsis pinicola* [102-108]

Chemical group	Compounds
Terpenoids	
alcohols	5α-Lanosta-7,9(11),24-triene-3β,21-diol; Pinicolol B, (+)-lanosta-7,9(11),24-trien-3β,15α,21-triol, 3α-acetyloxylanosta-8,24-diene-21-ol
acids	Pachymic acid, 16α-Hydroxyeburiconic acid; (þ)-Trametenolic acid B, Tsugaric acid A, Pinicolic acid A-E, Fomitopsic acid, 24-Methyl-3-oxo-Lanosta-8,25-dien-21-oic acid, Fomitopinic acid A-B, Polyporenic acid C, (16α)-16-(acetyloxy)-24-methylene-3-oxo-lanosta- 7,9(11)-dien-21-oic acid, Pinicolic acid D, Fomitopsic acid B, (-)-3,16-dioxolanosta-8,24-dien- 21-oic acid, (+)-3,4-secolanosta- 4(28),8,24-trien-3,21-dioic acid, (-)-3β-hydroxy-16-oxolanosta- 7,9(11),24-trien-21-oic acid, (+)-3a-acetoxylanosta-8,24-dien-21-oic acid, 3α-(4-carboxymethyl-3-hydroxy-3-methylbutanoyloxy)-lanosta-8,24-dien-21-oic acid
aldehydes	Pinicolol C
ketones	(+)-21-hydroxylanosta-8,24-dien-3-one; Lanosta-7,9(11),24-trien-3-one; (+)-21-(carboxyacetoxy)lanosta-7,9(11),24-trien-3-one
esters	(3α)-lanosta-8,24-diene-3,21-diol 3-acetate
glycosides	Fomitoside A-J
Steroids	
ergosterol and its derivates	Ergosterol D; Ergosta-7,22-dien-3β-ol, [3β(9Z,12Z),5α,22E]-ergosta-7,22-dien-3-ol, 9,12-octadecadienoate; [3β(9Z,12Z)]-ergosta-7,24(28)-dien-3-ol, 9,12-octadecadienoate; α-dihydroergosterol
other steroids	Fungisterollinoleate; Agnosterone; 21-hydroxy-agnosterone

The Pachymic acid isolated from this fungus has antitumor activity against cell lines of: breast cancer MDA-MB-231 and MCF-7, prostate cancer DU145, LNCaP, and gastric carcinoma HT-29 [111]. Dehydroeburicoic acid may cause apoptosis of HL60 cell line of human leukemia [112]. Fomitopinic acid A, fomitoside E and F (IC_{50} 0.15-1.15 μM) have shown significant anti-inflammatory activity against cylooxygenase-2 in comparison to control (indomethacin, IC_{50} of 0.60 μM) [107].

Fomitopsis betulina

F. betulinus (Bull.: Fr.) P. Karst. (Polyporaceae) is a bracket fungus, also named as the birch polypore, birch bracket, or razor strop conk (Figure 8). *F. betulina* grows both on living and on dead Birch trees, but can be artificially introduced to other types of tree. It has long been known that *F. betulina* has immunomodulating, anti-inflammatory, antiseptic, antifungal and styptic action. *F. betulina* fruiting bodies are used in European traditional medicine in cancer and stomach diseases treatment. Its active compounds belong to phenolic acids, indole compounds, steroids, and triterpenoids [113-125]. Terpenoid and steroid compounds of *F. betulina* are presented in Table 7.

Figure 8. *Fomitopsis pinicola.*

F. betulina extracts obtained with ethanol, diethyl ether, chloroform of dihormati contain mainly polyporic acids and exhibit bactericidal, fungicidal and acaricidal activity [120-122].

Table 7. Terpenoid and steroid compounds of *F. betulina* [113-119]

Chemical group	Compounds
Terpenoids	
acids	Piptolinic acids A-J; 3-epi-(30-hydroxy-30-methylglutaryloxyl)-dehydrotumulosic acid; Dehydroeburiconic acid; 6α-hydroxypolyporenic acid C; 3-epidehydropachymic acid; Polyporenic acid A-C; (3α,12α,25S)-3-(acetyloxy)-12-hydroxy-24-methylene-Lanost-8-en-26-oic acid;
	(3α,12α,25S)-3-[(carboxyacetyl)oxy]-12-hydroxy-24-methylene-Lanost-8-en-26-oic acid; (3α,12α,25S)-12-hydroxy-3-(3-methoxy-1,3-dioxopropoxy)-24-methylene-Lanost-8-en-26-oic acid; (3α,12α,25S)-3-[(3S)-4-carboxy-3-hydroxy-3-methyl-1-oxobutoxy]-12-hydroxy-24-methylene-Lanost-8-en-26-oic acid; (3α,12α,25S)-12-hydroxy-3-[[(3S)-3-hydroxy-5-methoxy-3-methyl-1,5-dioxopentyl]oxy]-24-methylene-Lanost-8-en-26-oic acid; (þ)-12α,28-dihydroxy-3α-(30-hydroxy-30-methylglutaryloxy)-24-methyllanosta-8,24(31)-dien-26-oic acid; 3β-acetoxy-16-hydroxy-24-oxo-5α-lanosta-8- ene-21-oic acid; Fomefficinic acid; 3-epi-dehydrotumulosic acid
esters	Methylpolyporenate C; 3α,12α-dihydroxy-24-methylene-lanost-8-en-26-oic acidmethyl ester; 3α,12α-dihydroxy-24-methylene-lanost-8-en-26-oic acid methyl ester 3-acetate
pentacyclic	Betulinic acid; Betulin
Steroids	
ergosterol and its derivates	5α,8α-epidioxy-22E-ergosta-6, 22-dien-3β-ol (ergosterol peroxide); 9,11-dehydroergosterol peroxide
other steroids	Cholecalciferol; Hexestrol

Methanolic extract of *Fomitopsis betulina* mycelium in concentration of 50 μg/ml exhibited significant cytotoxic activity against prostate cancer cells DU145 – 5% of viable cells remain, and a moderate effect on melanoma cell lines A375 – 21% of viable cells. At the same time, this extract did not have a cytotoxic effect on normal human skin fibroblasts and epithelial cells of the prostate gland.

Methanol extract from the fruit bodies of this fungus in a concentration of 50 μg/ml showed no cytotoxic effect on melanoma cells A375, prostate cancer DU145 and melanoma WM795 - 52 and 69% and 96% of viable cells, respectively. This extract was not toxic to human skin fibroblasts, but there was a strong cytotoxic effect on normal prostate epithelial cells – 5% of viable cells. The reason for differences in cytotoxicity *in vitro* and *in vivo* extracts obtained from fruit bodies and mycelial cultures of *F. betulina* may be due to the fact that the extract from the mycelium of the fungus contained more biologically active steroids and triterpenoids [115].

For ethanol extracts obtained from the cultured fungus *F. betulina* was demonstrated antiproliferative effect against A549 human lung carcinoma, HT-29 human colon adenocarcinoma and T47D human breast cancer cell lines with IC_{50} value in the range of 11.1-34.4 mg/ml. Ethanol extract of the fruiting bodies collected in the forest showed a less distinct effect on these cells – IC_{50} was about 48,1 μg/ml [123].

Ether extracts of *F. betulina* reveal the significant cytotoxic activity on FTC238 thyroid cancer, SK-N-AS neuroblastoma, T47D breast cancer, Hep-2 laryngeal cancer and HeLa cervical cancer [124].

Divers active components, isolated from *F. betulina*, act in different ways. For example, a 3β-acetoxy-16-hydroxy-24-oxo-5α-lanosta-8-ene-21-oic acid exhibits antimicrobial activity both against gram-positive bacterium (*Bacillus subtilis* and *Staphylococcus aureus*) and a little lower activity against gram-negative bacterium (*Pseudomonas aeruginosa* and *Escherichia coli*).

However, compounds of polyporenic acid C, ergosterol peroxide, 9,11-dehydroergosterol peroxide, polyporenic acid A from *F. betulina* exhibit slight antibacterial activity against *Bacillus subtilis* and *Staphylococcus aureus*, besides, do not act against gram-negative strains [114].

Polyporenic acids A and C, three derivatives of polyporenic acid A, 12R,28-dihydroxy-3R-(3'-hydroxy-3'methylglutaryloxy)-24-methyllanosta-8,24(31)-dien-26-oic acid suppressed TPA-induced edema to 49-86% in concentration of 400 mmol. Herewith, these components had higher activity than of control (glycyrrhetinic acid and indomethacin) [117].

Piptolinic acids A had an inhibitory effect on HL-60 human promyelocytic leukemia and THP-1 human acute monocytic leukemia with an IC_{50} of 1.77 mM and 8.21 mM, respectively.

And 25S,30S)-(+)-12α-hydroxy-3α-(30-hydroxy-40-methoxycarbonyl-30-ethyl-butyryloxy)-24-methyl-lanosta-8,24(31)-dien-26-oic acid exhibited a cytostatic effect on A-375 human melanoma with an IC_{50} of 42.8 mM and 786-O human renal carcinoma with an IC_{50} of 56.5 mM [114, 116, 118].

Thus, pharmacological studies of *F. betulina* have confirmed its potential indicating significant diverse therapeutic activity. The possibility of its successful cultivation in artificial conditions additionally promotes the applicability of this fungus.

Inonotus obliquus

Inonotus obliquus (Rers.) Pil. sterilis (Figure 9). This fungus, which is also called chaga, from a biological point of view, is a barren (sterile) stage of development of the birch fungus. According to the Botanical classification it's polypore. To date, more than 40 triterpenoids of the lanostane series have been isolated from this fungus (Table. 8).

In traditional medicine chaga and its water extracts have long been used for treating cancer and diseases of the gastrointestinal tract. With long-term use of medications produced from aqueous extracts of this fungus, health and general condition of patients with III-IV stages cancer, regardless of tumor location, significantly improve. After 3-4 weeks of these pharmaceuticals usage, patients pain decrease and cease [158-161]. Researchers have suggested that chaga doesn't possess a specific effect on the tumor, but can reveal a tonic effect on the central nervous system. And further, with a prolonged treatment, it normalizes impaired metabolic processes in the body, and thus has a retarding effect on tumor growth.

Figure 9. *Inonotus obliquus.*

Table 8. Terpenoids and steroids isolated from *Inonotus obliquus* [126-157]

Chemical group	Compounds
Terpenoids	
alcohols	Lanosterol (3β-hydroxylanosta-8,24-diene); Inotodiol (3β,22-dihydroxylanosta-8,24-diene); lanosta-8,24-dien-3β,11 β-diol (inonotsudiol A); 3β,22-dihydroxylanosta-7,9(11),24-triene; 3β,22,25-trihydroxylanosta-8-ene; 3β,22α,25-trihydroxy-lanosta-8,23E-diene (fuscoporianol C); 3β,22,24-trihydroxylanosta-8,25-diene; 3β,21-dihydroxylanosta-8,24-diene; 3β,22*R*,25-trihydroxylanosta-8,23*E*-diene; 3β,22*R*,25-trihydroxylanosta,7,9(11)23*E*-triene; Lanosta-8-en-3β,22R,24R-triol (inonotsutriol D); Lanosta-8-en-3β,24R,24S-triol (inonotsutriol E); Inoterpenes A-F; Inonotusanes A-C; lanosta-8-en-3β,22,24,25-tetraol-25-methyl oxide (inonotusane G); inonotusol A-C
acids	trametenolic (3β-hydroxylanosta-8,24-dien-21-oic); obliquinolic (3β-hydroxylanosta-8-en-21-oic); 3β-hydroxylanosta-8,25-dien-24-one-21-oic acid (inonotusane F); saponaceoic acid I; 3β-hydroxy-5α-lanosta-8,25-dien-21-oic acid (chagabusone A)

Table 8. (Continued)

Chemical group	Compounds
Terpenoids	
aldehydes	3β-hydroxylanosta-8,24-dien-21-al; 3β-hydroxy-25,26,27-trinorlanosta-8,22*E*-diene-24-al
ketones	(3S,7S,9R)-3,7-dihydroxy-7(8→9)abeo-lanost-24-en-8-one (spiroinonotsuoxodiol); (22R)-3β,22-dihydroxylanosta-8,24-dien-11-one (inonotsuoxodiol A); 21,24-cyclopenta-3,11,15,21,25-pentahydroxylanosta-8-ene-7-one; 21,24-cyclopenta-3,11,21,25-tetra-hydroxylanosta-8-ene-7-one; 3β,22-dihydroxylanosta-8,25-dien-24-one
lactones	3β-hydroxylanosta-8,24-diene-21,23-lactone; 24-methyl-3β-hydroxylanosta-8,24-diene-21,23-lactone (inotolactone B); inotolactone A; (20R,24S)-3β,25-dihydroxylanost-8-en-20,24-olide (inonotsulide A); (20R,24R)-3β,25-dihydroxylanost-8-en-20,24-olide (inonotsulide B); (20R,24S)-3β,25-dihydroxylanosta-7,9(11)-dien-20,24-olide (inonotsulide C)
peroxides	3β,22α-dihydroxylanosta-8,23*E*-dien-25-peroxide (fuscoporianol B), 3β,22α-dihydroxylanosta-8,24-dien-25-peroxide
with five-membered ring	21,24-cyclopentalanosta-3β,21,25-triol-8-ene; 25-methoxy-21,22-cyclolanosta-8-ene-3β,21α-diol (fuscoporianol A); (3β,21R,24S)-21,24-cyclolanost-8-ene-3,21,25-triol (inonotsutriol A); (3β,21R,24R)-21,24-cyclolanost-8-ene-3,21,25-triol (inonotsutriol B); (3β,21R,24S)-21,24-cyclolanosta-7,9(11)-diene-3,21,25-triol (inonotsutriol C); (21R,24S)-3β,12β,15α,21,25-pentahydroxy-21,24-cyclopentalanosta-7,9(11)-diene (inonotusane E)
epoxy	22*R*,25-epoxylanost-8-ene-3β,24*S*-diol (inonotsuoxide A); 22S,25-epoxylanost-8-ene-3β,24*S*-diol (inonotsuoxide B)
norlanostane	3β-hydroxy-24,25,26,27-tetranorlanosta-8-en-22-one (inonotusane D)
pentacyclic	Betulin; Upeol; Lupenon; Betulinic acid
Steroids	
ergosterol and its derivates	Ergosterol; Ergosterol peroxide; 3β-hydroxyergosta-7,22-diene; 3β-hydroxyergosta-7-ene (fungisterol); 9,11-dehydroergosterol peroxide
other steroids	Sitosterol; Stigmasterol; Sitostanol; Cholesterol

It is shown that the use of aqueous extracts of chaga reduces the size of tumors of MOP and S180 sarcoma, Lewis lung carcinoma, Ehrlich carcinoma, B16-F10 melanoma and U-87 MG glioblastoma. The antimetastatic effect of aqueous extracts of chaga was established *in vivo* on cells: sarcoma, HeLa cervical adenocarcinoma, Ehrlich carcinoma, and *in vitro* on cells: hepatoma, colon cancer, 180 sarcoma [162-174].

Melanin isolated from chaga exhibits moderate antitumor properties on the primary tumor and strongly expressed on metastases [175].

Biological activity of various organic solvents extracts of chaga: ethanol, methanol, petroleum, ethyl acetate, chloroform was investigated *in vitro*.

Ethanol chaga extract has an antiproliferative effect on B16F1 melanoma cells, inhibited the growth of NCI-H460 lung cancer cells by 60% [176-178]. Chloroform chaga extract reduced the proliferation of leukemic cells at a concentration of 20-40 μg/ml. The activity of this extract was significantly higher than that of aqueous chaga extract in relation to hepatoma cells and cervical cancer [179]. This indirectly indicates that steroid compounds of chaga in the chloroform extract were more active than melanin – the main active component of the aqueous extract of this fungus.

Ethanol extract also have been separated into fractions using petroleum ether and ethyl acetate. The corresponding petroleum and ethyl acetate extracts *in vitro* reduced the growth of PC3 prostate carcinoma cells and MDA-MB-231 breast carcinoma cells [180]. For these extracts IC_{50} have been determined: for PC3 of 29.57 ± 12.18 and 19.22 ± 0.46 μg/ml, for MDA-MB-231 of 57.39 ± 14.46 and 46.49 ± 13.21 μg/ml, respectively.

The activity of petroleum extract in relation to both cancer cells lines and the activity of ethyl acetate extract against MDA-MB-231 were at the level of known cytostatics-doxorubicin, and against cells PC3 – were three times lower compared with doxorubicin. The highest activity of petroleum extract was probably associated with the content in it of triterpenoid and steroid compounds.

Methanol extract of chaga (indicated as MeOH) was separated by organic solvents hexane, dichloromethane, ethyl acetate, butanol into 4 fractions – indicated as HX, MC, EA, BuOH, respectively. The extract itself and its fractions were tested for cytotoxic activity against A549, H1264, H1299 and Calu-6 human lung cancer cell lines. It was shown that the MeOH extract reduced the cell viability in all cell lines of lung cancer at IC_{50} of 2.03-2.40 mg/ml. Fractions containing triterpenoids, HX and MC showed significant cytotoxicity with respect to all cancer cell lines with IC_{50} values in the range from 95.3 μg/ml to 225.1 μg/ml [154].

Table 9. Antitumor activity of triterpenoid and steroid compounds of *Inonotus obliquus* [154, 181]

Compound	Cell line	Inhibition concentration (IC_{50})
Lanosterol	MCF-7 Breast adenocarcinoma	1 μg/ml; 250 μg/ml
	P388 leukemia	> 100 μg/ml
	HeLa Cervical adenocarcinoma	250 μg/ml
	A-549 Lung carcinoma	250 μg/ml
	AGS Gastric adenocarcinoma	250 μg/ml
	MDA-MB-231 Breast adenocarcinoma	> 40 μg/ml
	PC3 Prostate carcinoma	> 40 μg/ml
	L1210 Leukemia	> 80 μg/ml
	COLO205 Gastric adenocarcinoma	> 40 μg/ml
Inotodiol	MCF-7 Breast adenocarcinoma	1 μg/ml; 250 μg/ml
	256 Walker carcinosarcoma	10 μg/ml
	P388 Leukemia	6 μg/ml; > 40 μg/ml
	L1210 Leukemia	49 μg/ml
	COLO205 Gastric adenocarcinoma	75 μg/ml
	MDA-MB-231 Breast adenocarcinoma	> 40 μg/ml
	PC3 Prostate carcinoma	> 4μg/ml
Inotodiol	HeLa Cervical adenocarcinoma	250 μg/ml
	A-549 Lung carcinoma	250 μg/ml
	AGS Gastric adenocarcinoma	250 μg/ml
Trametenolic acid	MCF-7 Breast adenocarcinoma	5 and 10 μg/ml
	P388 Leukemia	12 мг/мл
	L1210 Leukemia	16 μg/ml
	MDA-MB-231 Breast adenocarcinoma	25 μg/ml
	PC3 Prostate carcinoma	29 μg/ml
	COLO205 Gastric adenocarcinoma	> 90 μg/ml
	A-549 Lung carcinoma	> 90 μg/ml. 227.4 μmol
	H1264 Lung adenocarcinoma	218.2 μmol
	H1299 Lung adenocarcinoma	210.3 μmol
	Calu-6 Lung adenocarcinoma	184.1 μmol

Compound	Cell line	Inhibition concentration (IC_{50})
3β-hydroxy-lanosta-8,24-dien-21-al	MCF-7 Breast adenocarcinoma	250 μg/ml
	HeLa Cervical adenocarcinoma	250 μg/ml
	P388 Leukemia	9 мг/мл
	MDA-MB-231 Breast adenocarcinoma	16 μg/ml
3β-hydroxy-lanosta-8,24-dien-21-al	P388 Leukemia	28 μg/ml
	PC3 Prostate carcinoma	33 μg/ml
	COLO205 Gastric adenocarcinoma	> 80 μg/ml
	AGS Gastric adenocarcinoma	250 μg/ml
	A-549 Lung carcinoma	250 μg/ml, 83,5 μmol
	H1264 Lung adenocarcinoma	90.9 μmol
	H1299 Lung adenocarcinoma	128.0 μmol
	Calu-6 Lung adenocarcinoma	75.1 μmol
3β,22*R*-dihydroxy-lanosta-8,24-dien-11-one	KB[1] Nasopharyngeal cancer	4,5 μg/ml
	HL-60 Leukemia	6,2 mg/ml
	P388 Leukemia	6,4 mg/ml
	L1210 Leukemia	9 mg/ml
	A-549 Lung carcinoma	> 4 μg/ml
	Bel-7402 Hepatoma	> 4 μg/ml
22S,25-epoxy-lanosta-8-ene-3β,24*S*-diol	P388 Leukemia	13 mg/ml
(+)-fusco-porianol C	H1264 Lung adenocarcinoma	203.7 μmol
	Calu-6 Lung adenocarcinoma	162.1 μmol
inonotsutriol A	H1299 Lung adenocarcinoma	239.3
chagabusone A	A-549 Lung carcinoma	141.2 μmol
	H1264 Lung adenocarcinoma	122.2 μmol
	H1299 Lung adenocarcinoma	82.0 μmol
	Calu-6 Lung adenocarcinoma	128.6 μmol
Ergosterol	PC3 Prostate carcinoma	3,8 μg/ml
	MDA-MB-231 Breast adenocarcinoma	> 40 μg/ml
Ergosterol peroxide	256 Walker carcinosarcoma	5 μg/ml
	MDA-MB-231 Breast adenocarcinoma	13 μg/ml
	PC3 Prostate carcinoma	16,3 μg/ml
	Gastric adenocarcinoma COLO205	> 80 μg/ml
	A-549 Lung carcinoma	> 80 μg/ml
	L1210 Leukemia	> 80 μg/ml
	MCF-7 Breast adenocarcinoma	> 80 μg/ml

In [127,128] the antitumor activity of triterpenoid compounds of chaga in experiments *in vitro* against ascitic cancer (Ehrlich and Crocker sarcoma) was shown for the first time. Inotodiol had a noticeable effect on tumor cells and lanosterol showed a very weak effect (primary change in cells). In Table 9 are presented the data of studies of individual compounds for different lines of cancer cells *in vitro*.

Currently, on the base of water extracts of *Inonotus obliquus* pharmaceutical companies produce drugs, for example, "Befungin", the therapeutic effectiveness of which is due to the presence of a chromogenic complex – melanin, which has a wide range of biological activity. In aqueous extracts triterpenoid and steroid compounds are bound or loosely bound to melanin. It was shown that in this drug the content of triterpenoids is much lower than in the raw material of chaga and was in the range from 0.01 to 0.02% equivalent to lanosterol [182].

Most of the hydrophobic compounds of chaga are bounded to melanin. Different chemical methods are used for their isolation. After acid hydrolysis of melanin, only betulin (0.01%) and retinol (0.005%) were found in it [183].

A wider range of biologically active compounds could be extracted from melanin with organic solvents. The butanol extract of melanin showed the presence in it of lanosterol, its derivative – inotodiol and unsaturated hydrocarbon squalene [184]. In petroleum extract of melanin the polycyclic and isoprenoid hydrocarbons (the fitan, cholestan, ergotan, campestan, stigmastane or cytoscan, gopan, oleanan), steroids (cholesterol and its 5 derivatives, derivative of ergosterol) and terpenoids (lanosterol and its 11 derivatives, betulin) compounds were identified [184].

Analysis of alcohol, chloroform and methyltretbutyl extracts of meal obtained after aqueous extraction of chaga showed the presence of unidentified substances of terpenoid nature, including iridoids, lactones and their glycosides, triterpenes, tetraterpenes and sesquiterpenes.

The composition of petroleum extract of meal was studied. It was shown that 65% of the extractives are lanostane compounds, steroid compounds occupy about 3.5%. Inotodiol, tramadolusa acid, lanosterol, 22,25 epoxylanost-8-en-3,24-diol, 3-hydroxylanosta-8,24-dien-21-al, 3-lanosta-8-en-3,22,24-triol, 25-methoxy-21,22-cyclopentalanosta 3,21,25-triol--8-ene, 3,22- dihydroxylanosta-8,23-dien-25-peroxide, 3,22-dihydro-xylanosta-8,24-dien-21,23-lactone, 3β,25-dihydroxylanosta-8-en20.24-

olid), derivative of cholesterol (3-hydroxy-5-cholan-24-oic acid) and steroid – stigmastane were identified in it.

The overall toxicity of this extract in the biotests on the *Stylonychia mytilus*, for white mice and rabbits was defined. It was established that petroleum extract of chaga meal does not cause the death and destruction of test organisms of *Stylonychia mytilus* – in a concentration of 10 mg/ml. It is not toxic when applied to the skin of rabbits at concentrations up to 40 mg/cm^2 and when the introduction of the probe to mice is at concentrations up to 750 mg/kg of body weight.

Antitumor activity of petroleum extract of chaga meal *in vitro* on human breast cancer MCF-7 cells (ATCC, USA) was determined. Earlier studies on antitumor activity of lanosterol, inotodiol, tramadolbuy acid, 3β-of hydrocyanate-8,24-Dien-21-al in respect of that cell line were performed (table 9). In the petroleum extract of chaga meal, these compounds were about 30% of all triterpenoid and steroid compounds. It was found that IC_{50} of meal extract against MCF-7 cancer cells was 7.87 μg/ml. This concentration is much lower than IC_{50} of triterpenoid-containing extracts obtained from chaga itself, as well as from other medical fungi (for example *G. lucidum*) and is comparable to the inhibitory concentration of individual compounds. For normal human skin fibroblast Hs-68 cells this extract showed low toxicity, cytotoxic concentration – CC_{50} exceeded 85 μg/ml [186].

Triterpenoid and steroid compounds from chaga were extracted with various organic solvents more often by hexane, petroleum ether, chloroform, methanol and ethanol. The use of various solvents and subsequent fractionation of the obtained extracts, their study using modern instrumental methods made it possible to obtain and identify a wide range of these compounds. However, the natural raw material of medicinal mushrooms is quite difficult to reproduce in nature as the source of terpenoid and steroid compounds. In addition, there may be differences in the content of these compounds in fruiting bodies collected in different natural areas, which are caused by genetic differences between populations, differences in habitual and environmental conditions affecting their growth and development. Therefore, scientists are developing technologies for the cultivation of medicinal mushrooms. Biomass obtained by biotechnological methods has advantages over forest fruit bodies, since the cultivation of fungi provides ecological purity of fungal preparations and independence of the raw material base from the harvest of fungi. Then again, opportunity strikes it is possible to influence the yield and quality of the final product by varying the conditions of cultivation of fungal cultures.

In China and Japan successfully grow mushrooms, which have long been used in traditional Eastern medicine. Today there are comprehensive literature reviews, patents on methods of growing mushrooms. Most researchers have managed to achieve increased yield of target compounds from Reishi, Shiitake and many other fungi. So, to increase production of ganoderma acid from the cultured mycelium of Ganoderma species different inducers have been used (phenobarbital and methyljasmonat, microbial polysaccharides, the two-stage cultures, etc.). There are studies on the metabolic engineering of some fungi, using the integration of transcriptom, proteomic tools, scientists are trying to find out the regulatory mechanism of the biosynthesis of ganoderic acids [187-190].

When growing I. obliquus in culture, triterpene compounds accumulate in a much smaller amount, but in the same composition as in natural chaga [151-152]. In the culture of chaga compared with natural chaga accumulates has a little more ergosterol and expands the range of its derivatives. Among them, sitosterol, stigmasterol, sitostanol and cholesterol are more common. In cultivated chaga in large quantities are found pentacyclic triterpenes and steroids, such as betulin, lupeol, lupenone in comparison with natural chaga.

Despite the huge amount of research, currently only 10 types of medical fungi are used to obtain medicines. In Japan, Russia, China, Korea, the United States, and Norway as drugs in tumor chemotherapy, based on biologically active substances released during deep cultivation of fungi are used. Currently in medical practice has been implemented the following medicinal preparations from some species of basidiomycetes as anticancer drugs, such as: flammulina (from *Flammulina velutipes*), lampterol (*Lampteromyces japonicus*), calvacin (*Calvatia gigantea*), lentinan (*Lentinus edodes*), kresin (*Trametes versicolor*), befungin (*Inonotus obliguus*) [191-192].

Pharmaceuticals based on medical fungi exhibit fewer side effects compared to synthetic analogues. It makes them promising to develop a reliable technology and base for the medical fungi cultures cultivation, obtaining extracts and active components isolation, including terpenoid and steroid compounds.

REFERENCES

[1] *State Pharmacopoeia of the Russian Federation* (2018). XIV-th Ed. 4:1833.

[2] *European pharmacopoeia* (2013). 8th edition, 1: 1456.

[3] *USP Dietary Supplements Compendium* (2015). 2: 1107.

[4] Wasser, P. S. (2011).Current findings, future trends, and unsolved problems in studies of medicinal mushrooms. *Appl. Microbiol. Biotechnol*, 89: 1323-1332.

[5] Fan, L., Pan, H., Soccol, A. T., Pandey, A. and Soccol, C. R. (2006). Advances in Mushroom Research. *Food Technol. Biotechnol*. 44 (3): 303-311.

[6] Lin, H. C., Hewage, R. T., Lu, Y. C. and Chooi, Y. H. (2019). Biosynthesis of bioactive natural products from Basidiomycota. *Org. Biomol. Chem*, 17: 1027-1036.

[7] Cheong, P. C. H., Tan C. S. and Fung, S. Y. (2018). Application of wild macrofungi as anticancer therapeutics. *Biology of Macrofungi*: 243-274.

[8] Lindequist, U., Niedermeyer, T. H. J. and Julich W. D. (2005). The pharmacological potential of mushrooms. *Evid Based Complement Alternat Med*, 2 (3): 285-299.

[9] De Silva, D.D., Rapior, S., Fons, F., Bahkali, A.H. and Hyde K.D. (2012). Medicinal mushrooms in supportive cancer therapies: an approach to anticancer effects and putative mechanisms of action. *Fungal Diversity*, 55: 1-35.

[10] Sánchez, C. (2017). Bioactives from Mushroom and Their Application. *Food Bioactives*: 23-57.

[11] Chaturvedi, V.K., Agarwal, S., Gupta, K. K., Ramteke, P. W. and Singh M. P. (2018). Medicinal mushroom: boon for therapeutic applications. 3 *Biotech*, 8 (8): 334.

[12] Gerasimenya, V. P., Gumargaliyeva, K. Z., Zakharov, S. V., Milevich, T. I. and Trezvova A. V. (2014). *Extracts of basidiomycetes and their multifunctional medico-biological activity*: 128 p.

[13] Ríos J. L. and Andújar I. (2017). Lanostanoids from Fungi as Potential Medicinal Agents. Fungal Metabolites. *Reference Series in Phytochemistry*: 931-964.

[14] Belova, N.V. (2016). Lanostane triterpenoids and steroids of higher mushrooms. *Advances in Biology & Earth Sciences*, 1 (1): 107-120.

[15] De las Heras, B., Rodríguez, B., Boscá, L., Villar A. M. (2003). Terpenoids: Sources, Structure Elucidation and Therapeutic Potential in Inflammation. *Current Topics in Medicinal Chemistry*, 3: 171-185.

[16] Moses T., Pollier, J., Thevelein, J. M. and Goossens A. (2013). Bioengineering of plant (tri)terpenoids: from metabolic engineering of plants to synthetic biology in vivo and in vitro. *New Phytologist*, 200: 27-43.

[17] Baby, S., Johnson, A. J., Govindan B. (2015). Secondary metabolites from Ganoderma. *Phytochemistry*, 114: 66-101.

[18] Schmidt-Dannert, C. (2014). Biosynthesis of Terpenoid Natural Products in Fungi. *Adv Biochem Eng Biotechnol*, 148: 19-61.

[19] Schmidt-Dannert, C. (2016). Biocatalytic portfolio of Basidiomycota. *Current Opinion in Chemical Biology*, 31: 40-49.

[20] Lu, M. Y., Fan, W. L., Wang, W. F., Chen, T., Tang, Y. C., Chu, F. H., Chang, T. T., Wang, S. Y., Li, M. Y., Chen, Y. H., Lin, Z. S., Yang, K. J., Chen, S. M., Teng, Y. C., Lin, Y. L., Shaw, J. F., Wang, T. F. and Li W. H. (2014). Genomic and transcriptomic analyses of the medicinal fungus Antrodia cinnamomea for its metabolite

biosynthesis and sexual development. *Proc. Natl. Acad.* Sci. U.S.A., 111 (44): 4744-4752.

[21] Sanodiya, B. S., Thakur, G. S., Baghel, R. K., Prasad, G. B. K. S. and Bisen, P. S. (2014). Ganoderma lucidum: A Potent Pharmacological Macrofungus. *Current Pharmaceutical Biotechnology*, 10: 717-742.

[22] Ahmad, M. F. (2018). Ganoderma lucidum: Persuasive biologically active constituents and their health endorsement. *Biomedicine & Pharmacotherapy*, 107: 507-519.

[23] Akihisa, T. (2005). Oxygenated lanostane-type triterpenoids from the fungus Ganoderma lucidum. *J. Nat. Prod*, 68: 559-563.

[24] Kubota, T., Asaka, Y., Miura, I. and Mori, H. (1982). Structures of Ganoderic acid A and B, two new lanostane type bitter triterpenes from Ganoderma lucidum (Fr.) Karst. *Helvetica Chimica Acta*, 65: 611-619.

[25] Hirotani, M., Furuya T. (1985). Ganoderic acid derivatives, highly oxygenated lanostane-type triterpenoids, from Ganoderma lucidum. *Phytochemistry*, 25: 1189-1193.

[26] Kohda, H., Tokumoto, W., Sakamoto, K., Fujii, M., Hirai, Y., Yamasaki K., Komoda, Y., Nakamura, H., Ishihara, S. and Uchida, M. (1985). The Biologically active constituents of Ganoderma lucidum (Fr.) Karst. histamine release-inhibitory triterpenes. *Chem. Pharm. Bull*, 33 (4): 1367-1374.

[27] Chen, S., Yong, T., Zhang, Y., Su, J., Jiao, C. and Xie, Y. (2017). Anti-tumor and anti-angiogenic ergosterols from Ganoderma lucidum. *Frontiers in Chemistry*, 5 (85): 1-12.

[28] Murata, C., Tran, Q.T., Onda, S. and Usuki, T. (2016). Extraction and isolation of ganoderic acid Σ from Ganoderma lucidum. *Tetrahedron Letters*, 57: 5368-5371.

[29] Gao, J. J., Min, B. S., Ahn, E. M., Nakamura, N., Lee, H. K. and Hattori, M. (2002). New triterpene aldehydes, lucialdehydes A-C, from Ganoderma lucidum and their cytotoxicity against murine and human tumor cells. *Chem. Pharm. Bull*, 50 (6): 837-840.

[30] Zhang, C. R., Yang, S. P. and Yue, J. M. (2008). Sterols and triterpenoids from the spores of Ganoderma lucidum. *Natural Product Research*, 22(13): 1137-1142.

[31] Min, B. S., Gao, J. J., Nakamura, N. and Hattori, M. (2000). Triterpenes from the spores of Ganoderma lucidum and their cytotoxicity against meth-A and LLC tumor cells. *Chem. Pharm. Bull*, 48 (7): 1026-1033.

[32] Shiao, M. S. (2003). Natural products of the medicinal fungus Ganoderma lucidum: occurrence, biological activities, and pharmacological functions. *The Chemical Record*, 3: 172-180.

[33] Wei, J. C., Wang, Y. X., Dai, R., Tian, X. G., Sun, C. P., Ma, X. C., Jia, J. M., Zhang, B. J., Huo, X. K. and Wang C. (2017). C27-Nor lanostane triterpenoids of the fungus Ganoderma lucidum and their inhibitory effects on acetylcholinesteras. *Phytochem. Lett*, 210: 263-268.

[34] Chen, B., Tian, J., Zhang, J., Wang, K., Liu, L., Yang, B., Bao, L. and Liu, H. (2017). Triterpenes and meroterpenes from Ganoderma lucidum with inhibitory activity against HMGs reductase, aldose reductase and α-glucosidase. *Fitoterapia*, 120: 6-16.

[35] Zhao, X. R., Zhang, B. J., Deng, S., Zhang, H. L., Huang, S. S., Huo, X. K., Wang, C., Liu, F. and Ma, X. C. (2016). Isolation and identification of oxygenated lanostane-type triterpenoids from the fungus Ganoderma lucidum. *Phytochemistry Letters*, 16: 87-91.

[36] Ma, J. Y., Ye, Q. Y., Zhang, D. C., Cooper, R., Chang, M. N. and Chan, J. Y. (2012). New lanostanoids from the mushroom Ganoderma lucidum. *Journal of Natural Products*, 65: 72-15.

[37] Feng, N., Wei, Y., Feng, J., Tang, Q., Zhang, Z., Zhang, J. and Han W. (2018). Preparative isolation of ganoderic acid S, ganoderic acid T and ganoderol B from Ganoderma lucidum mycelia by high-speed counter-current chromatography. *Biomedical Chromatography*, 32 (10): 4283.

[38] Oludemi, T., Barros, L., Prieto, M. A., Heleno, S. A., Barreiro, M. F. and Ferreira I. C. F. R. (2018). Extraction of triterpenoids and

phenolic compounds from Ganoderma lucidum: optimization study using the response surface methodology. *Food Funct*, 9: 209-226.

[39] Binh, P. T., Thao, N. P., Luyen, N. T., Trang, D. T., Binh, P. T. X., Nguyen, N. P. D., Hiep, N. T., Dang, N. H., Tri, T. M. and Dat, N. T. (2018). Lanostane-type triterpenoids from Ganoderma lucidum and G. multipileum fruiting bodies. *Natural Product Communications*, 13 (11): 1441-1444.

[40] Liang, C., Tian, D., Liu, Y., Li, H., Zhu, J., Li, M., Xin, M. and Xia, J. (2019). Review of the molecular mechanisms of Ganoderma lucidum triterpenoids: ganoderic acids A, C2, D, F, DM, X and Y. *European Journal of Medicinal Chemistry*, 32(10): 8.

[41] Sharma, C., Bhardwaj, N., Sharma, A., Tuli, H. S., Batra, P., Beniwal, V., Gupta, G. K. and Sharma, A. K. (2019). Bioactive metabolites of Ganoderma lucidum: Factors, mechanism and broad spectrum therapeutic potential. *Journal of Herbal Medicine*. https://doi.org/10.1016/j.hermed.2019.100268.

[42] Satria, D., Amen, Y., Niwa, Y., Ashour, A., Allam, A. E. and Shimizu, K. (2019). Lucidumol D, a new lanostane-type triterpene from fruiting bodies of Reishi (Ganoderma lingzhi). *Nat. Prod. Res*, 33 (2): 189-195.

[43] Barbieri, A. (2019). Anticancer and anti-inflammatory properties of Ganoderma lucidum extract effects on melanoma and triple-negative breast cancer treatment. *Nutrients*, 9 (210): 1-9.

[44] Feng, L., Yuan, L., Du, M., Chen, Y., Zhang, M. H., Gu, J. F., He, J. J., Wang, Y. and Cao, W. (2013). Anti-lung cancer activity through enhancement of immunomodulation and induction of cell apoptosis of total triterpenes extracted from Ganoderma lucidum (Leyss. ex Fr.) Karst. *Molecules*, 18 (8): 9966-9998.

[45] Sliva, D., Labarrere, C., Slivova, V., Sedlak, M., Lloyd F. P. Jr. and Ho, N.W. (2002). Ganoderma lucidum suppresses motility of highly invasive breast and prostate cancer cells. *Biochem Biophys Res Commun*, 298: 603-612.

[46] Cheung, W. M. W., Hui, W. S., Chu, P. W., Chiu, S. W. and Ip, N. Y. (2000). Ganoderma extract activates MAP kiases and induces the

neuronal differentiation of rat pheochromocytoma PC12 cells. *FEBS Letters*, 486 (3): 291-296.

[47] Lin, S. B., Li, C. H., Lee, S. S. and Kan, L. S. (2003).Triterpene-enriched extracts from Ganoderma lucidum inhibit growth of hepatoma cells via suppressing protein kinase C, activating mitogen-activated protein kinases and G2-phase cell cycle arrest. *Life Sci*, 72: 2381-2390.

[48] Yuen, J. W. and Gohel, M.D. (2005). Anticancer effects of Ganoderma lucidum: a review of scientific evidence. *Nutr. Cancer*, 53 (1): 11-17.

[49] Xie, J. T., Wang, C. Z., Wicks, S., Yin, J. J., Kong, J., Li, J., Li, Y. C. and Yuan, C. S. (2005). Ganoderma lucidum extract inhibits proliferation of SW 480 human colorectal cancer cells. *Exp. Oncol*, 28 (1): 25-29.

[50] Liu, J., Shimizu, K., Konishi, F., Noda, K., Kumamoto, S., Kurashiki, K. and Kondo R. (2005). Anti-androgenic activities of the triterpenoids fraction of Ganoderma lucidum. *Food Chemistry*, 100 (4): 1691-1696.

[51] Kimura, Y., Taniguchi, M. and Baba K. (2002). Antitumor and antimetastatic effects on liver triterpenoid fractions of Ganoderma lucidum: mechanism of action and isolation of an active substance. *Anticancer Research*, 22 (6A): 3309-3318.

[52] Toth, J. O., Luu, B. and Ourisson, G. (1983).Ganoderic acid T and Z: cytotoxic triterpenes from Ganoderma lucidum (Polyporaceae). *Tetrahedron Letters*, 24: 1081-1084.

[53] Liu, X., Yuan, J. P., Chung, C. K. and Chen X. (2002). Antitumor activity of the sporoderm-broken germinating spores of Ganoderma lucidum. *Cancer Letters*, 182 (2): 155-161.

[54] Zhong, L., Jiang, D. Z. and Wang Q. R. (1999). Effects of Ganoderma lucidum (Leyss ex Fr.) Karst. compound on the proliferation and differentiation of K652 leukemic cells. *Journal of Hunan Medical University*, 24: 521-524.

[55] Lee, S., Park, S., Oh, J. W. and Yang C. H. (1998).Natural inhibitors for protein prenyltransferase. *Planta medica*, 54: 303-308.

[56] Mizushina, Y. Hanashima, L., Yamaguchi, T., Takemura, M., Sugawara, F., Saneyoshi, M., Matsukage, A., Yoshida, S. and Sakaguchi, K. (1998). A mushroom fruiting body-including substance inhibits activities of replicative DNA polymerases. *Biochemical and biophysical research communications*, 249: 17-22.

[57] Mizushina, Y., Takahashi, N., Hanashima, L., Koshino, H., Esumi, Y., Uzawa, J., Sugawara, F. and Sakaguchi, K. (1999). Lucidenic acid O and lactone new terpene inhibitors of eukaryotic DNA polymerases from a basidiomycete Ganoderma lucidum. *Bioorg. Med. Chem*, 7: 2047-2052.

[58] Wu, G. S., Guo, J. J., Bao, J. L., Li, X. W., Chen, X. P., Lu, J. J. and Wang Y.T. (2013). Anti-cancer properties of triterpenoids isolated from Ganoderma lucidum – a review. *Expert Opin. Investig. Drugs*, 22 (8): 1-12.

[59] Suarez-Arroyo, I. J., Rosario-Acevedo, R., Aguilar-Perez, A., Clemente, P. L., Cubano, L. A., Serrano, J., Schneider, R. J. and Martínez-Montemayor, M. M. (2013). Anti-tumor effects of Ganoderma lucidum (Reishi) in inflammatory breast cancer in in vivo and in vitro models. *PLoS ONE*, 8 (2): e57431. doi: 10.1371/journal.pone.0057431.

[60] Cao, Y, Xu, X., Liu, S., Huang, L. and Gu, J. (2018). Ganoderma: A cancer immunotherapy review. *Front. Pharmacol*, 9: 1217. doi: 10.3389/fphar.2018.01217

[61] El-Mekkawy, S., Meselhy, M. R., Nakamura, N., Tezuka, Y., Hattori, M., Kakiuchi, N., Shimotohno, K., Kawahata, T. and Otake, T. (1998). Anti-HIV-1 and anti-HIV-1-protease substances from Ganoderma lucidum. *Phytochemistry*, 49: 1651-1657.

[62] Kim, D. H., Shim, S. B., Kim, N. J. and Jang, I. S. (1999). β-Glucuronidase-inhibitory activity and hepatoprotective effect of Ganoderma lucidum. *Biol. Pharm. Bull*, 22: 162-164.

[63] Berger, A., Rein, D., Kratky. E., Monnard, I., Hajjaj, H., Meirim, I., Piguet-Welsch, C., Hauser, J., Mace, K. and Niederberger P. (2004). Cholesterol-lowering properties of Ganoderma lucidum in vitro, in

vivo, and in hamsters and minipigs. *Lipids in health and disease*, 3: 2. doi: 10.1186/1476-511X-3-2.

[64] Rivera, A., Benavides, O.L. and Rios-Motta, J. (2009). (22E)-Ergosta-6,22-diene-3β,5α,8α-triol: A new polyhydroxysterol isolated from Lentinus edodes (Shiitake). *Natural Product Research*, 23(1): 293-300.

[65] Chen, J., Wei, S.L. and Gao, K. (2015). Chemical consituents and antibacterial activities of compounds from Lentinus edodes. *Chemistry of Natural Compounds*, 51 (3): 592-594.

[66] Sun, Y., Zhang, M. and Fang Z. (2019). Efficient physical extraction of active constituents from edible fungi and their potential bioactivities: A review. *Trends in Food Science & Technology*. https://doi.org/10.1016/j.tifs.2019.02.026.

[67] Wojtas, M., Bieñkowski, T., Tateyama, S., Sagami, H., Chojnacki, T., Danikiewicz, W. and Swiezewska, E. (2004). Polyisoprenoid alcohols from the mushroom Lentinus edodes. *Chemistry and Physics of Lipids*, 130: 109-115.

[68] Yaoita Y., Endo, M., Tani, Y., Machida, K., Amemiya, K., Furumura, K. and Kikuchi M. (1999). Sterol Constituents from Seven Mushrooms. *Chem. Pharm. Bull*, 47 (6): 847-851.

[69] Ohnuma, N., Amemiya, K., Kakuda, R., Yaoita, Y. and Machida K. (2000). Sterol constituents from two edible mushrooms, Lentinula edodes and Tricholoma matsutake. *Chem Pharm Bull*, 48 (5): 749-751.

[70] Wang, Q., Cheng, J., Wang, L., Yan, S., Wang, R., Zhang, H., Shao, H. and Yang, X. (2018). Valorization of spent shiitake substrate for recovery of antitumor fungal sterols by ultrasound-assisted extraction. *J Food Biochem*, 42. https://doi.org/

[71] 10.1111/jfbc.12602.

[72] Yaoita Y., Kikuchi, M. and Machida, K. (2014). Terpenoids and sterols from some Japanese mushrooms. *Natural Product Communications*, 9 (3): 419-426.

[73] Yaoita, Y., Amemiya, K., Ohnuma, H., Furumura, K., Masaki, A., Matsuki, T. and Kikuchi, M. (1998). Sterol constituents from five edible mushrooms. *Chem. Pharm. Bull*, 46: 944-950.

[74] Drori, A., Shabat, Y., Ya'acov, A. B., Danay, O., Levanon, D., Zolotarov, L. and Ilan, Y. (2016). Extracts from Lentinula edodes (Shiitake) edible mushrooms enriched with vitamin D exert an anti-inflammatory hepatoprotective effect. *J Med Food*, 19 (4): 383-389.

[75] Morales, D., Gil-Ramirez, A., Smiderle, F. R., Piris, A. J., Ruiz-Rodriguez, A. and Soler-Rivas, C. (2017). Vitamin D-enriched extracts obtained from shiitake mushrooms (Lentinula edodes) by supercritical fluid extraction and UV-irradiation. *Innov. Food Sci. Emerg. Technol*, 41: 330-336.

[76] Kitzberger, C. S. G., Smânia, A., Pedrosa, J. and Ferreira, S. R. S. (2007). Antioxidant and antimicrobial activities of shiitake (Lentinula edodes) extracts obtained by organic solvents and supercritical fluids. *Journal of Food Engineering*, 80: 631-638.

[77] Finimundy, T. C., Dillon, A. J. P., Henriques, J. A. P. and Ely, M.R. (2014). A Review on general nutritional compounds and pharmacological properties of the Lentinula edodes mushroom. *Food and Nutrition Sciences*, 5: 1095-1105.

[78] Muszynska, B., Pazdur, P., Lazur, J. and Sulkowska-Ziaja, K. (2017). Lentinula edodes (Shiitake) – biological activity. Medicina Internacia Revuo - *International Medicine Review*, 27 (108): 189-195.

[79] Rahman, M. A., Abdullah, N. and Aminudin, N. (2018). An assessment of in vitro anti-atherosclerotic bio-functionality. *Saudi Journal of Biological Sciences*, 25: 1515-1523.

[80] Zheng, Y. and Yang, X. W. (2008). Poriacosones A and B: two new lanostane triterpenoids from Poria cocos. *Journal of Asian Natural Products Research*, 10 (7): 640-646.

[81] Zheng, Y. and Yang, X. W. (2008). Two new lanostane triterpenoids from Poria cocos. *Journal of Asian Natural Products Research*, 10 (4): 289-292.

[82] Zou, Y. T., Long, F., Wu, C. Y., Zhou, J., Zhang, W., Xu, J., Zhang, Y. Q. and Li, S. L. (2019). A dereplication strategy for identifying triterpene acid analogues in Poria cocos by comparing predicted and acquired UPL-ESI-QTOF-MS/MS data. *Phytochemical Analysis*, 30: 292-310.

[83] Yang, P. F., Hua, T., Wang, D., Zhao, Z. W., Xi, G. L. and Chen, Z. F. (2019). Phytochemical and chemotaxonomic study of Poria cocos (Schw.) Wolf. *Biochemical Systematics and Ecology*, 83: 54-56.

[84] Lee, S., Lee, D., Lee, S. O., Ryu, J. Y., Choi, S. Z., Kang, K. S. and Kim, K. H. (2017). Anti-inflammatory activity of the sclerotia of edible fungus, Poria cocos Wolf and their active lanostane triterpenoids. *Journal of Functional Foods*, 32: 27-36.

[85] Chen, T., Hua, L., Chou, G., Mao, X. and Zou, X. (2018). A Unique naphthone derivative and a rare 4,5-seco-lanostane triterpenoid from Poria cocos. *Molecules*, 23 (10): 2508. doi:10.3390/molecules23102508.

[86] Zhu, L., Xu, J., Zhang, S., Wang, R., Huang, Q., Chen, H., Dong, X. and Zhao, Z. (2018). Qualitatively and quantitatively comparing secondary metabolites in three medicinal parts derived from Poria cocos (Schw.) Wolf using UHPLC-QTOF-MS/MS-based chemical profiling. *Journal of Pharmaceutical and Biomedical Analysis*, 150: 278-286.

[87] Zhang, G., Wang, H., Xie, W., Wang, Q., Wang, X., Wang, C., Du, Y., Huo, C. and Wang, Q. (2019). Comparison of triterpene compounds of four botanical parts from Poria cocos (Schw.) wolf using simultaneous qualitative and quantitative method and metabolomics approach. *Food Research International*, 121: 666-677.

[88] Zhu, L. X., Xu, J., Wang, R. J., Li, H. X., Tan, Y. Z., Chen, H. B., Dong, X. P. and Zhao, Z. Z. (2018). Correlation between quality and geographical origins of Poria cocos revealed by qualitative fingerprint profiling and quantitative determination of triterpenoid acids. *Molecules*, 23: 2200. doi:10.3390/molecules23092200.

[89] Dong, H. J., Xue, Z. Z., Geng, Y. L., Wang, X. and Yang, B. (2017). Lanostane triterpenes isolated from epidermis of Poria cocos. *Phytochemistry Letters*, 22: 102-106.

[90] Zhou, L., Zhang, Y., Gapter, L. A., Ling, H., Agarwal, R. and Ng, K. (2008). Cytotoxic and anti-oxidant activities of lanostane-type triterpenes isolated from Poria cocos. *Chem. Pharm. Bull*, 56 (10): 1459-1462.

[91] Lai, K. H. (2016). Cytotoxic lanostanoids from Poria cocos. *J. Nat. Prod*, 79: 2805-2813.

[92] Akihisa, T., Uchiyama, E., Kikuchi, T., Tokuda, H., Suzuki, T. and Kimura, Y. (2009). Anti-tumor-promoting effects of 25-methoxyporicoic acid A and other triterpene acids from Poria cocos. *J. Nat. Prod*, 72: 1786-1792.

[93] Ukiya, M., Akihisa, T., Tokuda, H., Hirano, M., Oshikubo, M., Nobukuni, Y., Kimura, Y., Tai, T., Kondo, S. and Nishino, H. (2002). Inhibition of tumor-promoting effects by poricoic acids G and H and other lanostane-type triterpenes and cytotoxic activity of poricoic acids A and G from Poria cocos. *J. Nat. Prod*, 65: 462-465.

[94] Scheepers, A., Joost, H. G. and Schurmann, A. (2004). The glucose transporter families SGLT and GLUT: molecular basis of normal and aberrant function. *J Parenter Enteral Nutr*, 28: 364-371.

[95] Sato, M., Tai, T., Nunoura, Y., Yajima, Y., Kawashima, S. and Tanaka, K. (2002). Dehydrotrametenolic acid induces preadipocyte differentiation and sensitizes animal models of noninsulindependent diabetes mellitus to insulin. *Biol Pharm Bull*, 25: 81-86.

[96] Akihisa, T., Nakamura, Y., Tokuda, H., Uchiyama, E. Suzuki, T., Kimura, Y., Uchikura, K. and Nishino, H. (2007). Triterpene acids from Poria cocos and their anti-tumor promoting effects. *J. Nat. Prod*, 70: 948-953.

[97] Ukiya, M., Akihisa, T., Tokuda, H., Hirano, M., Oshikubo, M., Nobukuni, Y., Kimura, Y., Tai, T., Kondo, S. and Nishino, H. (2002). Inhibition of tumor-promoting effects by poricoic acids G and H and other lanostane-type triterpenes and cytotoxic activity of

poricoic acids A and G from Poria cocos. *J Nat Prod*, 65 (4): 462-465.

[98] Gapter, L., Wang, Z., Glinski, J. and Ng, K. (2005). Induction of apoptosis in prostate cancer cells by pachymic acid from Poria cocos. *Biochemical and biophysical research communications*, 332 (4): 1153-1161.

[99] Sun, K. X. and Xia, H. W. (2018). Pachymic acid inhibits growth and induces cell cycle arrest and apoptosis in gastric cancer SGC-7901 cells. *Oncology letters*, 16 (2): 2517-2524.

[100] Cheng, S., Eliaz, I., Lin, J. and Sliva, D. (2013). Triterpenes from Poria cocos suppress growth and invasiveness of pancreatic cancer cells through the downregulation of MMP-7. *International journal of oncology*, 42 (6): 1869-1874.

[101] Kikuchi, T., Uchiyama, E., Ukiya, M., Tabata, K., Kimura, Y., Suzuki, T. and Akihisa, T. (2011). Cytotoxic and apoptosis-inducing activities of triterpene acids from Poria cocos. *J. Nat. Prod*, 74 (2): 137-144.

[102] Ling, H., Zhou, L., Jia, X., Gapter, L. A., Agarwal, R. and Ng, K. (2009). Polyporenic acid C induces caspase-8-mediated apoptosis in human lung cancer A549 cells. *Molecular carcinogenesis*, 48 (6): 498-507.

[103] Keller, A. C., Mallard, M. P. and Hostettmann, K. (1996). Antimicrobial steroids from the fungus Fomitopsis pinicola. *Phytochemistry*, 41 (4): 1041-1046.

[104] Rosecke, J. and Konig, W. A. (1999). Steroids from the fungus Fomitopsis pinicola. *Phytochemistry*, 52 (8): 1621-1627.

[105] Rosecke, J. and Konig, W. A. (2000). Constituents of various wood-rotting basidiomycetes. *Phytochemistry*, 54 (6): 603-610.

[106] Petrova, A., Popov, S., Gjosheva, M. and Bankova, V. (2007). A new triterpenic alcohol from Fomitopsis pinicola. *Natural Product Research*, 21(5): 401-405.

[107] Isaka, M., Chinthanom, P., Suvannakad, R. Thummaru-kcharoen, T., Feng, T. and Liu, J. K. (2019). Fomitopsins I and J, 24-methyl-

lanostane triterpenoids from fruiting bodies of the wood-rot basidiomycete Fomitopsis sp. *Phytochemistry Letters*, 29: 178-181.

[108] Yoshikawa, K., Inoue, M., Matsumoto, Y., Sakakibara, C., Miyataka, H., Matsumoto, H. and Arihara, S. (2005). Lanostane triterpenoids and triterpene glycosides from the fruit body of Fomitopsis pinicola and their inhibitory activity against COX-1 and COX-2. *J. Nat. Prod*, 68 (1): 69-73.

[109] Liu, X .T., Winkler, A. L., Schwan, W. R., Volk, T. J., Rott, M. and Monte, A. (2010). Antibacterial compounds from mushrooms II: lanostane triterpenoids and an ergostane steroid with activity against Bacillus cereus isolated from Fomitopsis pinicola. *Planta Medica*, 76 (5): 464-466.

[110] Ren, G., Liu, X. Y., Zhu, H. K., Yang, S. Z. and Fu, C. X. (2006). Evaluation of cytotoxic activities of some medicinal polypore fungi from China. *Fitoterapia*, 77 (5): 408-410.

[111] Gao, Y., Wang, P., Wang, Y., Wu, L., Wang, X., Zhang, K. and Liu, Q. (2017). In vitro and in vivo activity of Fomitopsis pinicola (Sw. Ex Fr.) Karst chloroform (Fpkc) extract against S180 tumor cells. *Cell Physiol Biochem*, 44 (5): 2042-2056.

[112] Hong, R., Shen, M. H., Xiem X.H. and Ruan, S. M. (2012). Inhibition of breast cancer metastasis via PITPNM3 by pachymic acid. *Asian Pac J Cancer Prev*,13 (5): 1877-1880.

[113] Du, Y. C., Chang, F. R., Wu, T. Y., Hsu, Y. M., El-Shazly, M., Chen, C. F., Sung, P. J., Lin, Y. Y., Lin, Y. H., Wu, Y. C. and Lu, M. C. (2012). Antileukemia component, dehydroeburicoic acid from Antrodia camphorata induces DNA damage and apoptosis in vitro and in vivo models. *Phytomedicine*, 19(8-9): 788-796.

[114] Popova, M., Trusheva, B., Gyosheva, M., Tsvetkova, I. and Bankova, V. (2009). Antibacterial triterpenes from the threatened wood-decay fungus Fomitopsis rosea. *Fitoterapia*, 80(5): 263-266.

[115] Alresly, Z., Lindequist, U., Lalk, M., Porzel, A., Arnold, N. and Wessjohann L. A. (2016). Bioactive triterpenes from the fungus Piptoporus betulinus. *Rec. Nat. Prod*, 10 (1): 103-108.

[116] Sułkowska-Ziaja, K., Szewczyk, A., Galanty, A., Gdula-Argasińska, J. and Muszyńska, B. (2018). Chemical composition and biological activity of extracts from fruiting bodies and mycelial cultures of Fomitopsis betulina. *Molecular Biology Reports*, 45 (6): 2535-2544.

[117] Tohtahon, Z, Xue, J., Han, J., Liu, Y., Hua, H. and Yuan, T. (2017). Cytotoxic lanostane triterpenoids from the fruiting bodies of Piptoporus betulinus. *Phytochemistry*, 143: 98-103.

[118] Kamo, T., Asanoma, M., Shibata, H. and Hirota, M. (2017). Anti-inflammatory lanostane-type triterpene acids from Piptoporus betulinus. *J. Nat. Prod*, V. 66 (8): 1104-1106.

[119] Khalilov, Q., Li, L., Liu, Y., Tohtahon, Z., Chen, X., Aisa, H. A. and Yuan, T. (2019). Piptolinic acids F-J, five new lanostane-type triterpenoids from Piptoporus betulinus. *Natural Product Research*, 33 (21): 3044-3051.

[120] Wangun, H. V., Berg, A., Hertel, W., Nkengfack, A. E. and Hertweck, C. (2004). Anti-inflammatory and antihyaluronate lyase activities of lanostanoids from Piptoporus betulinus. *J Antibiot*, 57 (11): 755-758.

[121] Kandefer-Szerszeń, M. and Kawecki, K. (1974). Ether extracts from the fruiting body of Piptoporus betulinus as interference inducers. Acta Microbiol Pol Series A: *Microbiologia Generalis*,.6 (2): 197-200.

[122] Keller, C., Maillard, M., Keller, J. and Hostettmann, K. (2002). Screening of European fungi for antibacterial, antifungal, larvicidal, molluscicidal, antioxidant and free-radical scavenging activities and subsequent isolation of bioactive compounds. *Pharm Biol*, 40 (7): 518-525.

[123] Pleszczyńska, M., Lemieszek, M. K., Siwulski, M., Wiater, A., Rzeski, W. and Szczodrak, J. (2017). Fomitopsis betulina (formerly Piptoporus betulinus): the Iceman's polypore fungus with modern biotechnological potential. *World J Microbiol Biotechnol*, 33:83.doi: 10.1007/s11274-017-2247-0.

[124] Pleszczyńska, M., Wiater, A., Siwulski, M., Lemieszek, M. K., Kunaszewska, J., Kaczor, J., Rzeski, W., Janusz, G. and Szczodrak,

J. (2016). Cultivation and utility of Piptoporus betulinus fruiting bodies as a source of anticancer agents. *World J Microbiol Biotechnol*, 32 (9): 151. doi: 10.1007/s11274-016-2114-4.

[125] Kaczor, J., Klecha, I. M., Rzeski, W. Paduch, R., Zdzisińska, B. Pożarowski, P. and Kandefer-Szerszeń, M. (2004). Extract from Piptoporus betulinus Bull. Fr. supress human tumor cell growth. *Postępy Fitoter*, 2: 62-66.

[126] Tomasi, S., Lohézic-Le Dévéhat, F., Sauleau, P., Bézivin, C. and Boustie, J. (2004). Cytotoxic activity of methanol extracts from Basidiomycete mushrooms on murine cancer cell lines. *Pharmacy*, 59 (4): 290-293.

[127] Ludwiczak, R .S. and Wrecino, U. (1962). Research on the chemical components of Inonotus obliquus. V. Further research on the structure of inotodiol /R.S. Ludwiczak, U. Wrecino// *Polish Journal of Chemistry*, 36: 497-502.

[128] Kier, L. B. (1961). Triterpenes of Poria oblique. *J. Pharm. Sci*, 50 (6): 471-474.

[129] Lovyagina, E. V. and Shivrina, A. H. (1962). About steroid compounds of chaga mushroom. *Biochemistry*, 27 (5): 794-800.

[130] Lovyagina, E. V. and Shivrina, A. H. (1965). *Characteristics of wood-destroying fungi on the content of steroid compounds in them. Feed proteins and physiologically active substances for livestock*: 59-64.

[131] Kahlos, K., Hiltunen, R. and Schantz, M. V. (1984). 3β-hydroxy-lanosta-8,24-dien-21-al,a new triterpene from Inonotus obliquus. *Planta Medica*, 50 (2): 197-198.

[132] Kahlos, K. and Hiltunen, R. (1986). 3β,22-Dihydroxylanosta-7,9(11),24-triene, a new minor compound from Inonotus obliquus. *Planta Medica*, 52: 495-496.

[133] Kahlos, K. Hiltunen, R. and Kangas, L. (1986). Antitumor activity of triterpenes in Inonotus obliquus. Planta Medica, 52: 554.

[134] Shin, Y., Tamai, Y. and Terazawa, M. (2000). Chemical constituents of Inonotus obliquus. I. A new triterpene, 3β-hydroxy-8,24-dien-

lanosta-21,23-lactone from sclerotium. *Euras. J. Forest Res*, 1: 43-50.

[135] Shin, Y., Tamai, Y. and Terazawa, M. (2000). Chemical constituents of Inonotus obliquus. III. A new triterpene, 3β,22,25-trihydroxy-lanosta-8-ene from sclerotium. *Int. J. of Med. Mush*, 2 (3): 201-207.

[136] Shin, Y., Tamai, Y. and Terazawa, M. (2001). Chemical constituents of Inonotus obliquus. II. A new triterpene, 21,24-cyclopentalanosta-3b,21,25-triene from sclerotium. *J. Wood Science*, 47 (4): 313-316.

[137] Shin, Y., Tamai, Y. and Terazawa, M. (2001). Triterpenoids, steroids and a new sesquiterpen from Inonotus obliquus (Pers.: Fr.) Bond. et Sing. *Int. J. of Med. Mush*, 3: 250.

[138] He, J., Feng, X., Zhao, B. and Lu, Y. (2000). Triterpenoids from Fuscoporia oblique. *Chin. Chem. Letters*, 11 (1): 45-58.

[139] He, J., Feng, X., Zhao, B. and Lu, Y. (2001). Three new triterpenoids from Fuscoporia oblique. *J. Asian Nat. Prod. Research*, 3: 55-61.

[140] Kim, E. J., Lee, Y. J., Shim, H. K. and Yoon J. H. (2006). A study on the mechanism by which the aqueous extract of Inonotus obliquus induces apoptosis and inhibits proliferation in HT-29 human colon cancer cells. *J. Korean Soc. Food Sci. Nutr*, 35 (5):516-523.

[141] Nakata, T., Taji, S. and Yamada T. (2007). Structure determination of inonotsuoxides A and B and in vivo anti-tumor promoting activity of inotodiol from the sclerotia of Inonotus obliquus. *Bioorg. Med. Chem*, 15 (1): 257-264.

[142] Taji, S. Yamada, T. and Tanaka, R. (2007). Three new lanostane triterpenoids from Inonotus obliquus. *Helvetica Chimica Acta*, 90: 2047-2057.

[143] Taji, S. Yamada, T. and Tanaka, R. (2008). Three new lanostane triterpenoids, Inonotsutriols A, B, and C, from Inonotus obliquus. *Helvetica Chimica Acta*, 91 (8): 1513-1524.

[144] Taji, S. Yamada, T., Wada, S., Tokuda, H., Sakuma, K. and Tanaka, R. (2008). Lanostane-type triterpenoids from the sclerotia of

Inonotus obliquus possessing antitumor promoting activity. *Eur. J. Med. Chem.* 43 (11): 2373-79.

[145] Nakata, T., Taji, S. and Yamada, T. (2009). New lanostane triterpenoids, inonotsutriols D and E from Inonotus obliquus. *Bulletin of Osaka University of Pharmaceutical Sciences*, 3: 53-64.

[146] Zhong, X., Ren, K., Lu, S. J., Yang, S. Y. and Sun, D. Z. (2009). Progress of research on Inonotus obliquus. *Chin. J. Integr. Med*, 15 (2): 156-60.

[147] Handa, N., Yamada, T. and Tanaka, R. (2010). An unusial lanostane-type triterpenoid, spiroinonotsuoxodiol, and other triterpenoids from Inonotus obliquus. *Phytochemistry*, 71 (14-15): 1774-1779.

[148] Handa, N., Yamada, T. and Tanaka, R. (2012). Four new lanostane-type triterpenoids from Inonotus obliquus. *Phytochemistry Letters*, 5 (3): 480-485.

[149] Kim, Y. J., Park, J., Min, B.S. and Shin, S. H. (2011). Chemical constituents from the sclerotia of Inonotus obliquus. *J. Korean Soc. Appl. Biol. Chem*, 54 (2): 287-294.

[150] Zheng, W., Miao, K., Liu, Y., Zhao, Y., Zhang, M., Pan, S. and Dai, Y. (2010). Chemical diversity of biologically active metabolites in the sclerotia of Inonotus obliquus and submerged culture strategies for up-regulating their production. *Appl. Microbiol. Biotechnol*, 87 (4): 1237-1254.

[151] Liu, C., Zhao, C., Pan, H. H., Kang, J. and Yu, X. (2014). Chemical constituents from Inonotus obliquus and their biological activities. *J. Nat. Prod*, 77 (1): 35-41.

[152] Shin, Y., Tamai, Y. and Terazawa, M. (2001). Chemical constituents of Inonotus obliquus IV. Triterpene and steroids from cultured mycelia. *Euras. J. Forest Res*, 2: 27-30.

[153] Gao, Y., Xu, H., Lu, Z. and Xu, Z. (2009). Quantitative determination of steroids in the fruiting bodies and submerged-cultured mycelia of Inonotus obliquus. *Chin. J. Chromatography*, 29 (6): 745-749.

[154] Nakamura, S., Iwami, J., Matsuda, H., Mizuno, S. and Yoshikawa M. (2009). Absolute stereostructures of inoterpenes A-F from sclerotia of Inonotus obliquus. *Tetrahedron*, 65 (12): 2443-2450.

[155] Baeka, J., Roh, H. S., Baek, K. H., Lee, S., Lee, S., Song, S. S. and Kim, K. H. (2018). Bioactivity-based analysis and chemical characterization of cytotoxic constituents from Chaga mushroom (Inonotus obliquus) that induce apoptosis in human lung adenocarcinoma cells. *Journal of Ethnopharmacology*, 224: 63-75.

[156] Zhao, F., Mai, Q., Ma, J., Xu, M., Wang, X., Cui, T., Qiu, F. and Han, G. (2015). Triterpenoids from Inonotus obliquus and their antitumor activities. *Fitoterapia*, 101: 34-40.

[157] Zhao, F., Xia, G., Chen, L., Zhao, J., Xie, Z., Qiu, F. and Han, G. (2016). Chemical constituents from Inonotus obliquus and their antitumor activities. *J Nat Med*, 70 (4): 721-730.

[158] Tanaka, R., Toyoshima, M. and Yamada, T. (2011). New lanostane-type triterpenoids, inonotsutriols D, and E, from Inonotus obliquus. *Phytochem. Lett*, 4 (3): 328-332.

[159] Berezina, M. P. (1959). Physiological studies of patients with stage IV cancer during treatment with chaga. *Chaga and its therapeutic use in stage IV cancer*: 143-159.

[160] Bulatov, P. K. (1959). Physiological studies of patients with stage IV cancer during treatment with chaga. *Chaga and its therapeutic use in stage IV cancer*: 261-270.

[161] Bulatov, P. K. and Martynova E. Y. (1961). Physiological studies of patients with stage IV cancer during treatment with chaga. *A comprehensive study of the physiologically active substances of lower plants*: 247-253

[162] Martynova, E. Y. (1959). Physiological studies of patients with stage IV cancer during treatment with chaga. *Chaga and its therapeutic use in stage IV cancer*: 271-278.

[163] Berezina, M. P., Bulatov, P. K. and Vandysheva F. Y. (1959). Influence of chaga on the development of MOP sarcoma in rats. *Chaga and its therapeutic use in stage IV cancer*: 119-126.

[164] Krotkina, N. A. (1959). Effect of chaga mushroom on transplantable tumors in rats. Chaga and its therapeutic use in stage IV cancer: 114-118.

[165] Rudakov, V. F. The effect of chaga on the metastasis of Ehrlich tumor in mice. Higher fungi and their physiologically active compounds: 52-54.

[166] Burczyk, J., Gawron, A., Slotwinska, M., Smietana, B. and Terminska K. (1996). Antimitotic activity of aqueous extracts of Inonotus obliquus. *Boll. Chim. Farm*, 135:. 306-309.

[167] Shashkina, M. Ja., Shashkin, P. N.and Sergeev, A. V. (2005). Chaga in oncology. *Russian therapeutic journal*, 4 (4): 59-72.

[168] Youn, M. J., Kim, J. K., Park, S. Y., Kim, Y., Kim, S. J., Lee, J. S., Chai, K. Y., Kim, H. J., Cui, M. X., So, H. S., Kim, K. Y. and Park, R. (2008). Chaga mushroom (Inonotus obliquus) induces G0/G1 arrest and apoptosis in human hepatoma HepG2 cells. *World J Gastroenterol*, 14 (4): 511-517.

[169] Kim, S. H., Shin, J. A., Park J. H. and Song K.Y. (2013). Inhibition effect of cell proliferation and apoptosis by Inonotus obliquus in human glioblastoma U-87 MG cells. *J. Korean Soc. Food Sci. and Nutr*, 42 (7): 1022-1028.

[170] Yoon, T. J., Moon, W. K. and Lee, K. H. (2007). Antitumor metastasis activity by extracts of Inonotus obliquus. *Korean J. Food and Nutr*, 20 (3): 253-258.

[171] Lee, S. H., Hwang, H. S. and Yun, J. W. (2009). Antitumor activity of water extract of a mushroom, Inonotus obliquus, against HT-29 human colon cancer cells. *Phytother. Res*, 23 (12): 1784-1789.

[172] Koyama, T., Gu, Y. and Taka, A. (2008). Fungal medicine, Fuscoporia obliqua, as a traditional herbal medicine: its bioactivities, in vivo testing and medicinal effects. *Asian Biomedicine*, 2(6): 459-469.

[173] Chen, C., Zheng, W., Ga, X. and Xiang, X. (2007). Aqueous extract of Inonotus obliquus (Fr.) Pilát (Hymenochaetaceae) significantly inhibits the growth of sarcoma 180 by inducing apoptosis. *American J. Pharm. and Tox*, 2(1): 10-17.

[174] Illana-Esteban, C. (2011). Interes medicinal del "Chaga" (Inonotus obliquus). *Bol. Soc. Micol*, 35: 175-185.

[175] Youn, M., Kim, J. K., Park, S. Y., Kim, Y., Park, C., Kim, E. S., Park, K. I., So, H. S. and Park, R. (2009). Potential anticancer properties of the water extract of Inonotus obliquus by induction of apoptosis in melanoma B16-F10 cells. *J. Ethnopharmacology*, 121 (2): 221-228.

[176] Ryzhova, G. L., Kravtsova, S. S., Matasova, S. A., Gribel', N. V., Pashynsky, V. G. and Duchko, K. A. (1997). Chemical and pharmacological properties of dry extract of chaga. *Chemical and pharmaceutical journal*, 10: 44-47.

[177] Park, E., Jeon, K. and Byun, B. H. (2005). Ethanol extract of Inonotus obliquus shows antigenotoxic effect on hydrogen peroxide induced DNA damage in human lymphocytes. *Cancer Prevention Research*, 10 (1): 54-59.

[178] Hackathorn, A., Boisvert, E., Masterjohn, K., Bellizzi, H. Flanders, T., Mitchell, D., Bryan, B. and Frink, D. (2013). Ethanol extracts from the mushroom Inonotus obliquus (chaga) alter melanoma cell viability and morphology. *BIOS*, 84 (12): 74-81.

[179] Sun, Y., Yin, T., Chen, X. H. and Zhang, G. (2013). In vitro antitumor activity and structure characterization of ethanol extracts from wild and cultivated chaga medicinal mushroom, Inonotus obliquus. *Int. J. Med. Mush*, 13 (2): 121-130.

[180] Nomura, M., Takahashi, T., Uesugi, A., Tanaka, R. and Kobayashi, S. (2013). Inotodiol, a lanostane triterpenoid, from Inonotus obliquus inhibits cell proliferation through caspase-3-dependent apoptosis. *Anticancer research*, 28 (5A): 2691-2696.

[181] Ma, L., Chen, H., Dong, P. and Lu, X. (2013). Anti-inflammatory and anticancer activities of extracts and compounds from the mushroom Inonotus obliquus. *Food Chemistry*, 139 (1-4): 503-508.

[182] Nikitina, S. A., Habibrakhmanova, V. R. and Sysoeva, M. A. (2016). Chemical composition and biological activity of triterpenes and steroids of chaga mushroom. *Biochemistry (Moscow) Supplement Series B: Biomedical Chemistry*, 10 (1): 63-69.

[183] Zhukovich, E. N., Semenova, M. Yu., Sharikova, L. A. and Pribytkova, T. F. (2010). Standardization of chaga tincture and befungin. *Pharmaceutical Chemistry Journal*, 44 (3): 144-146.

[184] Mazurkiewicz, W., Rydel, K., Pogocki, D., Lemieszek, M. K., Langner, E. and Rzeski, W. (2010). Separation of an aqueous extract Inonotus obliquus (chaga). A novel look at the efficiency of its influence on proliferation of A549 human lung carcinoma cells. *Acta Poloniae Pharmaceutica*, 67 (4): 397-406.

[185] Burmasova, M. A. and Sysoeva, M. A. (2017). Chemical composition and biological activity of the BuOH fraction from chaga melanin. *Pharmaceutical Chemistry Journal*, 51 (4): 292-294.

[186] Nikitina, S. A., Habibrakhmanova, V. R., Sysoeva, M. A. and Nosova, F. F. (2015). Investigation of chaga melanin. II. Composition of hydrocarbon fraction. *Pharmaceutical Chemistry Journal*, 49 (8): 543-546.

[187] Nikitina, S. A. (2015). The composition and properties of triterpenoids, steroid and related compounds of Inonotus obliquus: dis. cand. chemical of science. *Kazan*: 139.

[188] Sarnthima, R., Khammaung, S. and Sa-ard, P. (2017). Culture broth of Ganoderma lucidum exhibited antioxidant, antibacterial and a-amylase inhibitory activities. *J Food Sci Technol*, 54 (11): 3724-3730.

[189] Ye, L., Liu, S., Xie, F., Zhao, L. and Wu, X. (2018). Enhanced production of polysaccharides and triterpenoids in Ganoderma lucidum fruit bodies on induction with signal transduction during the fruiting stage. *PLoS ONE*, 13 (4): e0196287. doi: 10.1371/journal.pone.0196287.

[190] Cheong, P. C. H., Tan, C. S. and Fung, S. Y. (2019). Medicinal mushrooms: cultivation and pharmaceutical impact. *Biology of Macrofungi*: 287-304.

[191] Moses, T., Pollier, J., Thevelein, J. M. and Goossens, A. (2013). Bioengineering of plant (tri)terpenoids: from metabolic engineering of plants to synthetic biology in vivo and in vitro. *New Phytologist*, 200 (1): 27-43.

[192] Feofilova, E. P. (1998). Modern trends in the study of biologically active substances of basidiomycetes. *Applied Biochemistry and Microbiology*, 34 (6): 597-608.

[193] Korsun, V. F., Krasnopolskaya, L. M., Korsun, E. V. and Avhukova, M.A. (1998). Antitumor properties of fungi. Moscow: "*Meilyer*":210.

In: A Closer Look at Hormones
Editor: Alexander V. Knyazev
ISBN: 978-1-53617-254-6

Chapter 2

RED ALGAE: POTENTIAL SOURCE OF PHYTOHORMONES

R. Ukun M. S. Soedjanaatmadja*, Safri Ishmayana, Zakiyah Muntari and Saadah D. Rachman

Department of Chemistry, Faculty of Mathematics and Natural Sciences, Universitas Padjadjaran, Sumedang, West Java, Indonesia

ABSTRACT

Marine macroalgae, or seaweeds, are plant-like organisms that generally live attached to rock or other hard substrata in coastal areas. They belong to three different groups, empirically distinguished since the mid-nineteenth century on the basis of thallus color: brown algae (phylum Ochrophyta, class Phaeophyceae), red algae (phylum Rhodophyta), and green algae (phylum Chlorophyta, classes Bryopsidophyceae, Chlorophyceae, Dasycladophyceae, Prasinophyceae, and Ulvophyceae). Red algae are a low level form of thallus plants that have no leaves, roots and true stems, mostly growing on rocks in coral reefs. Red algae contain organic compounds such as polysaccharides, vitamins, as well as other

* Corresponding Author's Email: ukun@unpad.ac.id.

bioactive compounds. In addition, they contain phytohormones which potentially can be used as a biostimulant in fertilizer application. Although the exact physiological roles remain to be elucidated, the essential and bioactive forms of five "old timer phytohormones" such as auxin, ethylene, abscisic acid, cytokinins, and gibberellins have been detected in red algae. Our recent investigation identified auxin, cytokinins and gibberellins from red algae (*Gracilaria coronopifolia* and *Eucheuma cotonii*) and investigated their effects on the growth of rice (*Oryza sativa*). *Trans*-zeatin (cytokinin) from *G. coronopifolia* was 6.26×10^{-2} mg/g dry weight, while indole acetic acid (auxin) and GA3 (gibberellin) from *E. cotonii* was 5.34×10^{-3} and 1.32 mg/g dry weight, respectively.

Keywords: red algae, phytohormones, *G. coronopifolia*, *E. cotonii*

INTRODUCTION

Phytohormones (also known as plant hormones) are signaling molecules biosynthesized in plants which have regulatory roles in plant physiology, controlling architecture, developmental stage, and stress responses, and as such, are in principle, universal among plants. The active concentrations of phytohormones in plants are very low, ranging from 10^{-10} to 10^{-6} mol.kg^{-1} [1]. Traditionally, there are five major classes of so called "the classic" or "the old timer" phytohormones, including auxin, cytokinins, gibberellins (GAs), abscisic acid (ABA) and ethylene, each has different characteristics and physiological influences [2]. Recent studies have identified other phytohormones, including brasinosteroids, jasmonic acid (Jas), salicylic acid (SA) and strigolactones [1, 3]. A wide variety of plant hormones are important for the growth and development of plants, mutually interacting in diverse conditions to produce effects such as the fall fruit, growth plants, and so forth [4]. Typically, they are small compounds, synthesized from amino acids, mevalonate or nucleotides precursors, sometimes through several stages of the synthesis pathway (multi-step pathway), then deactivated through oxidation or conjugation, and interact with other molecules such as glucose and amino acids [5]. The chemical structure of the phytohormone is presented in Figure 1, while the synthesis location and physiological function is presented on Table 1.

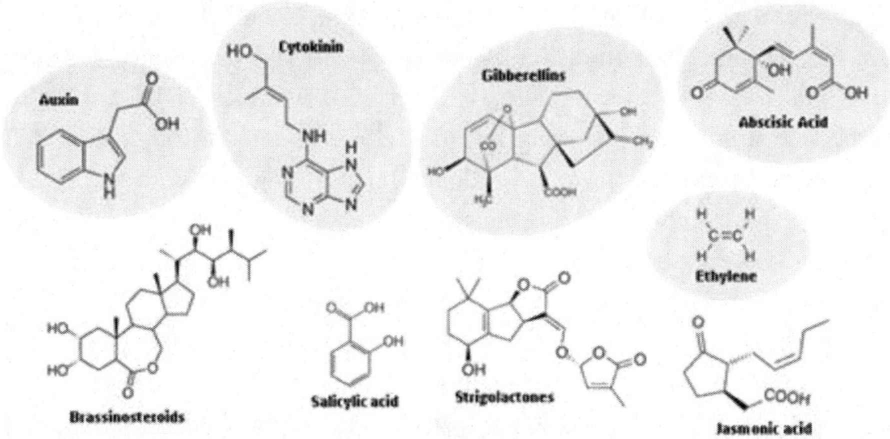

Figure 1. Chemical structure of phytohormone: blue shaded structures are the so called "old-timers" phytohormone, while the rest are "newcomers" [1, 3].

Like plants, seaweeds do not have a central nervous system and are controlled by a dispersed system [8]. They are a source of phytohormones, like their terrestrial plant relatives [9-11], however, the complete function of phytohormone in seaweed remains to be elucidated. This chapter focused on seaweed phytohormones, more specifically, describing three phytohormones isolated from red algae, namely *trans*-zeatin, indole acetic acid and gibberellin GA3.

SEAWEEDS

Marine macroalgae, or seaweeds, are plant-like organisms that generally live attached to rock or other hard substrata in coastal areas. They belong to three different groups, empirically distinguished since the mid-nineteenth century on the basis of thallus color: brown algae (phylum Ochrophyta, class Phaeophyceae); red algae (phylum Rhodophyta), and green algae (phylum Chlorophyta, classes Bryopsidophyceae, Chlorophyceae, Dasycladophyceae, Prasinophyceae, and Ulvophyceae). Distinguishing these three phyla, however, involves more substantial differences than color. In addition to pigmentation, they differ considerably in many ultrastructural and biochemical features, including photosynthetic pigments, storage compounds, composition of cell walls, presence/absence of flagella, ultra-structure of mitosis, connections between adjacent cells, and the fine structure of the chloroplasts. They originated through different evolutionary processes (primary endosymbiosis for green and red algae, secondary endosymbiosis for brown algae), and for this reason, they are now classified in different kingdoms; green algae and red algae in the Kingdom Plantae, and brown algae in the Kingdom Chromista [12].

Seaweeds form an integral part of marine coastal ecosystems, and include the macroscopic, multicellular marine algae that commonly inhabit the coastal regions of the world's oceans where suitable substrata exist. It has been estimated that there are about 9,000 species of macroalgae broadly classified into three main groups based on their pigmentation, for example, Phaeophyta, Rhodophyta, and Chlorophyta; or the brown, red, and green algae, respectively [12]. Brown seaweeds are the second most abundant group comprising about 2,000 species, which reach their maximum biomass levels on the rocky shores of temperate zones. They are most commonly used in agriculture [13], particularly, *Ascophyllum nodosum* (L.). These green, brown and red macroalgae constitute nearly 10,000 species distributed worldwide [14]. Red and brown algae are almost exclusively marine, while green algae are also common in freshwater (rivers and lakes), and even in terrestrial (rocks, walls, houses, and tree bark in damp places) habitats. Seaweed extracts can induce an array of physiological plant responses, such as the promotion of plant growth, improvement of flowering and yield, as well as enhancing the nutritional quality and shelf life of edible products, even at low concentrations [15]. Seaweed concentrates also contain aromatic cytokinins benzyl amino purine (BAP) and topolin (6-[3-hydroxybenzyl-amino] purine) derivatives) [16].

Seaweed products exhibit growth-stimulating activities, and the use of seaweed formulations as biostimulants in crop production is well established. Biostimulants are defined as "materials, other than fertilizers, that promote plant growth when applied in small quantities" and are also referred to as "metabolic enhancers" [17].

Table 1. Type of hormone, synthesis location, and the main physiological activity of hormones [6, 7]

Phytohormone	Synthesis location	Main physiological activity
Auxin (i.e., IAA)	Synthesized and found on the apical meristem of the buds, young leaves, in the embryo of seed and developing fruit	Induction of elongation growth; differentiation of phloem elements; apical dominance; tropisms; initiation of root formation; added long stem, affecting growth, differentiation and branching roots; fruit development; phototropism and geotropism
Cytokinins (i.e., Zeatin)	Synthesized and found on the root tips, embryo and fruit, young leaves, and developing seeds; moves from root to other organs	Control of cell division; root differentiation bud development; senescence retardation
Gibberellins (i.e., GA-3)	Synthesized and found on the apical meristem, young leaves, embryo, young shoot tissues and developing seeds	Encourages the development of seeds, growth of buds, stem and leaf growth of lengthening, encourages flowering and fruit development; affects growth and differentiation of the root
Abscisic acid (i.e., ABA)	Synthesized and found in the leaves; stems, root, and green fruit	Hinders growth; stimulates the closure of stomata in times of water shortage, maintains dormancy
Ethylene	Synthesized and found in the mature fruit and old leaf	Encourages fruit maturation; the opposite effect with some influences of auxin; encourages or inhibits the growth and development of roots, leaves, stems and flowers
Brassinosteroids (Brasinolida)	Synthesized and found in the pollen, seed of young vegetative tissue, and various tissues	Various effects on processes of plant growth and development, seed germination, rhizogenesis, flowering and the aging process; has an effect for plant resistance against some abiotic stress Control of division, growth by elongation, differentiation of the vascular system
Jasmonic acid	Synthesized and found in the pollen, seed and vegetative tissues; induced in response to injury (an open wound)	Inhibits the extension of the root; is necessary for pollen development; plant defense system against insects and pathogen viruses
Salicylic acid	Synthesized and found in various tissues	Plant defense system against pathogens and insects; induces complex defensive responses during pathogenesis
Strigolactones	Synthesized and found in the root	Acts against fungi mycorrhiza and supports plant germination, especially parasitic plant species of the genus of Striga

RED ALGAE (RHODOPHYTA)

The red color of these algae results from the pigments phycoerythrin and phycocyanin, which mask the other pigments, chlorophyll a (no Chlorophyll b), beta-carotene and several unique xanthophylls. The main reserves are typically floridean starch, and floridoside; true starch like that of higher plants and green algae is absent. The walls are made of cellulose, agars and carrageenan, long-chained polysaccharides in widespread commercial use. There are some unicellular representatives of diverse origin, with more complex thalli built up of filaments. A very important group of red algae is the coralline algae, which secrete calcium carbonate onto the surface of their cells. Some of these corallines are articulated and have been used in bone-replacement therapies. Coralline algae were used in ancient times as vermifuges, thus the binomial *Corallina officinalis* [18].

Most seaweed species are part of the division Rhodophyta, or red algae. They are found in a variety of physical forms, including simple and branched filaments, fleshy plants, and sheets. They can be single celled, while others can reach lengths of 2 or 3 meters, forming a distinct photosynthetic eukaryotic lineage that consists of around 6,000 unicellular to large multicellular taxa [19]. The red algae are unique among eukaryotes in lacking both flagella and centrioles during their entire life cycle [20, 21].

Red algae are ecologically significant as primary producers, providers of a structural habitat for other marine organisms, and their important role in the primary establishment and maintenance of coral reefs. Some red algae are economically important as providers of food and gels. For this reason, extensive farming and natural harvest of red algae occurs in numerous areas of the world [22].

As a storage product, the red algae produce granulated floridean starch in the cytoplasm that is different from green algal starch. In addition to these unique features, the monophyly of red algae is strongly supported by nuclear, plastid, and mitochondrial gene trees [23-28].

Bioactive extracts of red seaweeds *Pterocladiella capillacea* and *Osmundaria obtusiloba* (Floridophyceae: Rhodophyta) also have antioxidant and bacterial agglutination potential activity [29].

PHYTOHORMONES IN THE RHODOPHYTA (RED ALGAE)

The five "old timer" phytohormones, including auxin, abscisic acid (ABA), cytokinin (CK), ethylene (ET), and gibberellins (GAs), have been found in a broad spectrum of microalgal lineages. Although the functional role of microalgal endogenous phytohormones remains elusive, molecular evidence from the oleaginous microalga *Nannochloropsis oceanica* suggests that endogenous ABA and CK are functional and that their physiological effects are similar to those in higher plants. Phytohormone systems generally involve biosynthesis pathways that produce phytohormones and signal transduction pathways that mediate the effects of phytohormones. It has been suggested that higher plant hormone systems evolved from a pre-existing primary metabolic system in microalgae [30].

In the 1960-1970s, it was established that auxins and their inactive analogs were present in brown (Macrocystis and Laminaria), red (Botryocladia), and green (Enteromorpha, Chlorella, and Cladophora) algae as well as in cyanobacteria (Oscillatoria) [16]. Some data indicate that auxin controls thallus branching of red and brown macrophytes based on the principle of apical dominance [31, 32].

Gibberellic acid suppressed callus formation and organogenesis in tissue culture of the red alga *Grateloupia doryphora* [33]. Gibberellins may be involved in the control of brown macrophyte branching [34]. In brown and red macrophytes, exogenous gibberellins accelerate growth and increase the thallus length [35]. Thus, it seems likely that, like in higher plants, gibberellins control growth of axial structures.

Indole acetic acid (IAA) is found in the genera of Botryocladia and Porphyra [36] and cytokinin is found in genus Porphyra [16]. Cytokinins have been detected in fresh seaweeds [37] and seaweed extracts [38], including *trans*-zeatin, *trans*-zeatin riboside, and dihydro derivatives of these two forms [39]. Although the physiological roles remain largely unknown, the essential and bioactive forms of these phytohormones; auxin, ethylene, absisic acid, cytokinins, and gibberellins have been detected in red algae (Table 2).

Table 2. Discovery of phytohormones in red algae [40]

Phytohormone	Red Algae (Multicellular)
Auxin	*Prionitis lanceolata, Porphyra sp., Gelidium sp., Gracilaria sp., Gracilariopsis sp., Chondracanthus sp., Hypnea sp.* [41, 36, 42, 43], *E. cotonii* [44], *G. coronopifolia* [45]
Ethylene (ET)	*Porphyra tenera* [42]
Abscisic acid (ABA)	*Porphyra sp., Gelidium sp., Gracilaria sp., Gracilariopsis sp., Chondracanthus sp., Hypnea sp.* [43]
Cytokinins (CK)	*Porphyra sp., Gelidium sp., Gracilaria sp., Gracilariopsis sp., Chondracanthus sp., Hypnea sp., Gigartina clathrata, Hypnea sp.* [46, 47, 16], *G. coronopifolia* [48]
Gibberellins (GA)	*G. coronopifolia, E. cotonii* [49, 50]; *Hypnea musciformis* [47]

Endogenous cytokinins, auxins, and ABA were identified and quantified in eleven red algae collected from the Brazilian coast [43], confirming that cytokinins, auxins, and ABA are common constituents in red seaweeds, with this being the first report of the occurrence of ABA in Rhodophyta. The complexity of the hormone profiles suggests that plant hormones play a role in regulating physiological processes in Rhodophyta. Examples of cytokinins occurring in seaweeds include zeatin detected in *Macrocystis pyrifera* [51] and *Valoniopsis pachynema* [52], *cis*-zeatin riboside and isopentenyladenosine (iPR) found in *Sargassum muticum* [53] and *Laminaria japonica* [54], and iPR in *Porphyra perforata* [55]. Nineteen cytokinins, including both isoprenoid and aromatic groups, have been identified in five Chlorophyta, seven Phaeophyta, and nineteen Rhodophyta species from South Africa [16].

Phytohormones play an important role in the life of algae. Although the hormonal regulatory system in seaweeds has not been as extensively studied as in vascular plants, there is evidence that plant growth regulators are involved in seaweed developmental processes [56, 57], including the growth and morphogenesis of red algae. It has been indicated that the greatest number of phytohormones are characterized in the divisions Chlorophyta, Phaeophyta, and Rhodophyta [58].

Marine algae are also reportedly rich in auxins and auxin-like compounds [59]. *A. nodosum* extract was shown to contain as much as 50 mg IAA per gram of dry extract [60]. Similarly, an extract of *Ecklonia maxima* exhibited remarkable root-promoting activity on mung bean, an effect important of auxins [61]. Gas chromatography/mass spectroscopy (GC/MS) analysis of the extract revealed the presence of indole compounds, including IAA [62].

The benefits of seaweeds as sources of organic matter and fertilizer nutrients have led to their use as soil conditioners for centuries [13, 63]. Indeed, some 15 million metric tons of seaweed products are produced annually [64], with seaweed concentrates also containing aromatic cytokinins BAP and topolin (6-[3-hydroxybenzyl-amino] purine) derivatives) [65].

Application of seaweeds and seaweed extracts triggers the growth of beneficial soil microbes and secretion of soil-conditioning substances by these microbes. As mentioned, alginates affect soil properties, encouraging growth of beneficial fungi. Kuwada et al. [66] reported that organic fractions (25% methanol eluates) of red and green algae considerably improved *in vitro* hyphal growth of arbuscular mycorrhizal (AM) fungi. They showed that application of 25% methanol eluates of red and green algal extracts to roots of papaya (*Carica papaya* Linn.) and passion fruit (*Passiflora edulis* Sims.) improved mycorrhizal development more than the control treatment, concluding that both red and green algae have AM stimulatory compounds which play a part in mycorrhizal development in higher plants [66]. Auxins have been detected in other algal species like *Porphyra perforata*, but at low levels [55]. In higher plants, IAA occurs as an inactive conjugate with carboxyl groups, glycans, amino acids, and peptides, which, upon hydrolysis, are converted to free active IAA [67]. Stirk et al. [16] found four amino acid and three indole conjugates of IAA in extracts of two seaweeds, *E. maxima* and *Macrocystis pyrifera*. Biologically active auxin-like compounds other than IAA were reported in alkaline hydrolyzates of *A. nodosum*, *Fucus vesiculosus*, and other seaweeds [68]. Marine algae are also reportedly rich in auxins and auxin-like compounds [59]. Active gibberellins, GA1 and GA3, and inactive GA6 were isolated chromatographically from the tissue extracts of *F. vesiculosus* and *Fucus spiralis* [69].

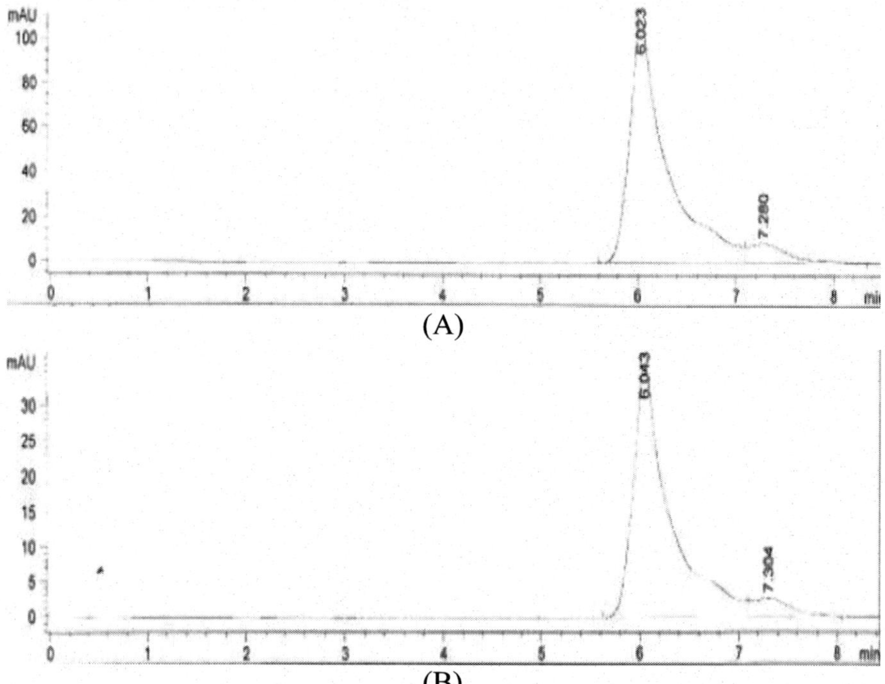

Figure 2. Reversed phase HPLC of (A) IAA standard and (B) preparative TLC of a purified *E. cotonii* extract. The separation was performed using C-18 Nucleosil ODS columns and detection using UV detector at 260 nm with 35% methanol in 20 mM acetate buffer (pH 3.5) as mobile phase at 0.7 mL/minute a flow-rate [44].

Our group isolated and identified auxin, cytokinin and gibberellin from two red algae species of *G. coronopifolia* and *E. cotonii* [49, 48, 50]. The cytokinin (*trans*-zeatin) content in *G. coronopifolia* was 6.26×10^{-2} mg/g dry weight, while auxin (IAA) and gibberellin (GA-3) content in *E. cotonii* were 5.34×10^{-3} mg/g and 1.32 mg/g dry weight, respectively. We also determined their effects on the growth of rice (*Oryza sativa*).

Isolation and purification of auxin and cytokinin was initiated by an extraction process using a maceration method with 80% methanol, followed by extraction with ethyl acetate and adsorption chromatography on Silica Gel-60 (70-230 mesh) using isopropanol-ammonium hydroxide-water (20:1:1) as the mobile phase. The fractions were then further purified on thin layer chromatography (TLC) silica gel GF-254, with the final step of the identification performed using RP-HPLC on Nucleosil C-18 column with 35% methanol in 20 mM acetate buffer (pH 3.5) as mobile phase with 0.7 mL/minute flow-rate [48, 44].

The reversed phase HPLC chromatogram of IAA spot on preparative TLC plate is presented on Figure 2, showing one major peak at 6.04 minutes which corresponds to the IAA standard (retention time 6.02). This result confirmed that *E. cotonii* contains auxin in the form of IAA. Comparison of the sample peak and standard indicated that *E. cotonii* samples contained 5.34×10^{-3} mg/g dry weight IAA.

Isolation and purification of gibberellin was performed using the method described by Hemphill et al. [70]. One hundred grams of dried algae sample were homogenized for three minutes at 4°C in 90% methanol, then stirred for 24 hours at 4°C and filtered under vacuum. The residue was then re-extracted as above and combined extracts were evaporated to dryness before the addition of 5% sodium bicarbonate. The bicarbonate solution was partitioned three times against equal volumes of ethyl acetate, then the pH was adjusted to 3.0 with dilute sulfuric acid and partitioned three times against equal volumes of ethyl acetate. Combined ethyl acetate layers were washed with distilled water, dried over anhydrous sodium sulfate, and evaporated to dryness. The gummy residue was dissolved in minimal ethanol or ethyl acetate and applied as a thin streak to ethanol-washed Whatman 3 MM paper (50 mg/chromatogram). The chromatograms (descending) were developed with isopropanol-ammonium hydroxide-water (10:1:1, v/v). Active compounds were eluted from strips of the chromatograms with 90% ethanol, the volume was reduced and the eluates were applied to Silica Gel G or Silica Gel F25 pre-coated preparative plates (Brinkmann Instruments, Inc.) for TLC. The TLC plate was developed with carbon tetrachloride-acetie acid-water (8:3:5, v/v), non-aqueous phase plus 10% ethyl acetate. Separation of gibberellin used adsorption column chromatography on Silica Gel G-60 (70-230 mesh). All column fractions were identified by TLC on Silica Gel GF-254 (analytic and preparative), with the identification and final step purification of gibberellin performed by reverse-phase HPLC on Nucleosil C-18 column, with mobile phase of 0.1% TFA (trifluoro acetic acid) in acetonitrile, at a flow-rate of 0.7 mL/minute [49].

Figure 3 shows the chromatogram of preparative TLC spot of the purified cytokinin (*trans*-zeatin). *Trans*-zeatin as standard was eluted at 8.905 minutes (Figure 3A). The chromatogram showed that the isolated cytokinin still contained some impurities, as indicated by the presence of peaks before the *trans*-zeatin peak, which eluted at earlier retention time, 5.315 and 5.678 minutes (Figure 3B).

The analysis of the purified gibberellin from *E. cotonii* extract is presented in Figure 4. GA-3 was used as standard (Figure 4A) and was eluted at 13.629 minutes. The isolated gibberellin was eluted at 13.615 and there were still some impurities detected in the sample.

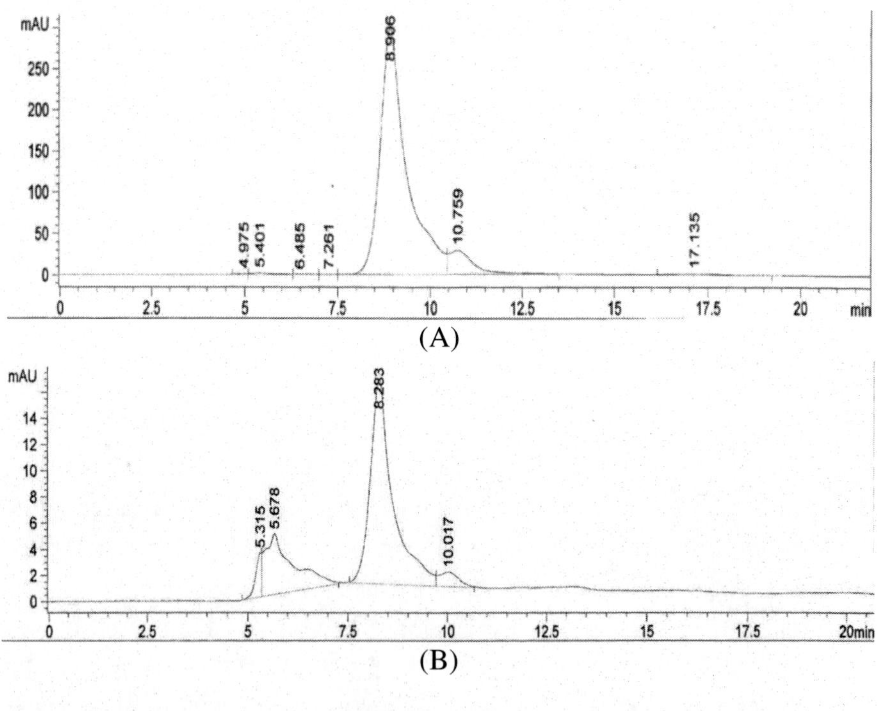

Figure 3. Reversed phase HPLC of (A) *trans*-zeatin standard and (B) preparative TLC result of the purified *G. coronopifolia* extract. The separation was performed using C-18 Nucleosil ODS columns and detection using UV detector 260 nm with 35% methanol in 20 mM acetate buffer (pH 3.5) as mobile phase at 0.7 mL/minute a flow-rate [48].

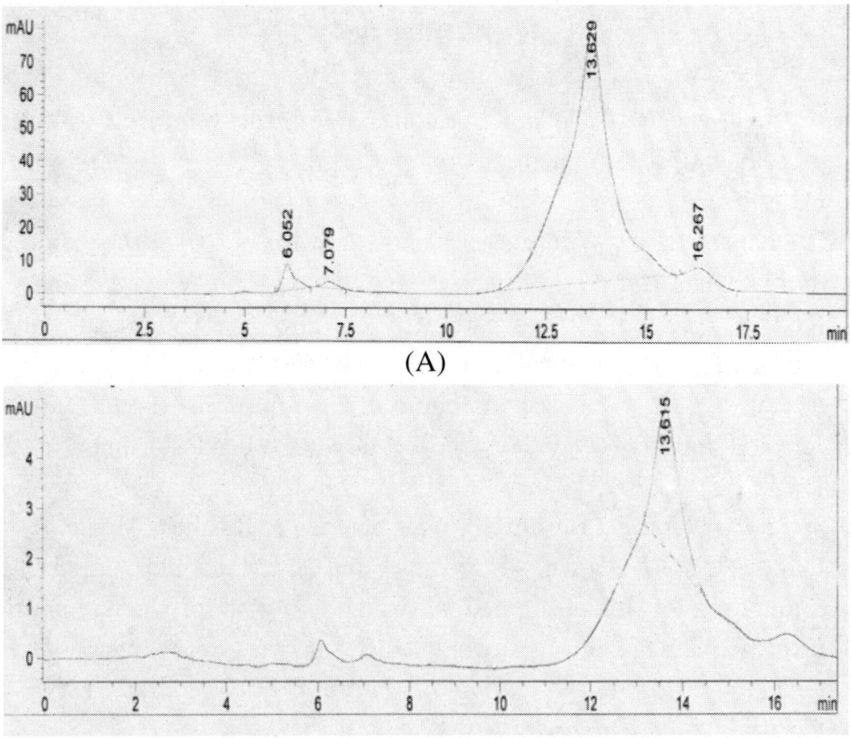

(A)

(B)

Figure 4. Reversed phase HPLC of (A) GA-3 standard and (B) preparative TLC of the purified *E. cotonii* extract. The separation was performed using C-18 Nucleosil ODS columns and detection using UV detector 254 nm with 0.1% TFA in acetonitrile as mobile phase at 0.7 mL/minute a flow-rate [48].

CONCLUSION

Seaweeds, especially red algae, i.e., *G. coronopifolia* and *E. cotonii*, are a potential source of phytohormones and can be used as biostimulant agents to improve the quality and quantity of crop production. The phytohormones in red algae, especially auxin, cytokinins, and gibberellins, can be applied to improve agriculture products, especially to increase productivity of horticultural crops.

REFERENCES

[1] Kamiya, Y. 2010. "Plant hormones: Versatile regulators of plant growth and development." *Annual Review of Plant Biology* no. 61 (1):null.

[2] Buchanan, B. B., W. Gruissem, and R. L. Jones. 2015. *Biochemistry and molecular biology of plants*. New Jersey: John Wiley & Sons.

[3] Kende, H., and J. Zeevaart. 1997. "The five" classical "plant hormones." *The Plant Cell* no. 9 (7):1197.

[4] Beck, W. S., K. F. Liem, and G. G. Simpson. 1965. *Life: an introduction to biology*. New York: Harper Collins Publishers.

[5] Davies, P. J. 2013. *Plant hormones: physiology, biochemistry and molecular biology*. Berlin: Springer Science & Business Media.

[6] Davis, J., M. Brown, C. Evans, and J. Mansfield. 2004. The integration of foliar applied seaweed and fish products into the fertility management of organically grown sweet peppers. In: *Organic Farming Research Foundation Project Report. North Carolina State University*.

[7] Polevoi, V. 1982. *Fitogormony (Phytohormones)*. Leningrad: Gos. Univ.

[8] Mori, I. C., Y. Ikeda, T. Matsuura, T. Hirayama, and K. Mikami. 2017. "Phytohormones in red seaweeds: a technical review of methods for analysis and a consideration of genomic data." *Botanica Marina* no. 60 (2):153-170.

[9] Basu, S., H. Sun, L. Brian, R. L. Quatrano, and G. K. Muday. 2002. "Early embryo development in *Fucus distichusIs* auxin sensitive." *Plant Physiology* no. 130 (1):292-302.

[10] Mikami, K., I. C. Mori, T. Matsuura, Y. Ikeda, M. Kojima, H. Sakakibara, and T. Hirayama. 2016. "Comprehensive quantification and genome survey reveal the presence of novel phytohormone action modes in red seaweeds." *Journal of Applied Phycology* no. 28 (4):2539-2548.

[11] Wang, C., Y. Liu, S. S. Li, and G. Z. Han. 2014. "Origin of plant auxin biosynthesis in charophyte algae." *Trends in Plant Science* no. 19 (12):741-743.

[12] Guiry, M. D. *What are seaweeds?* 2000. Available from http://www.seaweed.ie/algae/seaweeds.php.

[13] Blunden, G., and S. M. Gordon. 1986. "Betaines and their sulphonio analogues in marine algae." *Progress in Phycological Research* no. 4:39-80.

[14] Khan, W., U. P. Rayirath, S. Subramanian, M. N. Jithesh, P. Rayorath, D. M. Hodges, A. T. Critchley, J. S. Craigie, J. Norrie, and B. Prithiviraj. 2009. "Seaweed extracts as biostimulants of plant growth and development." *Journal of Plant Growth Regulation* no. 28 (4):386-399.

[15] El-Sharkawy, M., T. El-Beshsbeshy, R. Al-Shal, and A. Missaoui. 2017. "Effect of plant growth stimulants on alfalfa response to salt stress." *Agricultural Science* no. 8:267-291.

[16] Stirk, W., O. Novák, M. Strnad, and J. Van Staden. 2003. "Cytokinins in macroalgae." *Plant Growth Regulation* no. 41 (1):13-24.

[17] Xunzhong, Z., and R. E. Schmidt. 1997. "The impact of growth regulators on the alpha-tocopherol status in water stressed *Poa pratensis* L." *ITS Research Journal* no. 8:1364-1374.

[18] Guiry, M. D. *Rhodophyta: Red algae* 2000. Available from http://www.seaweed.ie/algae/rhodophyta.php.

[19] Guiry, M., W. Guiry, S. V. Miranda, D. Garbary, A. Langangen, P. Gabrielson, J. West, C. Schneider, M. Wynne, C. Carter, I. Bárbara, and R. Lavigne. Available from http://www.algaebase.org/.

[20] Gabrielson, P. W., D. J. Garbary, M. R. Sommerfeld, R. A. Townsend, and P. L. Tyler. 1990. "Phylum Rhodophyta." In: *Handbook of protoctista: The structure, cultivation, habitats and life histories of the eukayotic microorganisms and their descendants exclusive of animals, plants and fungi*, edited by L. Margulis, J. O. Corliss, M. Melkonian and D. J. Chapman. Boston: Jones & Bartlett Publishers.

[21] Graham, L. E., and L. W. Wilcox. 2000. *Algae*. Upper Saddle River: Prentice Hall.

[22] Freshwater, D. W. *Red Algae* 2000. Available from http://tolweb.org/Rhodophyta/2381/2000.03.24.

[23] Burger, G., D. Saint-Louis, M. W. Gray, and B. F. Lang. 1999. "Complete sequence of the mitochondrial DNA of the red alga *Porphyra purpurea*: cyanobacterial introns and shared ancestry of red and green algae." *The Plant Cell* no. 11 (9):1675-1694.

[24] Freshwater, D. W., S. Fredericq, B. S. Butler, M. H. Hommersand, and M. W. Chase. 1994. "A gene phylogeny of the red algae (Rhodophyta) based on plastid rbcL." *Proceedings of the National Academy of Sciences* no. 91 (15):7281-7285.

[25] Ragan, M. A., and R. R. Gutell. 1995. "Are red algae plants?" *Botanical Journal of the Linnean Society* no. 118 (2):81-105.

[26] Van de Peer, Y., and R. De Wachter. 1997. "Evolutionary relationships among the eukaryotic crown taxa taking into account site-to-site rate variation in 18S rRNA." *Journal of Molecular Evolution* no. 45 (6):619-630.

[27] Yoon, H. S., J. D. Hackett, and D. Bhattacharya. 2002. "A single origin of the peridinin-and fucoxanthin-containing plastids in dinoflagellates through tertiary endosymbiosis." *Proceedings of the National Academy of Sciences* no. 99 (18):11724-11729.

[28] Yoon, H. S., J. D. Hackett, C. Ciniglia, G. Pinto, and D. Bhattacharya. 2004. "A molecular timeline for the origin of photosynthetic eukaryotes." *Molecular Biology and Evolution* no. 21 (5):809-818.

[29] de Alencar, D. B., F. C. T. de Carvalho, R. H. Rebouças, D. R. dos Santos, K. M. dos Santos Pires-Cavalcante, R. L. de Lima, B. M. Baracho, R. M. Bezerra, F. A. Viana, and R. H. S. dos Fernandes Vieira. 2016. "Bioactive extracts of red seaweeds *Pterocladiella capillacea* and *Osmundaria obtusiloba* (Floridophyceae: Rhodophyta) with antioxidant and bacterial agglutination potential." *Asian Pacific Journal of Tropical Medicine* no. 9 (4):372-379.

[30] Kenrick, P., and P. R. Crane. 1997. "The origin and early evolution of plants on land." *Nature* no. 389 (6646):33.

[31] Moss, B. 1974. "Morphogenesis." In: *Algal Physiology and Biochemistry*, edited by W. D. P. Stewart, 788-813. Berkeley: University of California Press.

[32] Yokoya, N. S., and W. Handro. 1996. "Effects of auxins and cytokinins on tissue culture of *Grateloupia dichotoma* (Gigartinales, Rhodophyta)." *Hydrobiologia* no. 326 (1):393-400.

[33] Garcia-Jimenez, P., M. Rodrigo, and R. R. Robaina. 1998. "Influence of plant growth regulators, polyamines and glycerol interaction on growth and morphogenesis of carposporelings of Grateloupia cultured *in vitro*." *Journal of Applied Phycology* no. 10 (1):95-100.

[34] Dworetzky, B., R. M. Klein, and P. W. Cook. 1980. "Effect of growth substances on "apical dominance" in *Sphacelaria furcigera* (phaeophyta)." *Journal of Phycology* no. 16 (2):239-242.

[35] Jennings, R. 1968. "Gibberellins as endogenous growth regulators in green and brown algae." *Planta* no. 80 (1):34-42.

[36] Provasoli, L., and A. F. Carlucci. 1974. "Vitamins and growth regulators." In: *Algal Physiology and Biochemistry*, edited by W. D. P. Stewart, 741-787. Oxford: Blackwell Scientific Publications.

[37] Hussain, A., and A. Boney. 1969. "Isolation of kinin-like substances from *Laminaria digitata*." *Nature* no. 223 (5205):504.

[38] Brain, K., M. Chalopin, T. Turner, G. Blunden, and P. Wildgoose. 1973. "Cytokinin activity of commercial aqueous seaweed extract." *Plant Science Letters* no. 1 (6):241-245.

[39] Stirk, W., and J. Van Staden. 1996. "Comparison of cytokinin-and auxin-like activity in some commercially used seaweed extracts." *Journal of Applied Phycology* no. 8 (6):503-508.

[40] Lu, Y., and J. Xu. 2015. "Phytohormones in microalgae: a new opportunity for microalgal biotechnology?" *Trends in Plant Science* no. 20 (5):273-282.

[41] Ashen, J. B., J. D. Cohen, and L. J. Goff. 1999. "GC-SIM-MS detection and quantification of free indole-3-acetic acid in bacterial

galls on the marine alga Prionitis lanceolata (Rhodophyta)." *Journal of Phycology* no. 35 (3):493-500.

[42] Watanabe, T., and N. Kondo. 1976. "Ethylene evolution in marine algae and a proteinaceous inhibitor of ethylene biosynthesis from red alga." *Plant and Cell Physiology* no. 17 (6):1159-1166.

[43] Yokoya, N. S., W. A. Stirk, J. Van Staden, O. Novák, V. Turečková, A. Pěnčík, and M. Strnad. 2010. "Endogenous cytokinins, auxins, and abscisic acid in red algae from Brazil." *Journal of Phycology* no. 46 (6):1198-1205.

[44] Tampubolon, E., S. D. Rachman, and U. M. S. Soedjanaatmadja. 2008. Activity of auxin from *Eucheuma contonii* (red algae). Bandung: Universitas Padjadjaran.

[45] Ardiyani, D. E. P., S. D. Rachman, and U. M. S. Soedjanaatmadja. 2019. Isolasi dan Identifikasi Hormon Auksin (IBA) dari *Glacilaria coronopifolia* dan Uji Hayatinya pada Tanaman Padi *Oryza sativa*. Jatinangor: Universitas Padjadjaran.

[46] Jennings, R. 1969. "Cytokinins as endogenous growth regulators in the algae Ecklonia (Phaeophyta) and Hypnea (Rhodophyta)." *Australian Journal of Biological Sciences* no. 22 (3):621-628.

[47] Jennings, R., and A. McComb. 1967. "Gibberellins in the red alga *Hypnea musciformis* (Wulf.) lamour." *Nature* no. 215 (5103):872.

[48] Rachman, S. D., Z. Mukhtari, and R. U. M. Soedjanaatmadja. 2017. "Alga merah (*Gracilaria coronopifolia*) sebagai sumber fitohormon sitokinin yang potensial." *Chimica et Natura Acta* no. 5 (3):124-131.

[49] Florida, A., A. T. Hidayat, and U. M. S. Soedjanaatmadja. 2008. Activity of gibberellin from *Eucheuma contonii* (red algae). Jatinangor: Universitas Padjadjaran.

[50] Rahayu, M. U., S. D. Rachman, and U. M. S. Soedjanaatmadja. 2019. Isolasi dan identifikasi hormon gibberellin (GA3) dari *Glacilaria coronopifolia* dan uji hayatinya pada tanaman padi *Oryza sativa*. Jatinangor: Universitas Padjadjaran.

[51] De Nys, R., P. Jameson, N. Chin, M. Brown, and K. Sanderson. 1990. "The cytokinins as endogenous growth regulators in

Macrocystis pyrifera (L.) C. Ag.(Phaeophyceae)." *Botanica Marina* no. 33 (6):467-476.

[52] Farooqi, A. A., Y. N. Shukla, A. Shukla, and D. Bhakuni. 1990. "Cytokinins from marine organisms." *Phytochemistry* no. 29 (7):2061-2063.

[53] Zhang, W., D. J. Chapman, B. O. Phinney, C. R. Spray, H. Yamane, and N. Takahashi. 1991. "Identification of Cytokinins in *Sargassum muticum* (Phaeophyta) and *Porphyra perforata* (Rhodophyta)." *Journal of Phycology* no. 27 (1):87-91.

[54] Duan, D., X. Liu, F. Pan, H. Liu, N. Chen, and X. Fei. 1995. "Extraction and identification of cytokinin from *Laminaria japonica* Aresch." *Botanica Marina* no. 38 (1-6):409-412.

[55] Zhang, W., H. Yamane, and D. Chapman. 1993. "The phytohormone profile of the red alga *Porphyra perforata*." *Botanica Marina* no. 36 (3):257-266.

[56] Bradley, P. M. 1991. "Plant hormones do have a role in controlling growth and development of algae." *Journal of Phycology* no. 27 (3):317-321.

[57] Evans, L. V., and A. J. Trewavas. 1991. "Is algal development controlled by plant growth substances?" *Journal of Phycology* no. 27 (3):322-326.

[58] Tarakhovskaya, E., Y. I. Maslov, and M. Shishova. 2007. "Phytohormones in algae." *Russian Journal of Plant Physiology* no. 54 (2):163-170.

[59] Crouch, I. J., and J. Van Staden. 1993. "Evidence for the presence of plant growth regulators in commercial seaweed products." *Plant Growth Regulation* no. 13 (1):21-29.

[60] Kingman, A., and J. Moore. 1982. "Isolation, purification and quantitation of several growth regulating substances in *Ascophyllum nodosum* (Phaeophyta)." *Botanica Marina* no. 25 (4):149-154.

[61] Crouch, I., and J. Van Staden. 1991. "Evidence for rooting factors in a seaweed concentrate prepared from *Ecklonia maxima*." *Journal of Plant Physiology* no. 137 (3):319-322.

[62] Crouch, I., M. Smith, J. Van Staden, M. Lewis, and G. Hoad. 1992. "Identification of auxins in a commercial seaweed concentrate." *Journal of Plant Physiology* no. 139 (5):590-594.

[63] Temple, W., and A. Bomke. 1988. "Effects of kelp (*Macrocystis integrifolia*) on soil chemical properties and crop response." *Plant and Soil* no. 105 (2):213-222.

[64] FAO. 2006. Fishery and Aquaculture Statistics. Rome: FAO *Fisheries and Aquaculture Department*.

[65] Stirk, W. A., G. D. Arthur, A. F. Lourens, O. Novak, M. Strnad, and J. Van Staden. 2004. "Changes in cytokinin and auxin concentrations in seaweed concentrates when stored at an elevated temperature." *Journal of Applied Phycology* no. 16 (1):31.

[66] Kuwada, K., L. S. Wamocho, M. Utamura, I. Matsushita, and T. Ishii. 2006. "Effect of red and green algal extracts on hyphal growth of arbuscular mycorrhizal fungi, and on mycorrhizal development and growth of papaya and passionfruit." *Agronomy Journal* no. 98 (5):1340-1344.

[67] Bartel, B. 1997. "Auxin biosynthesis." *Annual Review of Plant Biology* no. 48 (1):51-66.

[68] Buggeln, R., and J. Craigie. 1971. "Evaluation of evidence for the presence of indole-3-acetic acid in marine algae." *Planta* no. 97 (2):173-178.

[69] Kato, J., W. K. Purves, and B. O. Phinney. 1962. "Gibberellin-like substances in plants." *Nature* no. 196 (4855):687-688.

[70] Hemphill, D. D., L. Baker, and H. Sell. 1972. "Isolation and identification of the gibberellins of *Cucumis sativus* and *Cucumis melo*." *Planta* no. 103 (3):241-248.

In: A Closer Look at Hormones
Editor: Alexander V. Knyazev
ISBN: 978-1-53617-254-6

Chapter 3

ANALYTICAL METHODS FOR THE DETERMINATION OF STEROID HORMONES IN ENVIRONMENTAL SYSTEMS

Rayco Guedes-Alonso*[*], *Zoraida Sosa-Ferrera* and *José Juan Santana-Rodríguez
Instituto UniversitaRíde Estudios Ambientales y Recursos Naturales (i-UNAT), Universidad de Las Palmas de Gran Canaria, Las Palmas de Gran Canaria, Las Palmas, Spain

ABSTRACT

Steroid hormones are an important group of biologically active compounds which are involved in many physiological functions of organisms with genomic and non-genomic effects. For this reason, many therapeutic treatments are based on the use of natural and synthetic hormones to treat disorders of the homeostasis of steroid hormones. After these compounds are used in organic systems, they are excreted at trace

[*] Corresponding Author's Email: rayco.guedes@ulpgc.es.

concentrations and reach wastewaters, which are the main source of these types of compounds in the environment. When purified wastewaters are discharged into the environment, steroid hormones are considerably diluted and for this reason are only detected at trace concentrations.

In recent decades, many scientific studies have revealed the potential toxicity of steroid hormones, even at $ng \cdot L^{-1}$ concentration levels. At trace concentrations, steroid hormones can produce changes in aquatic biota, including hermaphroditism, the development of unisexual populations and the malformation of sexual organs. Furthermore, steroid hormones not only interfere with the reproductive systems of exposed organisms, as has been seen in several studies which connect environmental concentrations of steroid hormones with alterations in growth, but can also produce changes in tissues and bones, and even behavioral changes in aquatic organisms. Some of these changes take time to manifest, very often only appearing in the following generations, and may be irremediable.

Because of the trace level concentrations of steroid hormones in the aquatic environment and the complexity of the matrices that have to be analyzed, appropriate analytical procedures are required. Efficient extraction/preconcentration methodologies are of vital importance, as are selective and sensitive determination methods. In this chapter, a general review is presented of the most commonly used and most recent methodologies for the identification and quantification of steroid hormones in environmental matrices.

Keywords: steroid hormones, analytical methods, environmental matrices

LIST OF ABBREVIATIONS

β-CD	β-cyclodextrin
ACN	Acetonitrile
APCI	Atmospheric pressure chemical ionization
BSTFA	Bis(trimethylsilyl) trifluoroacetamide
CO2	Carbon dioxide
DAD	Diode array detection
EC	European Commission
EDC	Endocrine disrupting compound
ELISA	Enzyme-linked immunosorbent assay
EPs	Emerging pollutants

ESI	Electrospray ionization
FD	Fluorescence detection
FPSE	Fabric phase sorptive extraction
FTIR	Fourier transform infrared spectroscopy
GC	Gas chromatography
HF-LPME	Hollow fiber liquid phase microextraction
HRMS	High resolution mass spectrometry
KOC	Soil adsorption coefficient
KOW	Octanol/water partition coefficient
LC	Liquid chromatography
LLE	Liquid – liquid extraction
LOD	Limit of detection
LOQ	Limit of quantification
LPME	Liquid-phase microextraction
MAE	Microwave assisted extraction
MIPs	Molecularly imprinted polymers
MS	Mass spectrometry
MS/MS	Triple quadrupole mass spectrometry
MSTFA	N-methyl-N-(trimethylsilyl)-trifluoroacetamide
NMR	Nuclear magnetic resonance
PAH	Polycyclic aromatic hydrocarbon
PDMS	Polydimethylsiloxane
PLE	Pressurized liquid extraction
PPCPs	Pharmaceuticals and personal care products
RIA	Radioimmunoassay
RSD	Relative standard deviation
SBSE	Stir-bar sorptive extraction
SD-LPME	Solid drop liquid phase microextraction
SFE	Supercritical fluid extraction
SPE	Solid phase extraction
SPME	Solid phase microextraction
TFME	Thin film microextraction
TMSI	Trimethylsilylimidazole
UAE	Ultrasound assisted extraction

UHPLC	Ultra-high-performance liquid chromatography
USEPA	United States Environmental Protection Agency
UV	Ultraviolet-visible
WWTP	Wastewater treatment plant

INTRODUCTION

The development of society in the last few centuries has been accompanied by the use of an ever-growing number of chemical products with different characteristics and purposes. This situation has resulted in severe damage to the environment, which has become the recipient body of the residues of these products used for human activities. Many thousands of tons of chemical compounds are discharged into the environment, with harmful effects that can persist for years and even decades. In principle, the chemical compounds that reach the various environmental compartments can simply be divided into inorganic and organic types. However, due to the enormous quantity of compounds and their extensive variety of structures, further subdivisions are convenient even though the subdivision of organic compounds can be extremely complicated. Some regulations, including European Directive 2000/60/EC which establishes a framework for Community action in the field of water policy, distinguish between priority and emerging pollutants (EPs). Priority substances are chemical compounds which present a significant risk to or via the aquatic environment. Consequently, priority substances are listed in the regulations which establish the maximum concentrations allowed in the environment. The inclusion of a chemical compound in such legislation means that it is has been determined to be harmful for the environment and for humans, and that the necessary measures need to be taken to eliminate the pollution associated to it. Some examples of priority pollutants are certain pesticides, polycyclic aromatic hydrocarbons (PAHs), herbicides and heavy metals.

For their part, EPs are a wide-ranging group of organic compounds with potentially adverse environmental effects, but which remain unregulated and whose fate and possible impact have been poorly studied.

The contamination associated to EPs is not new, as some types have been manufactured and used in households and commercial sectors for many years. However, due to their low concentrations, they had been ignored until improved or new analytical methods became available. Some examples of EPs are certain pharmaceuticals and personal care products (PPCPs), abused drugs, endocrine disrupting compounds (EDCs), flame retardants, perfluorinated compounds or plasticizers.

Steroid hormones form an interesting group of organic compounds because of their biological importance and physiological activities, as well as the extensive variety of the properties of many of them and, consequently, their pharmacological potential. Moreover, they form an important group among the EPs because of their potential harmful impact on aquatic biota (even at trace concentrations) and the lack of legislation about the maximum concentrations permitted in the environment. Steroid hormones are based on the cholesterol molecule, as are similar compounds named sterols (Lednicer 2011) and naturally-occurring steroid hormones are synthetized in sexual organs and adrenal glands. On this basis, steroid hormones can be divided into sex hormones and corticosteroids. In both these groups it is possible to find many of the synthetic steroid hormones that have been designed as pharmaceutical drugs to treat a wide variety of health disorders. The molecular structures and physicochemical properties of the most frequently studied steroid hormones are shown in Table 1.

The concentrations of steroid hormones in the human body are very low, from micromolar to picomolar levels. These concentrations are variable because they are present in the human body in cycles which depend on the moment of the day and the time of year, as well as the different external situations of the organisms (Del Río et al. 2018). However, these changes are not arbitrary and they are controlled by the endocrine system.

Because of the important activities in which steroid hormones are involved, these compounds are continuously being excreted via feces and urine (Kolodziej, Harter, and Sedlak 2004). This situation has a major impact on the environmental compartments which receive these residues, since the entrance of steroid hormones is constant. Steroid hormones can therefore also be considered pseudo-persistent pollutants. For this reason, it is absolutely necessary to control the emission sources of steroid hormones as well as their transport pathways into the environment. In this respect, wastewater treatment plants (WWTPs) are logically regarded to be the main source for the emission of steroid hormones into the environment (Yu et al. 2011; Zhou, Zha, and Wang 2012; Lapworth et al. 2012; Luo et al. 2014). This is because WWTPs are not designed to remove this type of pollutant using conventional treatments. Moreover, there are no legal regulations which establish concentration limits for wastewaters and environmental waters. Presently, in the European Union, only two naturally-occurring estrogens (estrone and 17β-estradiol) and one synthetic estrogen (17α-ethynylestradiol) are included in a "Watch List" in order to obtain "high-quality Union-wide monitoring data on potential water pollutants for the purpose of determining the risk they pose and thus whether Environmental Quality Standards (EQS) should be set for them at EU level" (Loos et al. 2018).

Apart from WWTPs, steroid hormones can be found in hospital area wastewaters, which present high concentrations because of their therapeutic uses, and also in livestock sewage. What is more, these compounds can reach the environment indirectly through uncontrolled spills or sewerage system losses (Pal et al. 2010). It is important to realize that, while both naturally-occurring and synthetic hormones are excreted as conjugated compounds (inactive molecules), the Escherichia coli bacteria present in feces and environmental waters can degrade significant amounts of the β-glucuronidase enzyme with the result that these conjugated hormones may enter the aquatic environment as biologically-active molecules (Routledge et al. 1998).

After reaching the aquatic environment, steroid hormones can be diffused by different transport mechanisms (Alda et al. 2002). From liquid to solid phase, the main transport mechanism is adsorption of the pollutants over the solid particles, the rate of which can be determined using the octanol/water partition coefficient (K_{OW}) and the soil adsorption coefficient (K_{OC}). The aqueous solubility and vapor pressure of a steroid hormone can also be used to indicate its affinity for the solid phase (Maes 2011). Table 1 summarizes these values for different steroid hormones.

Table 1. Physicochemical properties of the most frequently studied steroid hormones

		Log K_{OW}	Vapor pressure (mmHg)	Water solubility ($mg \cdot L^{-1}$)
Estrogens	17β-estradiol	4.01	$6.4 \cdot 10^{-9}$	3.0
	17α-ethynylestradiol	3.67	$1.9 \cdot 10^{-9}$	11.3
	Estriol	2.45	$9.9 \cdot 10^{-12}$	27.3
	Estrone	3.13	$2.5 \cdot 10^{-10}$	3.0
	Diethylstilbestrol	5.07	$3.3 \cdot 10^{-7}$	12.0
Androgens	Testosterone	3.32	$1.7 \cdot 10^{-8}$	23.4
	Nandrolone	2.62	$3.5 \cdot 10^{-8}$	24.0
	Boldenone	3.09	$2.0 \cdot 10^{-9}$	20.0
Progestogens	Progesterone	3.87	$3.54 \cdot 10^{-4}$	8.8
	Norethisterone	2.97	$3.1 \cdot 10^{-7}$	7.0
	Levonorgestrel	3.48	$3.9 \cdot 10^{-10}$	2.1
	Megestrol acetate	3.75	$2.1 \cdot 10^{-10}$	6.5
Glucocorticoids	Cortisone	1.43	$3.0 \cdot 10^{-15}$	140.0
	Prednisone	1.46	$3.8 \cdot 10^{-13}$	77.5
	Prednisolone	1.62	$1.2 \cdot 10^{-13}$	223.0

Because of their low vapor pressure, steroid hormones tend to remain in solid or liquid phase and are unlikely to be evaporated. Moreover, considering the solubility and soil adsorption coefficient of steroid hormones, these compounds will probably pass from liquid to solid phase. Furthermore, because of their lipophilicity, steroid hormones concentrate in sediments and produce bioaccumulation in adipose tissues of aquatic organisms (Lai, Scrimshaw, and Lester 2002; Huang et al. 2013). Steroid hormones can also appear in soils after the spreading of cattle manure over agricultural fields (Bartelt-Hunt et al. 2012).

With this in mind, it is logical to think that steroid hormones are most likely to be found in solid samples, but the presence of these pollutants in water bodies has been extensively evidenced in the literature. This is because the diffusion of steroid hormones to solid phases is not always possible with the result that steroid hormones can persist in environmental waters. The mechanisms which result in their remaining in water bodies include the formation of colloids or more soluble steroid hormone metabolites.

In light of all the above, analytical chemistry has faced the challenge of developing sensitive and selective analytical methods which permit the measurement of these EPs, even at trace concentrations, in order to be able to assess not only the quality of the environment but also the specific impact of EP sources. Although steroid hormones can be present in different environmental compartments, because of the complexity of solid and biological samples, far more studies have been carried out on the presence of steroid hormones in liquid samples. As a result, liquid sample extraction techniques for steroid hormones are more developed than those for solid samples. Broadly speaking, the highest concentrations of steroid hormones in water samples are found in those from WWTPs, with concentrations which range from $ng \cdot L^{-1}$ to $\mu g \cdot L^{-1}$. After the WWTP purification processes, steroid hormones can enter aquatic ecosystems. Because of the high dilution factors, the concentrations of steroid hormones in environmental waters tend to be at the $ng \cdot L^{-1}$ level (Vidal-Dorsch et al. 2012; Manickum and John 2014; Emnet et al. 2015; Patrolecco, Capri, and Ademollo 2015). Different studies have also found measurable concentrations of steroid hormones (from $pg \cdot L^{-1}$ to $ng \cdot L^{-1}$) in underground or surface waters, which can be indirectly polluted by these types of pollutants. Very few studies have been made on solid samples, but the concentration trends are similar. In fact, high steroid hormone concentrations have been found in WWTP sludge, with concentrations at the level of $\mu g \cdot g^{-1}$. In sediments of aquatic systems, the concentrations of steroid hormones are moderately lower, at the $ng \cdot g^{-1}$ level. As for organisms exposed to steroid hormones, both naturally-occurring and synthetic hormones have been found in aquatic organisms at trace and ultra-trace levels. Despite the lack of studies about the biomagnification of these contaminants, there have been some reports of moderate concentrations of these compounds in organisms at the highest levels of the food chain (Sosa-Ferrera, Mahúgo-Santana, and Santana-Rodríguez 2012; Sosa-Ferrera, Mahugo-Santana, and Santana-Rodríguez 2013).

Steroid hormones have been revealed as potent endocrine disrupting compounds (EDCs) for aquatic biota. The first studies about the potential toxicity of estrogens were carried out in the early 90s. Purdom et al. determined high levels of vitellogenin in river fishes exposed to WWTP effluents, which is an indication of estrogenic stimulation of liver in these organisms (Purdom et al. 1994). Subsequent studies focused on the determination of changes in the sexual organs or reproductive systems of aquatic organisms exposed to steroid hormones. On this basis, some studies have related the presence of steroid hormones at trace concentrations with the development of unisexual populations, hermaphroditism, lower fertility and the inhibition of egg production (Ankley et al. 2003; Zeilinger et al. 2009; McGree et al. 2010; Leet, Gall, and Sepúlveda 2011). Moreover, considering the wide spectrum of biological activities in which steroid hormones are involved, the adverse effects of steroid hormones in exposed organisms are not limited to reproductive systems. Studies by Pinto et al. and Hallgren et al. determined changes in fish biomass and mineralized tissues of organisms exposed to estrogens (Pinto, Estêvão, and Power 2014; Hallgren et al. 2014), and research carried out by Rivero-Wendt et al. and Zheng et al., revealed that androgens including 17α-methyltestosterone could produce malformations in zebrafish or oxidative stress in the liver of juvenile tilapia (Rivero-Wendt et al. 2016; Zheng et al. 2016). In addition, as shown in a study by Colman et al. which analyzed the effect of a synthetic estrogen, 17α-ethynylestradiol, on aggression and the courtship behavior of zebrafish (Colman et al. 2009), changes associated to the presence of steroid hormones in natural waters could even include the behavior of the organisms. As for glucocorticoids, because of their interaction in many biological systems including muscles, bones, lipids, blood and cardiovascular or nervous systems (McKay and Cidlowski 2003; Huetos Hidalgo 2004), it is reasonable to assume that they could affect the organic systems of exposed aquatic organisms. In this respect, Nakayama et al. determined an increase of free amino acids and glucose levels in blood serum of common carp (*Cyprinus carpio*) exposed to synthetic glucocorticoids, even at environmental concentrations (Nakayama et al.

2014). However, the same author determined in a subsequent study that the immune responses of common carp to bacterial infection are not affected by the exposure of environmental concentrations of these types of steroid hormones (Nakayama et al. 2016).

Given all of the above, there is a clear need for the proper and precise determination and control of steroid hormone compounds in environmental systems. For this, it is necessary to consider all the variables involved in the analytical process, including the kind of EP and matrix studied, the extraction process and the determination techniques to use. Research in this field needs to focus on the development of sensitive and selective analytical methodologies with low environmental impacts for the determination of these types of pollutants. Progress in this respect is underway and ongoing, following the trends in analytical chemistry of miniaturization, automation and the use of new materials and techniques.

In this chapter, an updated overview is presented of the different aspects of the analytical process used for the detection of steroid hormones in environmental samples, including sample pretreatment, extraction, preconcentration and determination methodologies, while focusing on the trends and studies of the last 10 years.

Sampling, Storage and Sample Pretreatments

Before the analysis of an environmental sample, the analytical chemist must first consider its complexity. Because of the presence of interferences in these types of samples, certain preparation steps such as filtration or purification must be applied before the extraction and determination processes. These preparation steps are essential for the selective and sensitive determination of steroid hormones, but in most cases are time-consuming processes which can affect the overall reproducibility and recovery of the analytical method. Before these preparation steps, it is important to collect and store samples in an appropriate way (Li and Migliaccio 2011; Fedorova et al. 2014). Depending on the extraction and subsequent preconcentration process, the quantity of the samples can range from a few milliliters to liters and from a few grams to hundreds of grams, and they are usually collected in clean glass containers. As a rule, for cleaner samples, the sample volume or quantity will be higher than for more complex matrixes. For example, the required volumes of WWTP influent samples are usually lower than those for effluent because of the large amount of interferences present in the former (Chang, Hu, and Shao 2007; Pedrouzo et al. 2009, 2011; Chang et al. 2011).

After collecting the sample, certain pretreatments are required to ensure its subsequent correct analysis and its storage in conditions which do not produce a degradation of the target analytes. In this regard, in samples with a high concentration of bacteria, such as wastewaters or WWTP sludge, the addition of reagents to prevent bacterial degradation is essential as the degradation of organic compounds like steroid hormones takes place at a faster rate than in cleaner samples such as freshwaters or sediments (Labadie and Budzinski 2005). Inhibition of bacterial degradation can be achieved by acidification of the liquid sample to pH values lower than 3.5 or by adding formaldehyde (Baronti et al. 2000; Labadie and Budzinski 2005; Patrolecco et al. 2013; Emnet et al. 2015; Avar et al. 2016). For solid samples, it is necessary to eliminate the water present in the sample because the humidity could produce undesired modifications of the matrix or partial dissolution of the target analytes. Dehydration of the sample can be performed by heating the sample in a laboratory oven at temperatures close to 100°C, though this might be problematic as the heating can result in volatilization losses. For this reason, many authors use a freeze-drying process to inhibit bacterial degradation.

After these pretreatments, environmental samples are stored in darkness to avoid photochemical degradation. Another important factor is the storage temperature of the sample. For liquid and solid samples which will be analyzed in the short or mid-term, samples can be kept refrigerated at low temperatures (below 5°C) until analysis, whereas for long-term storage the samples should be frozen at temperatures of -20°C.

Before the extraction and preconcentration processes, the homogenization of the sample is essential. In liquid samples with suspended particles, it is necessary to perform a procedure to remove them as they can produce interferences in the analytical methodology. To carry out the separation of suspended matter, two methods can be used: filtration or centrifugation. Filtration is recommended for large sample amounts, while centrifugation is better when the sample volume is small (Clara et al. 2004; Labadie and Budzinski 2005; Avar et al. 2016).

For solid samples, homogenization is necessary to ensure their representativeness and is achieved by pulverizing or grinding the samples. After pulverization, the samples must be sieved to obtain a fraction with similar particle size and thus ensure a homogenized system. Normally, the sieved samples should have a fraction size of 0.5-1.5 mm (Peysson and Vulliet 2013; Kumirska et al. 2015).

Extraction Techniques

Liquid Samples

The technological developments of the last decades in the field of analytical chemistry have permitted the development of more selective and sensitive detection techniques. Notwithstanding such advances, these techniques are not useful by themselves for the determination of steroid hormones because of the low concentrations found in environmental samples as well as the large amount of interferences present in these types of samples. Hence, selection of the extraction and preconcentration technique must take into account the characteristics of the matrix which is to be analyzed.

Liquid-liquid extraction (LLE) has traditionally been the technique employed. It is based on the transfer of analytes from an aqueous sample to an organic phase which is achieved after successive stages of shaking. Some studies have chosen this technique for environmental and biological samples (López de Alda and Barceló 2001; Li et al. 2015) due to its simplicity and acceptable recoveries. However, nowadays the use of LLE is discouraged because the large amounts of organic solvents it requires can be toxic and harmful for environment, and because the extraction times are usually very long.

Consequently, in recent years, extraction and preconcentration techniques have followed the trend of being sustainable and "greener" (Armenta, Garrigues, and de la Guardia 2008). One of these "greener" techniques is solid phase extraction (SPE), which has rapidly gained in popularity over the last few decades and has become a standardized extraction and preconcentration technique. In fact, SPE is the recommended extraction method in analysis protocols of the United States Environmental Protection Agency (USEPA) and the European Commission (EC) (United States Environmental Protection Agency 2010; Loos 2012). Solid phase extraction is based on the selective retention of analytes on an appropriate sorbent and a subsequent elution with an organic solvent. In a first step, the sorbent of the SPE device must be activated by passing a small quantity of organic solvent and a small quantity of water. After that, the sample is passed through the SPE device. In this step, the analytes in the sample are retained on the sorbent because they present a great affinity to it, while most of the interferences of the matrix pass through the SPE device. After the adsorption of the analytes, the sorbent is washed to eliminate the interferences that might have been retained and finally, in the last step, an organic solvent is used to elute the target analytes retained in the sorbent of the SPE device. Figure 1a shows the different steps involved in SPE. Depending on the sorbent and organic solvent used, the extraction can be highly selective, so the study and optimization of the SPE steps are crucial to obtain good extraction efficiencies. Elution volumes in SPE are usually in the range of a few milliliters, so this technique is not only an extraction method but also a preconcentration technique which produces more sensitive analytical procedures capable of extracting organic compounds at trace and sub-trace levels. The organic solvents used as eluents in SPE are usually methanol and acetonitrile, individually or combined and also in aqueous solution. More non-polar organic solvents, including ethyl acetate, methylene chloride or hexane, could be used as SPE extractants but present the disadvantage of the need to apply an evaporation step prior to the analysis (Tomšíková et al. 2012; Guedes-Alonso et al. 2014).

Because of the standardization of this extraction and preconcentration technique, nowadays we can find a wide variety of commercial polymeric sorbents, including reversed phase, normal phase or ionic exchange types. Hydrophilic-lipophilic balance polymers like OASIS HLB are commonly used for the determination of steroid hormones in wastewater and environmental samples (Pedrouzo et al. 2011; S. Liu et al. 2011; Tremblay, Gadd, and Northcott 2018; Chen et al. 2019). This type of polymer is extensively used because of its high capacity to retain compounds with different physicochemical properties. Another important type of sorbent used in the extraction of steroid hormones is silica-based and C18 end-capped, but it should be noted that these sorbents have a limited pH range and the silanol residues produced can be an important interference (Fontanals, Marcé, and Borrull 2005). As to the different devices used in SPE, cartridges are used more frequently for the determination of steroid hormones because of their lower cost than discs and because they can easily be prepared in the laboratory with different sorbents. Discs are available in only a limited range of sorbent chemistries because they are mainly produced in a manufacturing setting, but present a larger surface area per unit bed mass and so are appropriate for passive sampling (Poole 2003). Owing to the similarities between the structures and properties of the different families of steroid hormones, SPE is generally used for the combined extraction and preconcentration of estrogens, androgens, progestogens and glucocorticoids (Herrera-Melián et al. 2017; Jauković et al. 2017; Yao et al. 2018; Zhang and Fent 2018).

One of the current trends in analytical chemistry is its focus on the automation of analytical methodologies. In this regard, in the last 10 years, automated systems have been developed to perform SPE. Automation permits the elimination of handling errors and has even led to the development of a new SPE technique known as on-line SPE, which permits the coupling of an automated SPE system to a separation and determination instrument such as a liquid chromatograph. In these automated systems, the SPE device, which is similar to a chromatographic column, is installed after the injection valve and, by adjusting the valves of the system, it is possible to elute the cartridge with the chromatographic mobile phase. As a result, the elution volume can be tens or hundreds times lower than when using traditional SPE. For the extraction of steroid hormones, this SPE variant has been successfully used with similar or even higher extraction efficiencies in comparison with traditional SPE. Guedes-Alonso et al. developed two analytical methods based on offline and on-line SPE coupled to ultra-high performance liquid chromatography tandem mass spectrometry (SPE-UPLC-MS/MS), with both methods showing recoveries up to 90% and low ng·L^{-1} detection limits (Guedes-Alonso, Sosa-Ferrera, and Santana-Rodríguez 2013; Guedes-Alonso, Sosa-Ferrera, and Santana-Rodriguez 2015). Another advantage of on-line SPE methods is that they can be used to analyze small quantities of samples. For example, both Fayad et al. and Guedes-Alonso et al. have developed online SPE methodologies for the determination of different steroid hormones which use less than 10 mL of sample (Fayad, Prévost, and Sauvé 2013; Guedes-Alonso, Sosa-Ferrera, and Santana-Rodriguez 2015).

Another trend in analytical chemistry that has been introduced in SPE is the development of new and more selective sorbents, including molecularly imprinted polymers (MIPs). These new polymers are synthetic materials which are created using a molecule as a template. To create the material, one or two "functional monomers" are placed in the reactive positions of the template molecule. Next, the functional monomers are polymerized using a cross-linking agent. Then, the template molecule is removed using an appropriate solvent, leaving a cavity where only molecules with a similar structure to that of the template molecule fit. The variables that need to be considered in the design of an MIP are the choice of functional monomers depending on the molecule, the cross-linking polymer and the polymerization method (Le Noir et al. 2007). Once produced, the MIP presents a "molecular memory" which enables a high degree of selectivity as it permits the retention in the cavities of a limited number of molecules (Tamayo, Turiel, and Martín-Esteban 2007; Cheong, Yang, and Ali 2013; Cáceres Fuentes 2016). Moreover, this type of polymer has a high level of tolerance to wide ranges of pH, temperature and organic solvents (Lucci, Núñez, and Galceran 2011). For the extraction of steroid hormones, the template molecule should be a sterane or a cholesterol because all steroid hormones present this basic structure (Rachkov et al. 2000; Sadowski and Gadzała-Kopciuch 2013). Other studies have used a specific molecule as template to obtain materials with high specificity, a substantial reduction of extracted interferences and excellent reproducibility with relative standard deviations (RSDs) below 2% (Gadzała-Kopciuch, Ričanyová, and Buszewski 2009) and recoveries near 100% (Zacs, Perkons, and Bartkevics 2016).

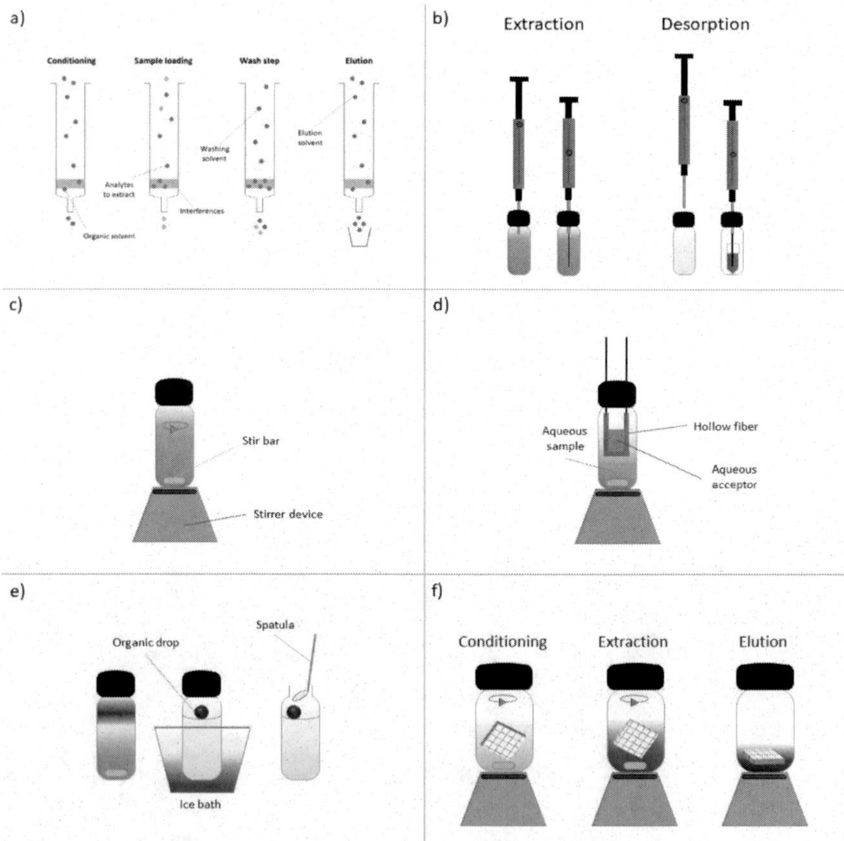

Figure 1. Different extraction techniques for environmental liquid samples. a) solid phase extraction; b) solid phase microextraction; c) stir bar sorptive extraction; d) hollow fiber liquid phase microextraction; e) solid-drop liquid phase microextraction; f) fabric phase sorptive extraction.

SPE is also a great technique to clean complex samples because of its capacity to remove interferences, divide extracted analytes depending on the elution steps applied, and even store analytes extracted from samples in which they can be degraded. As a clean-up methodology, the SPE cartridge retains the interferences and the clean extract is collected to be analyzed. Chang et al. used a combination of two sorbents to perform the clean-up of extracts of different families of steroid hormones (Chang et al. 2011), namely a silica gel cartridge for the purification of extracts to analyze androgens and progestogens and a Florisil cartridge for the purification of extracts to determine estrogens.

Apart from SPE and keeping the trends of miniaturization in mind, the solid phase microextraction (SPME) technique is now considered a good option. The SPME method was developed by Arthur and Pawliszyn in the early 90s, and the technique is based on the use of a fused silica fiber coated with a sorbent material with a thickness of a few hundred microns (Arthur and Pawliszyn 1990). The small size of the SPME devices allows the fiber to be placed in a syringe and transported, which means that extraction can be done *in situ*. After extraction of the target analytes, the fiber can be stored in the syringe until the time for desorption. This desorption can be liquid or thermal, depending on the detection system used (liquid or gas chromatography, respectively) and the physicochemical properties of the analytes (Figure 1b). The SPME technique presents linear results for a wide range of concentrations and shows great reproducibility even at low analyte concentrations. The standardization of SPME as a routine extraction method has resulted in the development of SPME fibers coated with many different polymers which are available commercially. For the extraction of steroid hormones, the most commonly used coatings are polyacrylate and carbowax/divinylbenzene (Peñalver et al. 2002; Yang, Luan, and Lan 2006; Zhang et al. 2009), but some authors like Liao et al. have developed specific coatings for SPME fibers based on monolithic fibers with similar efficiencies to commercial ones (Liao et al. 2016). The SPME method has been successfully used with liquid chromatography (LC) and mass spectrometry (MS). By way of example, Volmer et al. determined 13 corticosteroids in urine, and the SPME-LC-MS method has been shown to have similar efficiencies to those of traditional extraction techniques (Volmer and Hui 1997). Molecularly imprinted polymers have also been also as coating polymers in SPME, resulting in a higher selectivity of the method. Qiu et al. developed an SPME fiber using a testosterone-based MIP as coating polymer for the extraction of anabolic compounds. The fibers were placed in a homemade syringe and thermal desorption was applied in a gas chromatography (GC) system (Qiu et al. 2010). Automation of the SPME has also been developed by different authors, including in-tube SPME which uses as extraction device a fused silica capillary with an inner surface coated with a polymer (Eisert and Pawliszyn 1997; Kataoka 2002). The extraction and concentration of the analytes are directly performed in the stationary phase, thus presenting some advantages in comparison with conventional SPME including greater stability, a higher sorption capacity and the possibility of coupling the extraction technique with detection systems. Nevertheless, in-tube SPME presents an important drawback in the need to use very clean samples to avoid clogging of the capillary. In-tube SPME has been successfully used for the determination of steroid hormones. For example, Mitani et al. used an on-line in-tube SPME coupled to LC-MS to determine different estrogens in environmental waters, with high recoveries and detection limits in the $ng \cdot L^{-1}$ range (Mitani, Fujioka, and Kataoka 2005).

Another extraction technique used with steroid hormones has been the stir-bar sorptive extraction (SBSE) method which was developed by Baltussen et al. almost 10 years after SPME (Baltussen et al. 1999). In this technique, analyte extraction is performed on the surface of a stirring rod which is coated with a polymer (normally polydimethylsiloxane, PDMS). After extraction, as in SPME, desorption can be thermal or liquid depending on the properties of the compounds and the detection system, as can be seen in Figure 1c. The SBSE method presents some advantages in comparison with SPME, including a greater sorptive surface and the possibility of reusing the extraction device more than SPME fibers (Carpinteiro et al. 2012). In this respect, Kawaguchi et al. developed a method based on SBSE and GC-MS to determine estrogens in river water samples (Kawaguchi et al. 2004), while Almeida and Nogueira used an

SBSE-LC and diode array detection (DAD) to determine a group of estrogens and progestogens in water samples with detection limits that ranged from 0.3 to 1.0 $\mu g \cdot L^{-1}$ (Almeida and Nogueira 2006). Nevertheless, the available coatings of SBSE are limited, because only PDMS coated stir bars are commercially available and, as a result, the extractive capacity of SBSE is limited to non- or mid-polar compounds. To overcome this problem, some studies have focused on the development of new coatings. For example, Hu et al. tested a coating of polydimethylsiloxane and ß-cyclodextrin (PDMS/ß-CD) to determine four estrogens in environmental water samples, obtaining detection limits in the $ng \cdot L^{-1}$ range and high reproducibilities (RSDs lower than 10%) (Yuling Hu et al. 2007).

At the same time that SBSE was being developed, some studies started to evaluate the possibility of miniaturizing classic LLE. These techniques resolved some of the drawbacks of SPME and SBSE, especially in relation to the commercial availability of the extraction devices, offering greater methodological flexibility (Montesdeoca-Esponda 2013). Since the beginning of the XXI century, many liquid-liquid microextraction methods have been developed. The differences between these methods lie in the extraction device that is used or how the compounds are extracted from the sample. The first technique developed was liquid-phase microextraction (LPME), which keeps the amount of solvent to a minimum, as well as the exposure to toxic organic solvents. The LPME technique is performed using a small-quantity of water-inmiscible solvent (in the range of microliters or sub-microliters) and an aqueous matrix which contains the analytes. In the development of LPME, Pedersen-Bjedgaard and Rasmussen introduced a modification which they named hollow fiber liquid phase microextraction (HF-LPME) (Pedersen-Bjergaardand and Rasmussen 1999). In this technique, the solvent is placed in the porous hydrophobic hollow fiber and, as a result, the solvent in not in contact with the aqueous matrix and the samples can be stirred without losses in the extraction phase (Figure 1d). The HF-LPME technique has been successfully used for the determination of estrogens in reservoir and potable water samples (Basheer et al. 2005), as well as in tap water and wastewater samples (Zorita, Hallgren, and Mathiasson 2008), with detection limits in both studies in the range of a few ng·L^{-1}. Another modification that has been made of LPME is the solid drop liquid phase microextraction (SD-LPME) technique developed by Khalili-Zanjani et al. in 2007. In this technique, the glass vial which contains the sample and the extractant is immersed in an ice bath to obtain solidification of the organic extractant. As can be seen in Figure 1e, the solid drop of the extractant solvent located in the upper part of the liquid can be taken with a spatula for its subsequent analysis (Khalili Zanjani et al. 2007). The main problem with this technique is the limited variety of organic solvents with melting points near room temperature. Nevertheless, Chang et al. optimized the technique for the determination of four estrogens in river and tap water samples with high relative recoveries (between 87% and 116%) and detection limits that ranged from 0.8 to 2.7 µg·L^{-1} for spiked river water and 1.4 to 3.1 µg·L^{-1} for spiked tap water (Chang and Huang 2010).

Studies in recent decades have also focused on the miniaturization of solid phase techniques for the extraction of steroid hormones, including the thin film microextraction (TFME) method, which is an improvement on SPME, or, more recently, the fabric phase sorptive extraction (FPSE) method. The TFME technique uses chemically modified cellulose papers to perform the extraction of the analytes and overcomes one of the main disadvantages of microextraction techniques, the limited adsorption surface area. Saraji et al. used TFME to extract some synthetic and natural estrogenic hormones from aqueous samples. The detection limits for this extraction methodology ranged from 0.05 to 0.23 µg·L^{-1} (Saraji and Farajmand 2013).

The FPSE technique was developed by Kabir et al. in 2014 and is based on the use of a cellulose fabric coated with a polymeric sorbent where the analytes are extracted (Kabir and Furton 2014) (Figure 1f). This novel technique has been used for the extraction of different emerging pollutants, including surfactants, UV filters, pharmaceuticals and herbicides (Kumar et al. 2015; S Montesdeoca-Esponda et al. 2015; Racamonde et al. 2015; Roldán-Pijuán et al. 2015). Guedes-Alonso et al. 2016 used this new technique for the extraction of steroid hormones. In their work, the authors optimized a FPSE followed by UPLC-MS/MS to determine 10 steroid hormones (androgens and progestogens) in wastewater and urine samples. The detection limits, in the range of 1.7 to 267 $ng \cdot L^{-1}$, are very suitable for environmental samples, and relative recoveries higher than 70% were obtained (Guedes-Alonso et al. 2016). The main advantages of this extraction technique are its versatility (many different coatings can be used), the large sorptive surface (higher than SPME or SBSE), the short extraction times, and the easy handling required with the technique. Nevertheless, as with all the different extraction techniques, optimization of the variables involved in the extraction process is required, including the extraction and desorption times and volumes, to achieve high extraction efficiencies.

Table 2 summarizes the extraction methods and the analytical parameters for the analysis of steroid hormones in liquid samples.

Solid Samples

Considering the lipophilicity of the steroid hormones, with log KOW values ranging from 3 to 6 for most of them, it is logical to consider that solid samples like WWTP sludge or sediments could have considerable concentrations of steroid hormones. To confirm this, the development of analytical procedures is required that can extract these pollutants from solid samples for their subsequent analysis.

In the extraction of organic compounds from solid samples, an organic solvent with similar physicochemical properties to the analytes is placed in contact with the sample. By doing so, the adsorption forces which bind analyte and sample are broken and the analytes can then be dissolved in the organic solvent. Subsequently, it may be necessary to purify the extracts in order to eliminate interferences or matrix particles. The Soxhlet extraction technique was the first technique developed for solid samples. Nowadays, several procedures based on the Soxhlet technique are recommended by environmental quality institutions for the extraction of hormones from solid samples (USEPA 2007). Nevertheless, this technique presents some important drawbacks such as the long extraction times, which can be up to

48 hours, as well as the large volume of organic solvent used which has to be evaporated to concentrate the extracts.

Ultrasound assisted extraction (UAE) is one of the methods that has been developed as an alternative to Soxhlet extraction. This technique uses the energy produced by ultrasounds to favor the migration of the analytes adsorbed on the solid to the liquid phase. This takes place because the application of ultrasounds to the mixture of solid sample and organic extractant produces a process called cavitation, which consists of the formation and expansion of small bubbles until they burst and in this way generate local changes in temperature and pressure. The ultrasounds can be produced in a water bath, as well as through the use of probes, sonoreactors or microplate horns. Although UAE considerably reduces extraction times and organic solvent volumes, the local differences of temperatures and pressure produced by cavitation mean that that the results are less reproducible (Zuloaga et al. 2012). Nonetheless, the advantages of UAE are the minimization of the use of organic solvents, the possibility of performing extractions at different temperatures to that of the boiling point of the organic solvent, and the extremely low costs compared with more sophisticated techniques. After the extraction, the collected extractant must be cleaned by filtering or centrifugation. It should also be noted that, as with the other extraction techniques, the variables involved in UAE - extraction time, amount of sample or extractant volume- must be optimized to obtain high extraction efficiencies. This extraction technique has been widely used for the extraction of steroid hormones from solid samples. Lopez de Alda et al. used UAE coupled to LC-DAD-MS for the determination of estrogens and progestogens in river sediments exposed to wastewater effluents. The authors obtained appropriate detection limits that ranged from 0.04 to 1 $ng \cdot g^{-1}$ and extraction efficiencies from 64-100%, and found measurable concentrations of ethynylestradiol and estrone in real samples (Alda et al. 2002). The UAE technique is also a multiresidue extraction technique, as reported by Liu et al. In a study by these authors, a method was developed for the extraction of 28 steroid hormones from different families: estrogens, androgens, progestogens and glucocorticoids. The recoveries that were obtained, in a 63-138% range, were similar to those in other studies (Liu et al. 2011). In both the Lopez de Alda et al. and the Liu et el al. studies, a clean-up step was required to purify the obtained extract. For this, the former used octadecyl silica cartridges and the latter silica gel cartridges. In consequence, the overall analysis time with this method is considerably lengthened, while the implementation of clean-up steps can produce losses related to the handling of the samples.

As with liquid sample extraction techniques, current trends are focusing on the minimization of analysis times and reducing the use of organic solvents. In this regard, some efficient and robust extraction techniques have been developed in the last few decades, including pressurized liquid extraction (PLE), supercritical fluid extraction (SFE) and microwave assisted extraction (MAE).

The PLE method requires shorter times than UAE to perform the extraction of the analytes. The extraction is produced in a device where the sample and a small quantity of organic solvent are placed. This device can reach high pressures and temperatures to perform the extraction of the analytes, allowing the use of solvents at temperatures above their boiling points and, consequently, the breakup of the electromagnetic forces which bind the analytes to the sample (Mustafa and Turner 2011). Supercritical fluid extraction follows a similar principle to that of PLE, but in SFE a fluid in supercritical conditions is used as extractant agent. These supercritical conditions are that the extractant agent presents some properties of a fluid and some of a gas (Zougagh, Valcárcel, and Ríos 2004). Extraction using this technique is very efficient because of the high solubility and diffusivity of the fluids, and can be almost absolute because the extractant is continuously renewed in the system. Other advantages of this technique are the possibility of coupling it to detection systems, which is known as on-line mode, and the use of non-toxic, cheap and relatively environmental friendly supercritical fluids as carbon dioxide (CO2) (Zougagh, Valcárcel, and Ríos 2004; Herrero et al. 2010). Nevertheless, the main drawback of PLE and SFE is the high cost of the instrumentation required, and for this reason the use of these techniques has not been standardized. The PLE method has been used for the determination of steroid hormones with satisfactory results. In this respect, Gineys et al.

developed a PLE method with SPE as a clean-up step and LC-MS/MS as a separation and detection system for the determination of six sex hormones (estrogens, androgens and progestogens) and five glucocorticoids in soil samples with detection limits between 0.08 and 2.89 $ng \cdot g^{-1}$ (Gineys, Giroud, and Vulliet 2010). Hansen et al. also used PLE to extract ten steroid hormones from manure samples, which were then passed through two solid phase clean-up steps before derivatization and analysis by GC-MS. The recoveries of the PLE were between 67 and 105%, and appropriate detection limits were obtained for the measurement of these pollutants in manure and agriculture soil samples (Hansen et al. 2011).

As for SFE, Stolker et al. developed an SFE method for the determination of thirteen steroids in different animal tissues. In this work, CO2 was used as supercritical fluid and, after extraction, alkaline hydrolysis was performed to analyze the extracts by GC-MS. The method showed good detection limits, in the range of 2 to 5 $\mu g \cdot kg^{-1}$, and repeatability of 2 to 42% (Stolker, Zoontjes, and Van 1998).

Microwave assisted extraction was developed with a view to overcoming the drawback of the cost of PLE and SLE instrumentation. Microwave radiation is non-ionizing with a frequency between 0.3 and 300 GHz, although almost all microwave devices use a frequency of 2.45 GHz. This electromagnetic radiation can produce heating in two different ways depending on the properties of the molecules: by dipole rotation or by ionic conduction (Sanchez-Prado et al. 2015). In contrast to conventional heating, microwave heating of the samples takes place from the inside to the outside as the microwave energy directly affects the molecules of the sample. This is possible because the containers used in MAE are made from "transparent" materials like glass, quartz or Teflon which present very low dielectric constants. As a result, MAE is a very selective extraction methodology because the heating can be produced only in the sample or in the sample and solvent, depending on the properties of the target analytes. Another advantage is the relatively low cost of the instrumentation, comprising only a microwave heater and a carousel where the vessels are placed. The extraction can be performed at room temperature (in open-capped vessels) or under pressure and temperatures which can be higher than the boiling points of the organic solvents (using closed vessels) (Camel 2000). Several variables need to be optimized in the development of an MAE method, including the amount of sample, the organic solvent used and its volume, the power of the microwaves applied, and the extraction time. The MAE technique does not use large sample amounts, which tend to be in the range of a few grams, and a study of microwave power and extraction time is essential because of the possibility, if the conditions are extreme, of sample or target analyte degradation, solvent evaporation or multiple interference extraction (Madej 2009). The MAE technique also has its drawbacks, including the need for a clean-up step after extraction and the difficulty of coupling it with a detection system. For the extraction of steroid hormones from environmental solid samples, some authors have developed MAE-based analytical methods. Liu et al. used MAE to evaluate the presence of four estrogens in river sediments (Liu, Zhou, and Wilding 2004). The extracts were analyzed by GC-MS and the detection limits obtained were very good, in the range of 0.2 to 0.4 $ng \cdot g^{-1}$ and similar to those obtained in an MAE-GC-MS/MS method developed by Hibberd et al. (Hibberd et al. 2009). Even better detection limits were achieved when MAE extracts were analyzed by LC-MS/MS. In this case, Labadie et al. achieved detection limits in the range of 15 to 40 $pg \cdot g^{-1}$ for three estrogens (estriol, 17β-estradiol and 17α-ethynylestradiol) in river sediments and excellent overall recoveries, which ranged from 82 to 98% (Labadie and Hill 2007).

With respect to biological samples, there has been growing interest in the last few in the so-called QuEChERS technique. The name of this technique is the acronym of the words: quick, easy, cheap, effective, rugged and safe, which are the main properties of this extraction procedure. The QuEChERS extraction method was initially developed in 2003 by Anastassiades et al. for the determination of pesticides in complex edible matrices as in fruits and vegetables (Anastassiades et al. 2003). The method is based on micro-scale extraction with a few milliliters of organic solvent and different salts to favor the extraction of the analytes. To purify the sample, a dispersive solid phase extraction is then performed combining polymeric sorbents and magnesium sulphate (Cunha et al. 2007;

Wilkowska and Biziuk 2011). QuEChERS-based methods have been developed for the determination of steroid hormones not only in biological samples (Bergé and Vulliet 2015; Tan et al. 2016), but also in other matrices as sediments and soils (Camilleri and Vulliet 2015; Mulabagal, Wilson, and Hayworth 2017) In these studies, the authors achieved good detection limits, in the range of picograms, and recoveries in most cases higher than 75%. Moreover, in both studies, the authors detected different steroid hormones such as estrone, testosterone, cortisone or levonorgestrel in river and estuarine sediment samples.

Table 3 summarizes the extraction methods and analytical parameters for the analysis of steroid hormones in environmental solid samples.

Detection Techniques

After the extraction of steroid hormones from the matrix (liquid or solid), it is necessary to separate the extracted compounds for the subsequent determinations. The determination techniques can be divided in two major groups: biological and chromatographic.

The biological techniques are based on radioimmunoassays (RIA) or direct immunoassays such as ELISA (enzyme-linked immunosorbent assay). These biological techniques have been widely used in clinical samples, but they are seldom used to analyze environmental samples (López de Alda and Barceló 2001). Although they are easy to use and robust, the appropriateness of immunoassay methods has been questioned due to their low selectivity and strong dependence on matrix effects. As well as being time-consuming and often expensive, immunological methods, especially direct assays, tend to overestimate steroid hormone concentrations and, in most cases, it is not possible to measure many different steroid hormones in a single test. For these reasons, the use of immunoassays has progressively decreased and been replaced by chromatographic methodologies (Stanczyk and Clarke 2010), to the extent that chromatography-based techniques are now the mostly extensively used for steroid analysis in environmental matrices.

Table 2. Analytical techniques for the determination of steroid hormones in environmental liquid samples

Steroid hormones determined	Sample type	Extraction technique	Separation and determination technique	Analytical parameters	Reference
5 estrogens and 4 progestogens	Environmental waters and urine	SBSE	LC-DAD	LOD: 0.3-1.0 $\mu g \cdot L^{-1}$	Almeida and Nogueira, 2006
Drospirenone, levonorgestrel and progesterone	Freshwater samples	SPE	LC-MS/MS	LOQ: 0.03-0.11 $ng \cdot L^{-1}$	Avar et al. 2015
Estrone, 17β-estradiol, estriol and 17α-ethynylestradiol	Influent and effluent WWTP samples	SPE	LC-MS/MS	Recoveries: 86-90% LOQ: 0.2-0.9 $ng \cdot L^{-1}$	Baronti et al. 2000
Diethylstilbestrol, estrone, 17β-estradiol and 17α-ethynylestradiol	Reservoir and tap water	HF – LPME	GC-MS	LOD: 0.03-0.80 $ng \cdot L^{-1}$	Basheer et al. 2005
17β-estradiol, estrone, estriol, 17α-ethinyl estradiol, and diethylstilbestrol	River water	SPE	LC-MS	Recoveries: > 86% LOD: 3.2-10.6 $ng \cdot L^{-1}$	Benijts et al. 2002
Diethylstilbestrol, estrone, 17β-estradiol, 17α-ethynylestradiol and mestranol	Influent and effluent WWTP samples	SPME	GC-MS/MS	LOD: 0.2-3 $ng \cdot L^{-1}$	Carpinteiro et al. 2004
Estrone, 17β-estradiol, estriol and 17α-ethynylestradiol	River and tap water	SD – LPME	UHPLC-DAD	Recoveries: 87-116% LOD: 0.8-3.1 $\mu g \cdot L^{-1}$	Chang and Huang, 2010
six glucocorticoids (prednisone, predni-solone, cortisone, cortisol, dexamethasone, and 6α-methylprednisolone)	Influent and effluent WWTP samples and receiving river waters	SPE	UHPLC-MS/MS	Recoveries: 73-99% LOQ: 0.01-0.2 $ng \cdot L^{-1}$	Chang, Hu and Shao, 2007
24 steroid hormones (14 androgens, 5 progestogens and 5 glucocorticoids)	Natural WWTP water	SPE	UHPLC-MS/MS	Recoveries: 82-126%	Chen et al. 2019
Estrone, 17β-estradiol, estriol and 17α-ethynylestradiol	Seawater, sea ice and WWTP effluents	SPE	GC-MS	Recoveries: 94-142% LOQ: 0.4-7.0 $ng \cdot L^{-1}$	Emnet et al. 2015
Estrone, 17β-estradiol, estriol and 17α-ethynylestradiol, levonorgestrel, medroxyprogesterone, norethindrone and progesterone	WWTP influent and effluent samples	SPE	LC-MS/MS	Recoveries: 71-95% LOD: 8-60 $ng \cdot L^{-1}$	Fayad, Prévost and Sauvé, 2013
17β-estradiol, testosterone and progesterone	Urine	Molecularly imprinted SPE	LC-DAD	Recoveries: 77-96% LOD: 0.47-1.26 $\mu g \cdot L^{-1}$	Gadzała-Kopciuch, Ričanyová and Buszewski, 2009
5 estrogens, 1 androgen and 2 progestogens	WWTP influent and effluent samples	SPE	UHPLC-MS/MS	Recoveries: 73-90% LOD: 0.15-9.35 $ng \cdot L^{-1}$	Guedes. Alonso et al. 2013
4 estrogens, 3 androgens, 4 progestogens and 3 glucocorticoids	WWTP effluents	On-line SPE	UHPLC-MS/MS	Recoveries: 50-90% LOD: 0.5-13.2 $ng \cdot L^{-1}$	Guedes. Alonso et al. 2015

Table 2. (Continued)

Steroid hormones determined	Sample type	Extraction technique	Separation and determination technique	Analytical parameters	Reference
4 androgens and 6 progestogens	Hospital area wastewater, WWTP effluents and tap water	FPSE	UHPLC-MS/MS	LOD: 1.7-264 $ng \cdot L^{-1}$	Guedes- Alonso et al. 2016
Estrone, 17β-estradiol, 16α-hydroxyestrone and 17αethynylestradiol	River water	SPE	GC-MS/MS	Recoveries: 72-119% LOD: 0.22-0.28 $ng \cdot g^{-1}$	Hibberd et al. 2009
Estrone, 17β-estradiol, estriol, androsterone and progesterone	Surface water	SPE	LC-UV	Recoveries: 93.4-106.2% LOD: 0.07-0.61 μg·L-1	Hu et al. 2013
Estrone, 17β-estradiol, estriol and 17α-ethynylestradiol	River and lake water	SBSE	LC-UV	Recoveries: 84.6-124.1% LOD: 0.08-0.11 $\mu g \cdot L^{-1}$	Hu et al. 2007
Estrone, 17β-estradiol, estriol and 17α-ethynylestradiol	Wastewater and river and lake water	SPE	GC-MS	Recoveries: 83.8-94.7% LOD: 0.1-0.5 $ng \cdot L^{-1}$	Huang et al. 2013
33 androgens, 14 estrogens, 12 progestins and 11 corticosteroids	Seawater	SPE	UHPLC-HR-Q-Orbitrap-MS	Recoveries: 95-109% LOQ: 0.13-5.00 $ng \cdot L^{-1}$	Huysman et al. 2017
6 estrogens and 3 progestogens	Wastewater and river water	SPE	LC-MS/MS	Recoveries: 70-114% LOD: 1.7-15.0 $ng \cdot L^{-1}$	Jauković et al. 2017
Estrone, 17β-estradiol and 17α-ethynylestradiol	River water	SBSE	GC-MS	Recoveries: 90.3-105.7% LOD: 0.2-1 $ng \cdot L^{-1}$	Kawaguchi et al. 2004
Estriol, 17β-estradiol, testosterone, 17α-ethinylestradiol, estrone, progesterone and mestranol	Tap and lake water	SPME	LC-DAD	Recoveries: 77.7-117% LOD: 0.027-0.120 $\mu g \cdot L^{-1}$	Liao et al. 2015
17β-estradiol and 17α-ethynylestradiol	Wastewater, tap and river water	LPME	LC-FD	Recoveries: 98-106% LOD: 2.0-6.5 $ng \cdot L^{-1}$	Lima et al. 2013
4 estrogens, 14 androgens, 5 progestogens and 5 glucocorticoids	Influent and effluent wastewater and surface water	SPE	LC-MS/MS	Recoveries 44-200% LOD: 0.01-1.44 $ng \cdot L^{-1}$	Liu et al. 2011
Diethylstilbestrol, dienestrol, estrone, 17α-estradiol, 17β-estradiol, estriol and 17α-ethynylestradiol	River and tap water	Molecularly imprinted SPE	UHPLC-MS/MS	Recoveries: 42-106% LOD: 4.5-9.8 $ng \cdot L^{-1}$	Lucci, Nuñez and Galceran, 2011
Diethylstilbestrol, estrone, 17β-estradiol, estriol and 17α-ethynylestradiol	River water	In-tube SPME	LC-MS/MS	Recoveries: 86.1-106.8% LOD: 2.7- 11.7 $ng \cdot L^{-1}$	Mitani, Fujioka and Kataoka, 2005
Estrone, 17β-estradiol and 17α-ethynylestradiol	Influent and effluent wastewater and river water	SPE	LC-FD	Recoveries: 72-100% LOQ: 10-50 $ng \cdot L^{-1}$ (estrone: 800 - 1100 $ng \cdot L^{-1}$)	Patrolecco et al. 2013

Steroid hormones determined	Sample type	Extraction technique	Separation and determination technique	Analytical parameters	Reference
Estrone, 17β-estradiol and 17α-ethynylestradiol	Influent and effluent wastewater and river water	SPE	LC-ion-trap MS	Recoveries: > 65% LOQ: 4-150 $ng \cdot L^{-1}$	Patrolecco, Capri and Ademollo, 2015
11 estrogens and conjugates	Influent and effluent wastewater and river water	SPE	LC-MS/MS	Recoveries: 23-87% LOD: 10-70 $ng \cdot L^{-1}$	Pedrouzo et al. 2009
11 estrogens and conjugates	Influent and effluent wastewater	SPE	LC-MS/MS	Recoveries: 23-76% LOD: 10-100 $ng \cdot L^{-1}$	Pedrouzo et al. 2011
11 sex hormones	Mineral, underground and river waters	SPE	LC-UV-FD	LOD: 6-24 $ng \cdot L^{-1}$	Pérez and Escandar, 2016
Estrone, 17β-estradiol, mestranol, diethystilbestrol and 17α-ethynylestradiol	Lagoon waters	SPE	LC-MS	LOD: 0.1-2.6 $ng \cdot L^{-1}$	Pojana et al. 2007
Estrone, 17β-estradiol, estriol	Water	Molecularly imprinted SPE	LC-DAD	Recoveries: 74.3-82.2% LOD: 8.71-10.99 $\mu g \cdot L^{-1}$	Sadowski and Gadzala-Kopciuch, 2013
17α-ethynylestradiol	Pharmaceutical wastewaters	HF-LPME	LC-UV	Recoveries: 90-110% LOD: 10 $\mu g \cdot L^{-1}$	Seidi et al. 2019
16 sex hormones	Waters and wastewaters	On-line SPE	LC-MS/MS	Recoveries: 83.4-107.1%	Snow et al. 2013
6 corticosteroids, 4 androgens and progesterone	River and drinking waters	SPE	LC-MS/MS	Recoveries: 68-126% LOD: 0.06-0.21 $ng \cdot L^{-1}$	Tölgyesi et al. 2010
Estrone, 17β-estradiol, estriol, 17α-ethynylestradiol, progesterone, medroxyprogesterone, norethindrone and levonorgestrel	Influent and effluent wastewater	On-line SPE	LC-MS/MS	Recoveries: 92-110% (estriol: 85%) LOD: 2-50 $ng \cdot L^{-1}$	Viglino et al. 2008
27 steroid hormones	Surface water	SPE	LC-MS/MS	Recoveries: 60.2-100.5% LOD: 0.001-0.5 $ng \cdot L^{-1}$	Yao et al. 2018
17β-estradiol and 17α-ethynylestradiol	Influent and effluent wastewater	SPE	ELISA	LOD: 0.01-1 $ng \cdot L^{-1}$	Yien Fang et al. 2019
17β-estradiol and 17α-ethynylestradiol	Tap water	SPE	GC-HRMS	Recoveries: 99-113% LOQ: 0.08 $ng \cdot L^{-1}$	Zacs, Perkons and Bartkevics, 2016
33 steroid hormones	Influent and effluent wastewater and river water	SPE	LC-MS/MS	Recoveries: 70-120% LOD: 0.01-10 $ng \cdot L^{-1}$	Zhang and Fent, 2018
Estrone, 17β-estradiol and 17α-ethynylestradiol	Tap and sewage water	HF-LPME	GC-MS	LOD: 1.6-10 $ng \cdot L^{-1}$	Zorita, Hallgren, Mathiasson, 2008

Table 3. Analytical techniques for the determination of steroid hormones in environmental solid samples

Steroid hormones determined	Sample type	Extraction technique	Separation and determination technique	Analytical parameters	Reference
5 estrogens, 2 androgens and 4 progestogens	Earthworms	QuEChERS	HPLC-MS/MS	Recoveries: 58.0-98.7% LOD: 0.5-13.3 $ng \cdot g^{-1}$	Bergé and Vulliet, 2015
Estrone, 17β-estradiol, testosterone, megestrol acetate, progesterone	River sediments	QuEChERS	HPLC-MS/MS	Recoveries: > 74% LOD: 0.03-0.2 $ng \cdot g^{-1}$	Camilleri and Vulliet, 2015
24 steroid hormones (14 androgens, 5 progestogens and 5 glucocorticoids)	Natural WWTP solid samples	SPE	UHPLC-MS/MS	Recoveries: 68-107%	Chen et al. 2019
Estriol, estradiol, estrone and ethynylestradiol	Antarctic biota samples	ASE – SPE	GC-MS	Recoveries: 94-142% LOQ: 1.1-17.5 $ng \cdot g^{-1}$	Emnet et al. 2015
1 estrogen, 2 androgens, 3 progestogens and 5 corticosteroids	Soil samples	PLE – SPE	HPLC-MS/MS	Recoveries: 49-100% LOD: 0.08-2.84 $ng \cdot g^{-1}$	Gineys, Giroud and Vuillet, 2010
5 estrogens, 3 androgens, 4 progestogens and 3 glucocorticoids	WWTP sludge samples	MAE	UHPLC-MS/MS	Recoveries: 60-120% LOD: 2.1-192.8 $ng \cdot g^{-1}$	Guedes. Alonso et al. 2016
4 estrogens, 4 androgens and 2 progestogens	Manure and soil samples	PLE – SPE	GC-MS/MS	Recoveries: 67-107% LOD: 0.08-0.43 $ng \cdot g^{-1}$	Hansen et al. 2011
Estrone, 17β-estradiol, 16α-hydroxyestrone and 17αethynylestradiol	River sediments	MAE - SPE	GC-MS/MS	Recoveries: 90-102% LOD: 0.05-0.14 $ng \cdot g^{-1}$	Hibberd et al. 2009
Estrone, 17β-estradiol, estriol and 17α-ethynylestradiol	Sediments	MAE	GC-MS	Recoveries: 82.3-98.7% LOD: 0.2-0.5 $ng \cdot g^{-1}$	Huang et al. 2013
Diethylstilbestrol, estrone, 17β-estradiol, estriol and 17α-ethynylestradiol	Soils	MAE - SPE	GC-MS	Recoveries: 75.9-86.3% (estrone: 38.7%) LOD: 0.4-0.9 $ng \cdot g^{-1}$	Kumirska et al. 2015
Estrone, 17β-estradiol and 17α-ethynylestradiol	River sediments	MAE - SPE	HPLC-MS/MS	Recoveries: 82-98% LOD: 15-40 $pg \cdot g^{-1}$	Labadie and Hill, 2007
4 estrogens, 14 androgens, 5 progestogens and 5 glucocorticoids	WWTP sludge	UAE - SPE	HPLC-MS/MS	Recoveries 62.6-138% LOD: 0.08-2.06 $ng \cdot g^{-1}$	Liu et al. 2011

Steroid hormones determined	Sample type	Extraction technique	Separation and determination technique	Analytical parameters	Reference
7 estrogens, 4 androgens, 2 progestogens and 4 glucocorticoids	Marine sediments	QuEChERS – dispersive- SPE	UHPLC-MS/MS	Recoveries: 64.0-123.3%	Mulabagal, Wilson and Hayworth, 2016
11 sex hormones	River sediments	UAE	HPLC-UV-FD	LOD: 0.1-0.9 $ng \cdot g^{-1}$	Pérez and Escandar, 2016
6 estrogens, 4 androgens, 3 progestogens	Sewage sludge	QuEChERS – dispersive- SPE	HPLC-TOF MS	Recoveries: 49-135% LOD: 200-2500 $ng \cdot g^{-1}$	Peysson and Vulliet, 2013
Estrone, 17β-estradiol, mestranol, diethystilbestrol and 17α-ethynylestradiol	Lagoon sediments and biota	UAE	HPLC-MS	Recoveries: 44-82% LOD: 0.2-0.5 $ng \cdot g^{-1}$	Pojana et al. 2007
16 sex hormones	Sediments and soils	MAE	HPLC-MS/MS	Recoveries: 66.5-128.1% LOD: 0.2 $ng \cdot g^{-1}$	Snow et al. 2013

Chromatographic Techniques Gas Chromatography (GC)

Gas chromatography is a separation technique appropriate for the separation and determination of volatile and thermally stable compounds. The resolutions and sensitivities achieved with GC are very adequate for the analysis of environmental samples, but its application field is limited to thermally stable compounds and matrices. Because of the poor thermal stability and volatility of steroid hormones, derivatisation is required. Using GC, the molecules become separated by their distribution in a stationary phase which is usually an immobilized solid or liquid located in a chromatographic column and a mobile phase which is a carrier gas. The first chromatographic separations using GC were done in packed columns, but nowadays this type of column is in disuse and the separations are carried out in capillary columns. Coupling of GC is possible with many different detection techniques, including Fourier transform infrared spectroscopy (FTIR), nuclear magnetic resonance (NMR) and mass spectrometry. In this latter case, an appropriate interface is necessary because the mass spectrometry detection must be performed in a vacuum. Gas chromatography was the main chromatographic technique for the determination of steroid hormones until the late 90s (López de Alda and Barceló 2001). This technique coupled to mass spectrometry (GC-MS) or triple quadrupole mass spectrometry (GC-MS/MS) were the most developed types as these systems provided low detection limits, very appropriate for the determination of steroid hormones in environmental samples. The advantages of GC-MS and GC-MS/MS include better chromatographic resolution and sensitivity than liquid chromatography for certain specific compounds, including sterol derivatives and steroid hormone metabolites. For this reason, these techniques are very useful in metabolomics (Krone et al. 2010). Nevertheless, GC also has its drawbacks, including long sample preparation times, the impossibility of full automation of the process or the analysis of low-volatility compounds such as conjugated steroid hormones (Krone et al. 2010). Other aspects to be considered include the need to derivatize steroid hormones to increase their volatility and thermal stability, and, consequently, the sensitivity of the method. For example, bis(trimethylsilyl) trifluoroactamide (BSTFA), trimethylsilylimidazole (TMSI) or (N-methyl-N-(trimethylsilyl)-trifluoroacetamide (MSTFA) have all been used for the silylation of steroid hormones. Derivatization is usually a time-consuming step, potentially increasing the overall analysis time and reducing accuracy, as has been reported in some studies (Abdel-Khalik, Björklund, and Hansen 2013). Furthermore, for conjugated hormones, it necessary to apply a hydrolysis step before derivatization which reduces the sensitivity and reproducibility of the method because of the errors associated with these extra steps (Gomes, Scrimshaw, and Lester 2003). In light of the above, Croley et al. compared the sensitivity of analytical methods based on gas and liquid chromatography coupled to mass spectrometry, and reported the following sensitivity order: LC-MS/MS > GC-MS/MS > LC-MS (Croley et al. 2000).

Several authors have published studies on the use of GC for the determination of steroid hormones. Rutishauser et al. applied GC coupled to high resolution mass spectrometry (HRMS) for the determination of estrogens in wastewater samples. The derivatization agent was a mixture of MSTFA, TMSI and dithierythriol and the method showed appropriate detection limits for environmental samples which ranged from 0.5 to 1.0 $ng \cdot L^{-1}$ (Rutishauser et al. 2004).

The derivatization step requires a long heating time of between 10 and 60 min. In one study, Zuo et al. overcame the problem of heating during derivatization by using a microwave to perform this step. The authors showed that microwave irradiation at a power of 800 W for 60 seconds provided similar results to those of conventional heating in terms of linearity and precision, for all the steroid hormones that were tested. By combining SPE with microwave-accelerated derivatization, the authors achieved detection limits of 0.02-0.1 ng L−1 (Zuo, Zhang, and Lin 2007).

As for the extraction techniques applied before the chromatographic methods, apart from standardized techniques such as SPE, some authors have used GC as a separation and detection technique to evaluate the efficiency of microextraction techniques. In this respect, Carpinteiro et al. used SPME to extract estrogens from water samples and perform on-fiber derivatization of the target analytes (Carpinteiro et al. 2004), obtaining detection limits from 0.2 to 3 ng·L^{-1}, while Zorita et al. developed an analytical method for the determination of three estrogens in tap water and wastewater using an HF-LPME-GC-MS method, achieving detection limits that ranged from 1.6 to 10 ng·L^{-1} (Zorita, Hallgren, and Mathiasson 2008).

Liquid Chromatography (LC)

Analytical methods based on liquid chromatography have seen an exponential growth in the last few decades because of their versatility and sensitivity. The use of LC for the determination of organic compounds has progressively increased, and nowadays this technique is one of the most widely used separation techniques in analytical laboratories.

The popularity of LC is primarily due to its versatility and ability to analyze a large number of organic compounds, as well as to the development of different detectors which can be coupled to this technique. One of the most recent developments is that of ultra-high-performance liquid chromatography (UHPLC). This evolution of LC uses short columns with particle sizes even lower than HPLC (1.7 μm or less) and higher pressures, in the range of 5000 to 10000 psi, which produce very short chromatograms of narrow peaks. The analysis time, which involves the separation and detection of compounds, is usually just a few minutes. The use of LC to determine organic compounds is gaining in popularity, and this trend has also been noticeable in the analysis of steroid hormones. This family of compounds have traditionally been determined by GC despite the low volatility of steroids, but in recent years LC coupled to different detection systems has been gaining favor over GC (Guedes-Alonso et al. 2014). Both optical and mass spectrometry detectors have been used as detection systems. In view of the similar structures of the different steroid hormones, it is important to achieve a good chromatographic separation, not only when optical detectors are used, but also with mass spectrometry detectors because the parent and daughter ions can be, in some cases, very similar (Lucci, Núñez, and Galceran 2011). In fact, the improvement in MS and MS/MS systems over recent years has resulted in these detectors, coupled to LC or UHPLC, being used for the determination of steroid hormones in environmental samples.

Regarding the chromatographic columns used to perform the separation of steroid hormones, reversed phase columns are the preferred option in the analysis of these compounds. Octadecyl carbon chain (C_{18}) bonded silica columns are the most frequently used (Guedes-Alonso et al. 2014), while some studies have used phenyl-bonded silica or alkyl-amide columns to analyze steroid hormone families in environmental water samples (Chang and Huang 2010; Lucci, Núñez, and Galceran 2011; Tölgyesi et al. 2010) or C_8 columns to analyze estrogens in mussel samples (Pojana et al. 2007). The mobile phase used in these columns is usually formed by water and organic solvents such as methanol, widely used to separate androgens, progestogens or corticoids, or acetonitrile which is generally used in the analysis of estrogens (Guedes-Alonso et al. 2014). Many authors add modifiers to the mobile phase to improve the selectivity of the method, using in most cases acetic and formic acid, and less frequently, acid or basic buffers or ammonia. Using MS detectors, the composition of the mobile phase is a key factor in ionization efficiency. Some authors state that the addition of modifiers to the mobile phase can produce important ion suppression in the ionization of estrogens (Benijts et al. 2002), but others have obtained excellent results using acid or basic modifiers (Guedes-Alonso, Sosa-Ferrera, and Santana-Rodríguez 2013; Guedes-Alonso, Sosa-Ferrera, and Santana-Rodriguez 2015).

Optical detectors are commonly used in the analysis of steroid hormones by LC because of their lower cost in comparison with mass spectrometry detectors. The sensitivity of this type of detector is lower than with MS, but they are an excellent choice for routine analyses or to optimize extraction methods (Chang and Huang 2010; Yinfen Hu et al. 2013; Pérez and Escandar 2016; Seidi et al. 2019). Both ultraviolet-visible (UV) and diode array detector (DAD) have been used, with the most frequent determination wavelength being 240 nm, although some authors have reported excellent results using different wavelengths between 190 and 280 nm. Another optical technique used is fluorescence detection (FD). The main drawback of FD is the low number of steroid compounds which show this property. Some estrogens, including 17β-estradiol, estriol or 17α-ethynylestradiol, can be analyzed using FD as they present a phenolic ring in their structure which produces fluorescence when the molecule is excited (Fedorov, Cherkasova, and Samoilova 2004). For example, Vega-Morales et al. and Guedes-Alonso et al. developed SPE-HPLC-FD and SPE-UHPLC-FD methods respectively, for estrogen determination in wastewater samples (Vega-Morales, Sosa-Ferrera, and Santana-Rodríguez 2010; Guedes-Alonso, Sosa-Ferrera, and Santana-Rodríguez 2012). According to the authors, the sensitivity of the methods was similar to that obtained with chromatographic methods using mass spectrometry detection. Also, Lima et al. developed a low-cost methodology based on dispersive liquid–liquid microextraction and LC with fluorescence detection to determine 17β-estradiol and 17α-ethynylestradiol in water samples. The authors obtained detection limits between 2.0 and 6.5 $ng \cdot L^{-1}$, which are similar to those obtained using mass spectrometry detection (Lima et al. 2013).

Nevertheless, most studies on the determination of steroid hormones in environmental samples use MS or MS/MS detectors because of their sensitivity and selectivity. The most commonly-used interfaces are electrospray ionization (ESI) and atmospheric pressure chemical ionization (APCI) both in positive and negative modes (López de Alda and Barceló 2000). Estrogens are usually determined in ESI-negative mode, using as parent ions the negative forms of molecular ions $[M-H]^-$, although some authors have analyzed estrogens in positive mode using dehydrated molecular ions as precursor ions (Viglino et al. 2008; Chang et al. 2011). These same ions are also used in positive mode when an APCI interface is used (Lucci, Núñez, and Galceran 2011; Huysman et al. 2017). With respect to the other steroid families, they are commonly analyzed in positive mode using molecular ions as precursor ions $[M-H]^+$ (Guedes-Alonso et al. 2014). Triple quadrupole mass spectrometry is the most commonly used MS-based detection system because it provides good detection limits and selectivity. Croley et al. used different systems of chromatography (GC and LC) with mass spectrometry detection (MS and MS/MS) and showed that, for the determination of steroid hormones, detection limits using LC-MS/MS and GC-MS/MS were similar, in the range of a few ng·L^{-1}, while for LC-MS, the detection limits were 40 times higher (Croley et al. 2000). Comparing GC-MS/MS and LC-MS/MS, the first separation and detection technique has the drawback of derivatization so, despite the similar detection limits, the analysis procedure is easier when LC in used instead of GC.

In light of the above, many studies have used LC to monitor the presence of steroid hormones in environmental samples (Vidal-Dorsch et al. 2012; Emnet et al. 2015; Patrolecco, Capri, and Ademollo 2015; Zhang, Zhao, and Fent 2017). With respect to liquid samples, the highest concentrations of steroid hormones have been found in influent samples of WWTPs, at concentrations up to µg·L^{-1}, while the concentrations after a purification system has been applied are usually lower (Manickum and John 2014; Yarahmadi et al. 2018; Yien Fang et al. 2019). In recipient waters such as river or surface waters, the concentrations are usually low, in the range of a few ng·L^{-1} (Pal et al. 2010; Aris, Shamsuddin, and Praveena 2014; Luo et al. 2014; Barbosa et al. 2016). Studies for solid samples are scarcer, although Snow et al. determined sixteen steroid hormones in soils and sediments obtaining detection limits near 0.2 ng·g^{-1} (Snow et al. 2013). For their part, Guedes-Alonso et al. developed an MAE-UHPLC-MS/MS for the determination of estrogens, androgens, progestogens and glucocorticoids in wastewater sludge samples that showed recoveries over 70% and detection limits in the range of 1.1 to 7.9 ng·g^{-1} (Guedes-Alonso et al. 2016).

CONCLUSION

Steroid hormones constitute an important group of emerging pollutants which have attracted the attention of the international scientific community. This is because steroid hormones are essential in many body functions and can produce harmful effects in aquatic organisms at trace and ultra-trace levels. In fact, several studies have reported adverse effects in aquatic biota, including hermaphroditism, unisexual populations and infertility, as well as malformations, problems in different organic systems and behavioral changes of exposed organisms. Moreover, steroid hormones are constantly being introduced into the environment since these compounds are continuously excreted by humans and animals, and wastewater treatments are not designed to remove them. The concern about the growing problem of steroid hormones has resulted in some of them being included in water quality regulations as potential pollutants. Although the study of the presence of steroid hormones in solid samples is not as broad as for liquid samples, there has been a steady increase in such studies in recent years.

As steroid hormones are generally present in different environmental compartments at trace concentrations, it is a real challenge for analytical scientists to develop sufficiently sensitive and selective analytical methods for their correct determination. The first requirement is for extraction and preconcentration techniques which allow separation of the steroid hormones from the matrix in which they are present and their concentration to measurable levels. In this regard, solid phase extraction is now one of the most widely used routine extraction methods. Nonetheless, the trends in analytical chemistry towards miniaturization, automation and the use of new materials have resulted in the development of many new techniques that have been shown to be excellent alternatives to traditional extraction procedures.

After separation, extraction and concentration, sensitive detection techniques are required. In this respect, given certain drawbacks with immunological methods in the analysis of steroid hormones in environmental samples, biological techniques such as immunoassays have gradually been replaced by chromatographic techniques. Of these techniques, gas chromatography has commonly been used for the determination of steroid hormones because of its sensitivity and the possibility of coupling it to many different detection systems. Nevertheless, the unavoidable step of derivatization in gas chromatography to increase sensitivity for the analysis of these compounds considerably increases the analysis times and the possible accumulation of errors. As a result, liquid chromatography, which has undergone important methodological developments in recent decades, is now the most extensively used separation and determination technique for steroid hormones. Its success is due to the low volatility of steroid hormones as well as the possibility of coupling it to optical and mass spectrometry detectors. The combination of liquid chromatography and mass spectrometry has enabled the development of analytical procedures to identify and quantify hormones with very similar or identical mass weight, and even isomer compounds. For all of these reasons, LC techniques are considered reference methods for the analysis of steroid hormones in environmental samples.

If the present trends of automation, miniaturization and the use of new materials for the analysis of steroid hormones continue in the future, even more sensitive and selective methods will assuredly be elaborated to not only better analyze the presence of steroid hormones in environmental compartments, but also to determine the quality of these environments.

References

Abdel-Khalik, J., Erland, B. and Martin, H. (2013). Simultaneous Determination of Endogenous Steroid Hormones in Human and Animal Plasma and Serum by Liquid or Gas Chromatography Coupled to Tandem Mass Spectrometry. *Journal of Chromatography B*, 928: 58–77.

Alda, M. J., Gil, A., Paz, E., and Barceló, D. (2002). Occurrence and Analysis of Estrogens and Progestogens in River Sediments by Liquid Chromatography-Electrospray-Mass Spectrometry. *Analyst*, 127 (10): 1299–1304.

Almeida, C. and Nogueira, J. M. F. (2006). Determination of Steroid Sex Hormones in Water and Urine Matrices by Stir Bar Sorptive Extraction and Liquid Chromatography with Diode Array Detection. *Journal of Pharmaceutical and Biomedical Analysis*, 41 (4): 1303–1311.

Anastassiades, M., Lehotay, S.J., Štajnbaher, D., and Schenck, F. J. (2003). Fast and Easy Multiresidue Method Employing Acetonitrile

Extraction/Partitioning and 'Dispersive Solid-Phase Extraction' for the Determination of Pesticide Residues in Produce. *Journal of AOAC International*, 86 (2): 412–431.

Ankley, G. T., Jensen, K. M., Makynen, E. A., Kahl, M. D., Korte, J. J., Hornung, M. W., Henry, T. R., et al. (2003). Effects of the Androgenic Growth Promoter 17-β-Trenbolone on Fecundity and Reproductive Endocrinology of the Fathead Minnow." *Environmental Toxicology and Chemistry*, 22 (6): 1350–1360.

Aris, A. Z., Aida, S. S. and Sarva, M. P. (2014). Occurrence of 17α-Ethynylestradiol (EE2) in the Environment and Effect on Exposed Biota: A Review. *Environment International*, 69: 104–119.

Armenta, S., Garrigues, S. and de, la Guardia, M. (2008). Green Analytical Chemistry. *TrAC Trends in Analytical Chemistry*, 27 (6): 497–511.

Arthur, C. L. and Pawliszyn, J. (1990). Solid Phase Microextraction with Thermal Desorption Using Fused Silica Optical Fibers. *Analytical Chemistry*, 62 (19): 2145–2148.

Avar, P., Maasz, G., Takács, P., Lovas, S., Zrinyi, Z., Svigruha, R., Takátsy, A., Tóth, L. G. and Pirger, Z. (2016). HPLC-MS/MS Analysis of Steroid Hormones in Environmental Water Samples. *Drug Test. Analysis*, 8 (1): 123–127.

Baltussen, E., Pat, S., Frank, D. and Carel, C. (1999). Stir Bar Sorptive Extraction (SBSE), a Novel Extraction Technique for Aqueous Samples: Theory and Principles. *Journal of Microcolumn Separations*, 11 (10): 737–747.

Barbosa, M. O., Moreira, N. F. F., Ribeiro, A. R., Pereira, M. F. R. and Silva, A. M. T. (2016). Occurrence and Removal of Organic Micropollutants: An Overview of the Watch List of EU Decision 2015/495. *Water Research*, 94: 257–279.

Baronti, C., Curini, R., D'Ascenzo, G., Di Corcia, A., Gentili, A., and Samperi, R., (2000). Monitoring Natural and Synthetic Estrogens at Activated Sludge Sewage Treatment Plants and in a Receiving River Water. *Environmental Science & Technology*, 34 (24): 5059–5066.

Bartelt-Hunt, S. L., Snow, D. D., Kranz, W. L., Mader, T. L. Shapiro, C. A., van Donk, S. J., Shelton, D. P., Tarkalson, D. D., and Zhang, T. C.

(2012). Effect of Growth Promotants on the Occurrence of Endogenous and Synthetic Steroid Hormones on Feedlot Soils and in Runoff from Beef Cattle Feeding Operations. *Environmental Science & Technology*, 46 (3): 1352–1360.

Basheer, C., Jayaraman, A., Kee, M. K., Valiyaveettil, S. and Lee, H. K. (2005). Polymer-Coated Hollow-Fiber Microextraction of Estrogens in Water Samples with Analysis by Gas Chromatography-Mass Spectrometry. *Journal of Chromatography A*, 1100 (2): 137–143.

Benijts, T., Dams, R., Günther, W., Lambert, W., and De Leenheer. A. (2002). Analysis of Estrogenic Contaminants in River Water Using Liquid Chromatography Coupled to Ion Trap Based Mass Spectrometry. *Rapid Communications in Mass Spectrometry*, 16 (14): 1358–1364.

Bergé, A. and Vulliet. E. (2015). Development of a Method for the Analysis of Hormones and Pharmaceuticals in Earthworms by Quick, Easy, Cheap, Effective, Rugged and Safe (QuEChERS) Extraction Followed by Liquid Chromatography-Tandem Mass Spectrometry (LC-MS/MS). *Analytical and Bioanalytical Chemistry*, 407 (26): 7995–8008.

Cáceres Fuentes, C. A. (2016). Síntesis de Polímeros Impresos Molecularmente (MIPs) En Formato Micro y Nano Métrico, Para Preconcentrar Difenilamina y Para Ser Utilizados Como Receptores En El Test de ELISA. [Synthesis of Molecularly Imprinted Polymers (MIPs) in micro and nano metric format to preconcentrate diphenylamine and to be used as receptors in ELISA test.] *Universidad de Concepción. Facultad de Ciencias de la Ingeniería. Departamento de Ingeniería Química.* [*University of Concepción. Engeeniering Sciences Faculty. Chemistry Engeeniering Department*].

Camel, V. (2000). Microwave-Assisted Solvent Extraction of Environmental Samples. *TrAC Trends in Analytical Chemistry*, 19 (4): 229–248.

Camilleri, J. and Vulliet, E. (2015). Determination of Steroid Hormones in Sediments Based on Quick, Easy, Cheap, Effective, Rugged, and Safe (Modified-QuEChERS) Extraction Followed by

Liquid Chromatography-Tandem Mass Spectrometry (LC-MS/MS). *Analytical Methods*, 7 (22): 9577–9586.

Carpinteiro, I., Ramil, M., Rodríguez, I., and Nogueira, J. M. F. (2012). Combining Stir-Bar Sorptive Extraction and Large Volume Injection-Gas Chromatography-Mass Spectrometry for the Determination of Benzotriazole UV Stabilizers in Wastewater Matrices. *Journal of Separation Science*, 35 (3): 459–467.

Carpinteiro, J., Quintana, J. B., Rodríguez, I., Carro, A. M., Lorenzo, R. A. and Cela, R. (2004). Applicability of Solid-Phase Microextraction Followed by on-Fiber Silylation for the Determination of Estrogens in Water Samples by Gas Chromatography–tandem Mass Spectrometry. *Journal of Chromatography A*, 1056 (1–2): 179–185.

Chang, C. C. and Huang, S. D. (2010). Determination of the Steroid Hormone Levels in Water Samples by Dispersive Liquid–liquid Microextraction with Solidification of a Floating Organic Drop Followed by High-Performance Liquid Chromatography. *Analytica Chimica Acta*, 662 (1): 39–43.

Chang, H., Hu, J., and Shao, B. (2007). Occurrence of Natural and Synthetic Glucocorticoids in Sewage Treatment Plants and Receiving River Waters. *Environmental Science and Technology*, 41 (10): 3462–3468.

Chang, H., Wan, Y., Wu, S., Fan, Z., and Hu, J. (2011). Occurrence of Androgens and Progestogens in Wastewater Treatment Plants and Receiving River Waters: Comparison to Estrogens. *Water Research*, 45 (2): 732–740.

Chen, J., Liu, Y-S. Deng, W-J. and Ying, G-G. (2019). Removal of Steroid Hormones and Biocides from Rural Wastewater by an Integrated Constructed Wetland. *Science of The Total Environment*, 660: 358–365.

Cheong, W. J., Yang, S. H. and Ali, F. (2013). Molecular Imprinted Polymers for Separation Science: A Review of Reviews. *Journal of Separation Science*, 36 (3): 609–628.

Clara, M., Strenn, B., Saracevic, E. and Kreuzinger, N. (2004). Adsorption of Bisphenol-A, 17β-Estradiole and 17α-Ethinylestradiole to Sewage Sludge. *Chemosphere*, 56 (9): 843–851.

Colman, J. R., Baldwin, D., Johnson, L. L. and Scholz, N. L. (2009). Effects of the Synthetic Estrogen, 17α-Ethinylestradiol, on Aggression and Courtship Behavior in Male Zebrafish (Danio Rerio). *Aquatic Toxicology*, 91 (4): 346–354.

Croley, T. R., Hughes, R. J., Koenig, B. G., Metcalfe, C. D. and March, R. E. (2000). Mass Spectrometry Applied to the Analysis of Estrogens in the Environment. *Rapid Communications in Mass Spectrometry*, 14 (13): 1087–1093.

Cunha, S. C., Lehotay, S. J., Mastovska, K., Fernandes, J. O., Beatriz, M., and Oliveira, P. P. (2007). Evaluation of the QuEChERS Sample Preparation Approach for the Analysis of Pesticide Residues in Olives. *Journal of Separation Science*, 30 (4): 620–632.

Del Río, J. P., Alliende, M. I., Molina, N., Serrano, F. G., Molina, S. and Vigil, P. (2018). Steroid Hormones and Their Action in Women's Brains: The Importance of Hormonal Balance. *Frontiers in Public Health*, 6: 141.

Eisert, R. and Pawliszyn, J. (1997). Automated In-Tube Solid-Phase Microextraction Coupled to High-Performance Liquid Chromatography. *Analytical Chemistry,* 69 (16): 3140-3147.

Emnet, P., Gaw, S., Northcott, G., Storey, B. and Graham, L. (2015). Personal Care Products and Steroid Hormones in the Antarctic Coastal Environment Associated with Two Antarctic Research Stations, McMurdo Station and Scott Base. *Environmental Research*, 136: 331–342.

Fayad, P. B., Prévost, M. and Sauvé, S. (2013). On-Line Solid-Phase Extraction Coupled to Liquid Chromatography Tandem Mass Spectrometry Optimized for the Analysis of Steroid Hormones in Urban Wastewaters. *Talanta*, 115: 349–360.

Fedorov, V. I., Cherkasova, O. P. and Samoilova, E. S. (2004). Native Fluorescence of Steroid Hormones Excited by UV Laser Radiation. In

Saratov Fall Meeting 2003: Optical Technologies in Biophysics and Medicine, edited by Valery V. Tuchin, 5474: 80–87.

Fedorova, G., Golovko, O., Randak, T. and Grabic, R. (2014). Storage Effect on the Analysis of Pharmaceuticals and Personal Care Products in Wastewater. *Chemosphere*, 111: 55–60.

Fontanals, N., Marcé, R. M. and Borrull. F. (2005). New Hydrophilic Materials for Solid-Phase Extraction. *TrAC Trends in Analytical Chemistry*, 24 (5): 394–406.

Gadzała-Kopciuch, R., Ričanyová, J. and Buszewski, B. (2009). Isolation and Detection of Steroids from Human Urine by Molecularly Imprinted Solid-Phase Extraction and Liquid Chromatography. *Journal of Chromatography B*, 877 (11–12): 1177–84.

Gineys, N., Giroud, B. and Vulliet. E. (2010). Analytical Method for the Determination of Trace Levels of Steroid Hormones and Corticosteroids in Soil, Based on PLE/SPE/LC-MS/MS. *Analytical and Bioanalytical Chemistry*, 397 (6): 2295–2302.

Gomes, R. L., Scrimshaw, M. D. and Lester, J. N. (2003). Determination of Endocrine Disrupters in Sewage Treatment and Receiving Waters. *TrAC Trends in Analytical Chemistry*, 22 (10): 697–707.

Guedes-Alonso, R., Ciofi, L., Sosa-Ferrera, Z., Santana-Rodríguez, J. J., Del Bubba, M., Kabir, A. and Furton. K. G. (2016). Determination of Androgens and Progestogens in Environmental and Biological Samples Using Fabric Phase Sorptive Extraction Coupled to Ultra-High Performance Liquid Chromatography Tandem Mass Spectrometry. *Journal of Chromatography A*, 1437: 116–126.

Guedes-Alonso, R., Montesdeoca-Esponda, S., Sosa-Ferrera, Z. and Santana-Rodríguez, J. J. (2014). Liquid Chromatography Methodologies for the Determination of Steroid Hormones in Aquatic Environmental Systems. *Trends in Environmental Analytical Chemistry*, 3–4: 14–27.

Guedes-Alonso, R., Santana-Viera, S., Montesdeoca-Esponda, S., Afonso-Olivares, C., Sosa-Ferrera, Z. and Santana-Rodríguez. J. J. (2016). Application of Microwave-Assisted Extraction and Ultra-High Performance Liquid Chromatography–tandem Mass Spectrometry for the Analysis of Sex Hormones and Corticosteroids in Sewage Sludge Samples. *Analytical and Bioanalytical Chemistry*, 408 (24): 6833–6844.

Guedes-Alonso, R., Sosa-Ferrera, Z. and Santana-Rodríguez, J. J (2012). Optimization of a Solid-Phase Extraction Procedure Combined with Ultra-High Performance Liquid Chromatography and Fluorescence Detection to Determine Estrogens in Wastewater Samples. *Luminescence*, 27 (6): 560–61.

Guedes-Alonso, R., Sosa-Ferrera, Z. and Santana-Rodríguez, J. J. (2013). Simultaneous Determination of Hormonal Residues in Treated Waters Using Ultrahigh Performance Liquid Chromatography-Tandem Mass Spectrometry. *Journal of Analytical Methods in Chemistry*, 2013: e210653.

Guedes-Alonso, R., Sosa-Ferrera, Z. and Santana-Rodríguez, J. J. (2015). An On-Line Solid Phase Extraction Method Coupled with UHPLC-MS/MS for the Determination of Steroid Hormone Compounds in Treated Water Samples from Waste Water Treatment Plants. *Analytical Methods*, 7: 5996–6005.

Hallgren, P., Nicolle, A., Hansson, L. A., Brönmark, C., Nikoleris, L., Hyder, M. and Persson, A. (2014). Synthetic Estrogen Directly Affects Fish Biomass and May Indirectly Disrupt Aquatic Food Webs. *Environmental Toxicology and Chemistry*, 33 (4): 930–36.

Hansen, M., Krogh, K. A., Halling-Sørensen, B. and Björklund, E. (2011). Determination of Ten Steroid Hormones in Animal Waste Manure and Agricultural Soil Using Inverse and Integrated Clean-up Pressurized Liquid Extraction and Gas Chromatography-Tandem Mass Spectrometry. *Analytical Methods*, 3 (5): 1087–95.

Herrera-Melián, J. A., Guedes-Alonso, R., Borreguero-Fabelo, A., Santana-Rodríguez, J. J. and Sosa-Ferrera. Z. (2017). Study on the Removal of Hormones from Domestic Wastewaters with Lab-Scale Constructed Wetlands with Different Substrates and Flow Directions. *Environmental Science and Pollution Research*, 25 (21): 20374–20384.

Herrero, M., Mendiola, J. A., Cifuentes, A. and Ibáñez, E. (2010). Supercritical Fluid Extraction: Recent Advances and Applications. *Journal of Chromatography A*, 1217 (16): 2495–2511.

Hibberd, A., Maskaoui, K., Zhang, Z. & Zhou. J. L. (2009). An Improved Method for the Simultaneous Analysis of Phenolic and Steroidal Estrogens in Water and Sediment. *Talanta*, 77 (4): 1315–1321.

Hu, Y., Zhang, M., Tong, C., Wu, J. and Liu, W. (2013). Enrichment of Steroid Hormones in Water with Porous and Hydrophobic Polymer-Based SPE Followed by HPLC-UV Determination. *Journal of Separation Science*, 36 (20): 3321–3329

Hu, Y., Zheng, Y., Zhu, F. and Li, G. (2007). Sol–gel Coated Polydimethylsiloxane/β-Cyclodextrin as Novel Stationary Phase for Stir Bar Sorptive Extraction and Its Application to Analysis of Estrogens and Bisphenol A. *Journal of Chromatography A*, 1148 (1): 16–22.

Huang, B., Wang, B., Ren, D., Jin, W., Liu, J., Peng, J. and Pan, X. (2013). Occurrence, Removal and Bioaccumulation of Steroid Estrogens in Dianchi Lake Catchment, China. *Environment International*, 59: 262–273.

Huetos Hidalgo, O. (2004). "*Estudio Compararivo y Evaluación de Diferentes Técnicas Cromatográficas En El Análisis de Residuos de Corticosteroides En Muestras Biológicas*." [*Comparative Study and Evaluation of Different Chromatographic Techniques in the Analysis of Corticosteriod Residues in Biological Samples*.] Madrid: Universidad Complutense de Madrid. Madrid: Complutense University.

Huysman, S., Van Meulebroek, L., Vanryckeghem, F., Van Langenhove, H., Demeestere, K. and Vanhaecke, L. (2017). Development and Validation of an Ultra-High Performance Liquid Chromatographic High Resolution Q-Orbitrap Mass Spectrometric Method for the Simultaneous Determination of Steroidal Endocrine Disrupting Compounds in Aquatic Matrices. *Analytica Chimica Acta*, 984: 140–150.

Jauković, Z. D., Grujić, S. D. Matić Bujagić, I. V. and Laušević, M. D. (2017). Determination of Sterols and Steroid Hormones in Surface Water and Wastewater Using Liquid Chromatography-Atmospheric Pressure Chemical Ionization-Mass Spectrometry. *Microchemical Journal*, 135: 39–47.

Kabir, A. and Furton, K. G. (2014). Fabric phase sorptive extractors (FPSE), issued 2014.

Kataoka, H. (2002). Automated Sample Preparation Using In-Tube Solid-Phase Microextraction and Its Application – a Review. *Analytical and Bioanalytical Chemistry*, 373 (1–2): 31–45.

Kawaguchi, M., Ishii, Y., Sakui, N., Okanouchi, N., Ito, R., Inoue, K., Saito, K. and Nakazawa, H. (2004). Stir Bar Sorptive Extraction with *in Situ* Derivatization and Thermal Desorption–gas Chromatography–mass Spectrometry in the Multi-Shot Mode for Determination of Estrogens in River Water Samples."*Journal of Chromatography A*, 1049 (1–2): 1–8.

Khalili Zanjani, MR., Yamini, Y., Shariati, S. and Jönsson, J. A. (2007). A New Liquid-Phase Microextraction Method Based on Solidification of Floating Organic Drop. *Analytica Chimica Acta*, *585*, (2): 286–293.

Kolodziej, E. P., Harter, T. and Sedlak, D. L. (2004). Dairy Wastewater, Aquaculture, and Spawning Fish as Sources of Steroid Hormones in the Aquatic Environment. *Environmental Science and Technology*, 38 (23): 6377–6384.

Krone, N., Hughes, B. A., Lavery, G. G., Stewart, P. M., Arlt, W. and Shackleton, C. H. (2010). Gas Chromatography/Mass Spectrometry (GC/MS) Remains a Pre-Eminent Discovery Tool in Clinical Steroid Investigations Even in the Era of Fast Liquid Chromatography Tandem Mass Spectrometry (LC/MS/MS). *The Journal of Steroid Biochemistry and Molecular Biology*, 121 (3–5): 496–504.

Kumar, R., Gaurav, D., Kabir, A., Furton, K. G. and Malik, A. K. (2015). Development of a Fabric Phase Sorptive Extraction with High-Performance Liquid Chromatography and Ultraviolet Detection Method for the Analysis of Alkyl Phenols in Environmental Samples. *Journal of Separation Science*, 38 (18): 3228–3238.

Kumirska, J., Migowska, N., Caban, M., Łukaszewicz, P. and Stepnowski, P. (2015). Simultaneous Determination of Non-Steroidal Anti-Inflammatory Drugs and Oestrogenic Hormones in Environmental Solid Samples. *Science of the Total Environment*, 508: 498–505.

Labadie, P. and Budzinski, H. (2005). Development of an Analytical Procedure for Determination of Selected Estrogens and Progestagens in Water Samples. *Analytical and Bioanalytical Chemistry*, 381 (6): 1199–1205.

Labadie, P. and Hill, E. M. (2007). Analysis of Estrogens in River Sediments by Liquid Chromatography–electrospray Ionisation Mass Spectrometry: Comparison of Tandem Mass Spectrometry and Time-of-Flight Mass Spectrometry. *Journal of Chromatography A*, 1141 (2): 174–181.

Lai, K M., Scrimshaw, M. D. and Lester, J. N. (2002). Prediction of the Bioaccumulation Factors and Body Burden of Natural and Synthetic Estrogens in Aquatic Organisms in the River Systems. *Science of the Total Environment*, *289* (1–3): 159–168.

Lapworth, D. J., Baran, N., Stuart, M. E. and Ward. R. S. (2012). Emerging Organic Contaminants in Groundwater: A Review of Sources, Fate and Occurrence. *Environmental Pollution*, 163: 287–303.

Le Noir, M., Lepeuple, A. S., Guieysse, B. and Mattiasson, B. (2007). Selective Removal of 17β-Estradiol at Trace Concentration Using a Molecularly Imprinted Polymer. *Water Research*, 41 (12): 2825–2831.

Lednicer, D. (2011). *Steroid Chemistry at a Glance*. John Wiley & Sons.

Leet, J. K., Gall, H. E. and Sepúlveda, M. S. (2011). A Review of Studies on Androgen and Estrogen Exposure in Fish Early Life Stages: Effects on Gene and Hormonal Control of Sexual Differentiation. *Journal of Applied Toxicology*, 31 (5): 379–398.

Li, N., Yang, X., Nian, L., Wang, Z., Lei, L., Wang, K., Zhang, H., Yu, A. and Zhang, Z. (2015). Determination of Steroid Hormones in Milk Using Aqueous Two-Phase Extraction Coupled to Liquid Chromatography. *Analytical Methods*, 7 (6): 2514–22.

Li, Y. and Migliaccio, K. W. (2011). *Water Quality Concepts, Sampling, and Analyses*. CRC Press.

Liao, K., Mei, M., Li, H., Huang, X. and Wu, C. (2016). Multiple Monolithic Fiber Solid-Phase Microextraction Based on a Polymeric Ionic Liquid with High-Performance Liquid Chromatography for the Determination of Steroid Sex Hormones in Water and Urine. *Journal of Separation Science*, *39* (3): 566–575.

Lima, D.L., Silva, C.P., Otero, M. and Esteves, V. I. (2013). Low Cost Methodology for Estrogens Monitoring in Water Samples Using Dispersive Liquid–liquid Microextraction and HPLC with Fluorescence Detection. *Talanta*, 115: 980–985.

Liu, R., Zhou, J. L. and Wilding, A. (2004). Microwave-Assisted Extraction Followed by Gas Chromatography–mass Spectrometry for the Determination of Endocrine Disrupting Chemicals in River Sediments. *Journal of Chromatography A*, 1038 (1–2): 19–26.

Liu, S., Ying, G. G., Zhao, J.L., Chen, F., Yang, B., Zhou, L. J. and Lai, H. J. (2011). Trace Analysis of 28 Steroids in Surface Water, Wastewater and Sludge Samples by Rapid Resolution Liquid Chromatography–electrospray Ionization Tandem Mass Spectrometry. *Journal of Chromatography A*, 1218 (10): 1367–1378.

Loos, R. (2012). Analytical Methods Relevant to the European Commission's 2012 Proposal on Priority Substances under the Water Framework Directive. *JCR Scientific and Policy Reports*

Loos, R., Marinov, D., Sanseverino, I., Napierska, D. and Lettieri, T. (2018). Review of the 1st Watch List under the Water Framework Directive and Recommendations for the 2nd Watch List - European Commission. Luxembourg.

López de Alda, M. J. and Barceló, D. (2000). Determination of Steroid Sex Hormones and Related Synthetic Compounds Considered as Endocrine Disrupters in Water by Liquid Chromatography–diode Array Detection–mass Spectrometry. *Journal of Chromatography A*, *892* (1–2): 391–406.

López de Alda, M. J. and Barceló, D. (2001). Review of Analytical Methods for the Determination of Estrogens and Progestogens in

Waste Waters. *Fresenius' Journal of Analytical Chemistry*, 371 (4), 437–447.

Lucci, P., Núñez, O. and Galceran, M. T. (2011). Solid-Phase Extraction Using Molecularly Imprinted Polymer for Selective Extraction of Natural and Synthetic Estrogens from Aqueous Samples. *Journal of Chromatography A*, *1218* (30): 4828–4833.

Luo, Y., Guo, W., Ngo, H. H., Nghiem, L. D., Hai, F. I., Zhang, J., Liang S. and Wang X. C. (2014). A Review on the Occurrence of Micropollutants in the Aquatic Environment and Their Fate and Removal during Wastewater Treatment. *Science of the Total Environment*, 473–474: 619–641.

Madej, K. (2009). Microwave-Assisted and Cloud-Point Extraction in Determination of Drugs and Other Bioactive Compounds. *TrAC Trends in Analytical Chemistry*, 28 (4): 436–446.

Maes, H. (2011). *Fate of ethinylestradiol in the aquatic environment and the associated effects on organisms of different trophic levels*. Aachen: Aacen University.

Manickum, T. and John, W. (2014). Occurrence, Fate and Environmental Risk Assessment of Endocrine Disrupting Compounds at the Wastewater Treatment Works in Pietermaritzburg (South Africa). *Science of the Total Environment*, 468–469: 584–97.

McGree, M. M., Winkelman, D. L., Vieira, N. K. M. and Vajda, A. M. (2010). Reproductive Failure of the Red Shiner (Cyprinella Lutrensis) after Exposure to an Exogenous Estrogen. *Canadian Journal of Fisheries and Aquatic Sciences*, 67 (11), 1730–1743.

McKay, L. I. and Cidlowski, J. A. (2003). "Physiologic and Pharmacologic Effects of Corticosteroids," In: Kufe, D. W., Pollock, R. E., Weichselbaum, R.R., et al., editors. *Holland-Frei Cancer Medicine*. 6th edition.

Mitani, K., Fujioka, M. and Kataoka, H. (2005). Fully Automated Analysis of Estrogens in Environmental Waters by In-Tube Solid-Phase Microextraction Coupled with Liquid Chromatography–tandem Mass Spectrometry. *Journal of Chromatography A*, 1081 (2): 218–224.

Montesdeoca-Esponda, S., Sosa-Ferrera, Z., Kabir, A., Furton, K. G. and Santana-Rodríguez. J. J. (2015). Fabric Phase Sorptive Extraction Followed by UHPLC-MS/MS for the Analysis of Benzotriazole UV Stabilizers in Sewage Samples. *Analytical and Bioanalytical Chemistry,* 407 (26): 8137–8150.

Montesdeoca-Esponda, Sarah. (2013). *Desarrollo de Metodologías de Extracción y Microextracción Combinadas Con Sistemas Cromatográficos Para La Determinación de Fármacos y Productos de Cuidado Personal En Muestras Ambientales.* [*Development of extraction and microextraction methodologies combined with chromatographic systems for the determination of pharmaceuticals and personal care products in environmental.*] Universidad de Las Palmas de Gran Canaria. Las Palmas de Gran Canaria, España. [University of Las Palmas de Gran Canaria. Las Palmas de Gran Canaria, Spain.]

Mulabagal, V., Wilson, C. and Hayworth, J. S. (2017). An Ultrahigh-Performance Chromatography/Tandem Mass Spectrometry Quantitative Method for Trace Analysis of Potential Endocrine Disrupting Steroid Hormones in Estuarine Sediments. *Rapid Communications in Mass Spectrometry*, 31 (5): 419–429.

Mustafa, A. and Turner, C. (2011). Pressurized Liquid Extraction as a Green Approach in Food and Herbal Plants Extraction: A Review. *Analytica Chimica Acta*, 703 (1): 8–18.

Nakayama, K., Inoue, Y., Ikeda, N., Hashizume, N., Murakami, H., Ishibashi, T., Ikeda, H., Isobe, T., Kitamura, S. and Suzuki, G. (2014). Uptake and Biological Effects of Synthetic Glucocorticoids in Common Carp (Cyprinus Carpio). *Marine Pollution Bulletin, 85,* (2): 370–375.

Nakayama, K., Sato, K., Shibano, T., Isobe, T., Suzuki, G. and Kitamura S. (2016). Occurrence of Glucocorticoids Discharged from a Sewage Treatment Plant in Japan and the Effects of Clobetasol Propionate Exposure on the Immune Responses of Common Carp (Cyprinus Carpio) to Bacterial Infection. *Environmental Toxicology and Chemistry*, 35 (4): 946–952.

Pal, A., Gin, K. Y., Lin, A. Y. and Reinhard, M. (2010). Impacts of Emerging Organic Contaminants on Freshwater Resources: Review of Recent Occurrences, Sources, Fate and Effects. *Science of The Total Environment*, 408 (24): 6062–6069.

Patrolecco, L., Ademollo, N., Grenni, P. Tolomei, A. Barra Caracciolo, A. and Capri, S. (2013). Simultaneous Determination of Human Pharmaceuticals in Water Samples by Solid Phase Extraction and HPLC with UV-Fluorescence Detection. *Microchemical Journal*, 107: 165–171.

Patrolecco, L., Capri, S. and Ademollo, N. (2015). Occurrence of Selected Pharmaceuticals in the Principal Sewage Treatment Plants in Rome (Italy) and in the Receiving Surface Waters. *Environmental Science and Pollution Research*, 22 (8): 5864–5876.

Pedersen-Bjergaardand, S. and Rasmussen, K. E. (1999). Liquid−Liquid−Liquid Microextraction for Sample Preparation of Biological Fluids Prior to Capillary Electrophoresis. *Analytical Chemistry*, 71 (14): 2650–2656.

Pedrouzo, M., Borrull, F., Pocurull, E. and Marcé, R. M. (2009). Estrogens and Their Conjugates: Determination in Water Samples by Solid-Phase Extraction and Liquid Chromatography–tandem Mass Spectrometry. *Talanta*, 78 (4–5): 1327–31.

Pedrouzo, M., Borrull, F., Pocurull, E. and Marcé, R. M. (2011). Presence of Pharmaceuticals and Hormones in Waters from Sewage Treatment Plants. *Water, Air, and Soil Pollution, 217* (1–4): 267–281.

Peñalver, A., Pocurull, E., Borrull, F. and Marcé, R. M. (2002). Method Based on Solid-Phase Microextraction–high-Performance Liquid Chromatography with UV and Electrochemical Detection to Determine Estrogenic Compounds in Water Samples. *Journal of Chromatography A*, 964 (1–2): 153–160.

Pérez, R. L. and Escandar, G. M. (2016). Multivariate Calibration-Assisted High-Performance Liquid Chromatography with Dual UV and Fluorimetric Detection for the Analysis of Natural and Synthetic Sex Hormones in Environmental Waters and Sediments. *Environmental Pollution*, 209: 114–122.

Peysson, W. and Vulliet, E. (2013). Determination of 136 Pharmaceuticals and Hormones in Sewage Sludge Using Quick, Easy, Cheap, Effective, Rugged and Safe Extraction Followed by Analysis with Liquid Chromatography–time-of-Flight-Mass Spectrometry. *Journal of Chromatography A*, 1290: 46–61.

Pinto, P. I. S., Estêvão, M. D. and Power. D. M. (2014). Effects of Estrogens and Estrogenic Disrupting Compounds on Fish Mineralized Tissues. *Marine Drugs*, 12 (8): 4474–4494.

Pojana, G., Gomiero, A., Jonkers, N. and Marcomini, A. (2007). Natural and Synthetic Endocrine Disrupting Compounds (EDCs) in Water, Sediment and Biota of a Coastal Lagoon. *Environment International*, 33 (7): 929–936.

Poole, C. F. (2003). New Trends in Solid-Phase Extraction. *TrAC Trends in Analytical Chemistry*, 22 (6): 362–373.

Purdom, C. E., Hardiman, P. A., Bye, V. V. J., Eno, N. C., Tyler, C. R. and Sumpter, J. P. (1994). Estrogenic Effects of Effluents from Sewage Treatment Works. *Chemistry and Ecology*, 8 (4): 275–285.

Qiu, L., Liu, W., Huang, M. and Zhang, L. (2010). Preparation and Application of Solid-Phase Microextraction Fiber Based on Molecularly Imprinted Polymer for Determination of Anabolic Steroids in Complicated Samples. *Journal of Chromatography A*, 1217 (48): 7461–7470.

Racamonde, I., Rodil, R., Quintana, J. B., Sieira, B. J., Kabir, A., Furton, K. G. and Cela, R. (2015). Fabric Phase Sorptive Extraction: A New Sorptive Microextraction Technique for the Determination of Non-Steroidal Anti-Inflammatory Drugs from Environmental Water Samples. *Analytica Chimica Acta, 865* (1): 22–30.

Rachkov, A., McNiven, S., El'skaya, A., Yano, K. and Karube, I. (2000). Fluorescence Detection of β-Estradiol Using a Molecularly Imprinted Polymer. *Analytica Chimica Acta*, 405 (1–2): 23–29.

Rivero-Wendt, C. L., Oliveira, R., Monteiro, M. S., Domingues, I., Soares, A. M. and Grisolia, C. K. (2016). Steroid Androgen 17α-Methyltestosterone Induces Malformations and Biochemical

Alterations in Zebrafish Embryos. *Environmental Toxicology and Pharmacology*, 44: 107–113.

Roldán-Pijuán, M., Lucena, R., Cárdenas, S., Valcárcel, M., Kabir, A. and Furton. K. G. (2015). Stir Fabric Phase Sorptive Extraction for the Determination of Triazine Herbicides in Environmental Waters by Liquid Chromatography. *Journal of Chromatography A*, 1376: 35–45.

Routledge, E. J., Sheahan, D., Desbrow, C., Brighty, G. C., Waldock, M. and Sumpter, J. P. (1998). Identification of Estrogenic Chemicals in STW Effluent. 2. *In Vivo* Responses in Trout and Roach. *Environmental Science & Technology*, 32 (11): 1559–1565.

Rutishauser, B. V., Pesonen, M., Escher, B. I., Ackermann, G. E., Aerni, H. R., Suter, M. J. and Eggen, R. I. (2004). Comparative Analysis of Estrogenic Activity in Sewage Treatment Plant Effluents Involving Three *in Vitro* Assays and Chemical Analysis of Steroids. *Environmental Toxicology and Chemistry*, 23 (4): 857–864.

Sadowski, R. and Gadzała-Kopciuch, R. (2013). Isolation and Determination of Estrogens in Water Samples by Solid-Phase Extraction Using Molecularly Imprinted Polymers and HPLC. *Journal of Separation Science*, 36 (14): 2299–2305.

Sanchez-Prado, L., Garcia-Jares, C. Dagnac, T. and Llompart, M. (2015). Microwave-Assisted Extraction of Emerging Pollutants in Environmental and Biological Samples before Chromatographic Determination. *TrAC Trends in Analytical Chemistry*, 71: 119–143.

Saraji, M., and Farajmand, B. (2013). Chemically Modified Cellulose Paper as a Thin Film Microextraction Phase. *Journal of Chromatography A*, 1314: 24–30.

Seidi, S., Alavi, L., Jabbari, A. and Shanehsaz, M. (2019). Three-Phase Carrier-Mediated Hollow Fiber Microextraction Based on Deep Eutectic Solvent Followed by HPLC–UV for Determination of Raloxifene and Ethinylestradiol in Pharmaceutical Wastewater Treatment Plants. *Journal of the Iranian Chemical Society*, 16 (5): 1007–1018

Snow, D. D., Damon-Powell, T., Onanong, S. amd Cassada, D. A. (2013). Sensitive and Simplified Analysis of Natural and Synthetic Steroids in

Water and Solids Using On-Line Solid-Phase Extraction and Microwave-Assisted Solvent Extraction Coupled to Liquid Chromatography Tandem Mass Spectrometry Atmospheric Pressure Photoioni. *Analytical and Bioanalytical Chemistry*, 405 (5): 1759–1771.

Sosa-Ferrera, Z., Mahugo-Santana, C. and Santana-Rodríguez, J. J. (2012). Steroid Hormones in Biological and Environmental Samples: Extraction and Determination Techniques. In *Steroids: Biosynthesis, Functions and Health Implications*, edited by Carlos Amada Figueiredo, 83–128. New York: Nova Science Publishers.

Sosa-Ferrera, Z., Mahugo-Santana, C. and Santana-Rodríguez, J. J. (2013). Analytical Methodologies for the Determination of Endocrine Disrupting Compounds in Biological and Environmental Samples. *BioMed Research International*, 2013: 674838. 1–23

Stanczyk, F. Z. and Clarke, N. J. (2010). Advantages and Challenges of Mass Spectrometry Assays for Steroid Hormones. *The Journal of Steroid Biochemistry and Molecular Biology*, 121 (3–5): 491–495.

Stolker, A. A. M., Zoontjes, P. W. and van Ginkel, L. A. (1998). The Use of Supercritical Fluid Extraction for the Determination of Steroids in Animal Tissues. *Analyst*, 123 (12), 2671–2676.

Tamayo, F. G., Turiel, E. and Martín-Esteban. A. (2007). Molecularly Imprinted Polymers for Solid-Phase Extraction and Solid-Phase Microextraction: Recent Developments and Future Trends. *Journal of Chromatography A*, 1152 (1–2): 32–40.

Tan, X. T., Li, Z. M., Deng, L. G., Zhao, S. C. and Wang. M. L. (2016). Analysis of 13 Kinds of Steroid Hormones in Raw Milk Using Modified QuEChERS Method Combined with UPLC-QTOF-MS. *Journal of Integrative Agriculture*, 15 (9): 2163–2174.

Tölgyesi, A., Verebey, Z., Sharma, V. K., Kovacsics, L. and Fekete, J. (2010). Simultaneous Determination of Corticosteroids, Androgens, and Progesterone in River Water by Liquid Chromatography–tandem Mass Spectrometry. *Chemosphere*, 78 (8): 972–979.

Tomšíková, H., Aufartová, J., Solich, P., Nováková, L., Sosa-Ferrera, Z. and Santana-Rodríguez, J. J. (2012). High-Sensitivity Analysis of

Female-Steroid Hormones in Environmental Samples. *TrAC Trends in Analytical Chemistry*, 34: 35–58.

Tremblay, L. A., Gadd, J. B. and Northcott, G. L. (2018). Steroid Estrogens and Estrogenic Activity Are Ubiquitous in Dairy Farm Watersheds Regardless of Effluent Management Practices. *Agriculture, Ecosystems & Environment*, 253: 48–54.

United States Environmental Protection Agency. (2007). *Method 1698: Steroids and Hormones in Water, Soil, Sediment, and Biosolids by HRGC/HRMS.*

United States Environmental Protection Agency. (2010). *Method 539: Determination of Hormones in Drinking Water by Solid Phase Extraction (SPE) and Liquid Chromatography Electrospray Ionization Tandem Mass Spectrometry (LC-ESI-MS/MS).*

Vega-Morales, T., Sosa-Ferrera, Z. and Santana-Rodríguez, J. J. (2010). Simultaneous Determination of Endocrine Disrupting Chemicals in Wastewater Samples by High Performance Liquid Chromatography with Fluorescence Detection. *Luminescence*, 25 (3): 246–247.

Vidal-Dorsch, D. E., Bay, S.M., Maruya, K., Snyder, S.A., Trenholm, R.A., and Vanderford, B.J. (2012). Contaminants of Emerging Concern in Municipal Wastewater Effluents and Marine Receiving Water. *Environmental Toxicology and Chemistry*, 31 (12): 2674–2682

Viglino, L., Aboulfadl, K., Prévost, M. and Sauvé, S. (2008). Analysis of Natural and Synthetic Estrogenic Endocrine Disruptors in Environmental Waters Using Online Preconcentration Coupled with LC-APPI-MS/MS. *Talanta*, 76, (5): 1088–1096.

Volmer, D. A. and Hui, J. P. M. (1997). "Rapid Determination of Corticosteroids in Urine by Combined Solid Phase Microextraction/Liquid Chromatography/Mass Spectrometry." *Rapid Communications in Mass Spectrometry*, *11* (17): 1926–1934.

Wilkowska, A. and Biziuk, M. (2011). Determination of Pesticide Residues in Food Matrices Using the QuEChERS Methodology. *Food Chemistry*, 125 (3): 803–812.

Yang, L., Luan, T. and Lan, C. (2006). Solid-Phase Microextraction with on-Fiber Silylation for Simultaneous Determinations of Endocrine

Disrupting Chemicals and Steroid Hormones by Gas Chromatography-Mass Spectrometry. *Journal of Chromatography A*, 1104 (1–2): 23–32.

Yao, B., Li, R., Yan, S., Chan, S. A. and Song, W. (2018). Occurrence and Estrogenic Activity of Steroid Hormones in Chinese Streams: A Nationwide Study Based on a Combination of Chemical and Biological Tools. *Environment International*, 118: 1–8.

Yarahmadi, H., Duy, S. V., Hachad, M., Dorner, S., Sauvé, S. and Prévost M. (2018). Seasonal Variations of Steroid Hormones Released by Wastewater Treatment Plants to River Water and Sediments: Distribution between Particulate and Dissolved Phases. *Science of the Total Environment*, *635*: 144–155.

Yien Fang, T., Praveena, S. M., Aris, A. Z., Syed Ismail, S. N. and Rasdi, I.. (2019). Quantification of Selected Steroid Hormones (17β-Estradiol and 17α-Ethynylestradiol) in Wastewater Treatment Plants in Klang Valley (Malaysia). *Chemosphere*, 215: 153–162.

Yu, Y., Huang, Q., Wang, Z., Zhang, K., Tang, C., Cui, J., Feng, J. and Peng, X. (2011). Occurrence and Behavior of Pharmaceuticals, Steroid Hormones, and Endocrine-Disrupting Personal Care Products in Wastewater and the Recipient River Water of the Pearl River Delta, South China. *Journal of Environmental Monitoring*, 13 (4): 871–878.

Zacs, D., Perkons, I. and Bartkevics. V. (2016). Determination of Steroidal Oestrogens in Tap Water Samples Using Solid-Phase Extraction on a Molecularly Imprinted Polymer Sorbent and Quantification with Gas Chromatography-Mass Spectrometry (GC-MS). *Environmental Monitoring and Assessment*, *188* (7): 433.

Zeilinger, J., Steger-Hartmann, T., Maser, E., Goller, S., Vonk, R. and Länge, R. (2009). Effects of Synthetic Gestagens on Fish Reproduction. *Environmental Toxicology and Chemistry*, 28 (12): 2663–2670.

Zhang, K. and Fent, K. (2018). Determination of Two Progestin Metabolites (17α-Hydroxypregnanolone and Pregnanediol) and Different Classes of Steroids (Androgens, Estrogens, Corticosteroids, Progestins) in Rivers and Wastewaters by High-Performance Liquid

Chromatography-Tandem Mass Spectrometry (HPLC-MS/MS). *Science of the Total Environment*, 610–611: 1164–72.

Zhang, K., Zhao, Y. and Fent, K. (2017). Occurrence and Ecotoxicological Effects of Free, Conjugated, and Halogenated Steroids Including 17α-Hydroxypregnanolone and Pregnanediol in Swiss Wastewater and Surface Water. *Environmental Science & Technology*, 51 (11): 6498–6506.

Zhang, Z., Duan, H., Zhang, L., Chen, X., Liu, W. and Chen, G. (2009). Direct Determination of Anabolic Steroids in Pig Urine by a New SPME–GC–MS Method. *Talanta*, 78 (3), 1083–1089.

Zheng, Y., Qu, J., Qiu, L., Fan, L., Meng, S., Song, C., Bing, X. and Chen, J. (2016). Effect of 17α-Methyltestosterone (MT) on Oxidation Stress in the Liver of Juvenile GIFT Tilapia, Oreochromis Niloticus. *SpringerPlus*, 5: 338

Zhou, Y., Zha, J. and Wang, Z. (2012). Occurrence and Fate of Steroid Estrogens in the Largest Wastewater Treatment Plant in Beijing, China. *Environmental Monitoring and Assessment*, 184 (11): 6799–6813.

Zorita, S., Hallgren, P. and Mathiasson, L. (2008). Steroid Hormone Determination in Water Using an Environmentally Friendly Membrane Based Extraction Technique. *Journal of Chromatography A*, 1192 (1): 1–8.

Zougagh, M., Valcárcel, M. and Ríos, A. (2004). Supercritical Fluid Extraction: A Critical Review of Its Analytical Usefulness. *TrAC Trends in Analytical Chemistry*, 23 (5): 399–405.

Zuloaga, O., Navarro, P., Bizkarguenaga, E., Iparraguirre, A., Vallejo, A., Olivares, M. and Prieto. A. (2012). Overview of Extraction, Clean-up and Detection Techniques for the Determination of Organic Pollutants in Sewage Sludge: A Review. *Analytica Chimica Acta*, 736: 7–29.

Zuo, Y., Zhang, K. and Lin, Y. (2007). Microwave-Accelerated Derivatization for the Simultaneous Gas Chromatographic–mass Spectrometric Analysis of Natural and Synthetic Estrogenic Steroids. *Journal of Chromatography A*, 1148 (2): 211–218.

In: A Closer Look at Hormones
Editor: Alexander V. Knyazev
ISBN: 978-1-53617-254-6

Chapter 4

CHEMICAL THERMODYNAMICS AND POLYMORPHISM IN STEROID HORMONES

Alexander Knyazev*[*], *Anastasiya Shipilova*, *Ekaterina Gusarova*, *Svetlana Knyazeva* and *Alexey Amosov
Lobachevsky University, Nizhni Novgorod, Russia

ABSTRACT

For the first time, the crystallographic and thermodynamic characteristics of steroid hormones were analyzed. A complex polymorphism for crystalline modifications of steroid hormones has been shown. The structural patterns in the formation of molecular crystals of steroid hormones are revealed.

Keywords: steroid hormone, crystallographic characteristics, polymorphic forms, thermodynamic characteristics

[*] Corresponding Author's Email: knyazevav@gmail.com.

INTRODUCTION

Steroids are large group of chemical substances that share a 17-carbon-atom skeleton composed of four fused rings (three six-membered rings and one five-membered ring) (Figure 1). Steroids vary from one another in the nature of attached groups, their position, and the configuration of the steroid nucleus. The smallest modifications in structures of steroids produce significant differences in their biological activities. Many, but not all, steroids are hormones. Naturally occurring steroid hormones in humans include glucocorticoids, mineralocorticoids, androgens, estrogens, and progestins.

The goals of this chapter include systematization of thermodynamic (standard enthalpies, entropies and Gibbs functions of compounds, as well as characteristics of phase transitions) and structural (results of X-ray structural analysis, symmetry of crystal structures) data on steroid hormones.

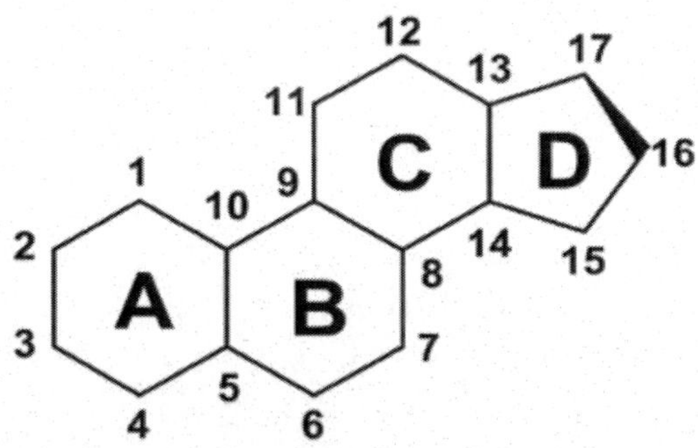

Figure 1. Steroids skeleton.

For the first time, we will generalize the experimental data on the enthalpies of combustion of steroid hormones. The standard molar enthalpies of formation in the crystalline state at 298.15 K were derived from the combustion experiments. The temperature dependences of heat capacity of the some steroid hormones have been measured in the range from 5 to 350 K by precision adiabatic vacuum calorimetry. These data allowed us to calculate the absolute entropies and standard entropies of formation of these substances. Using combination of the adiabatic and combustion calorimetry results the thermodynamic functions of formation of the steroid hormones at T = 298.15 K and P = 0.1 MPa have been calculated. In addition, we plan to compare the experimental data on thermodynamic functions with model calculations using different methods.

An important task of our chapter will be the analysis of the diversity of polymorphic forms for a whole series of steroid hormones. This is extremely important information for the identification of steroid hormones, because the biological activity of polymorphic forms is significantly different. In the pharmaceutical industry, monitoring the crystal structure of the drug is also a mandatory procedure, as well as studying the composition, purity and biological activity of the compound.

Crystallographic Characteristics and Polymorphism

The systematization of structural information on steroid hormones was carried out on the basis of literature data and structural databases for the first time (Table 1). 198 crystal structures of steroid hormones were determined by X-ray diffraction analysis. We found that most often (62.6%) steroid hormones crystallize in orthorhombic syngony (Figure 2). An interesting fact is that all orthorhombic structures are represented by only one space group $P2_12_12_1$. Also, quite a few (33.3%) crystals of steroid hormones crystallize in monoclinic syngony. Monoclinic steroid hormones crystallize in 5 space groups, but the most common (84.8%) is $P2_1$ (Figure 3). Some confusion is caused by articles in which the space group I2 is presented, which is not in the list of possible space groups. Thus, more than 90 percent (90.9%) of steroid hormones structures crystallize in two hierarchically related space groups $P2_1 \rightarrow P2_12_12_1$.

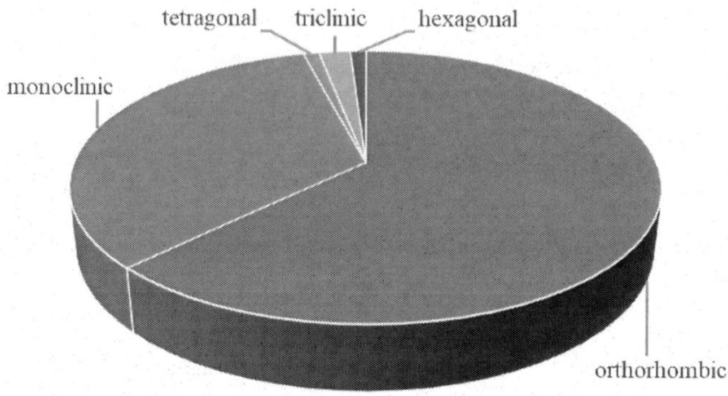

Figure 2. Distribution of syngonies in crystals of steroid hormones.

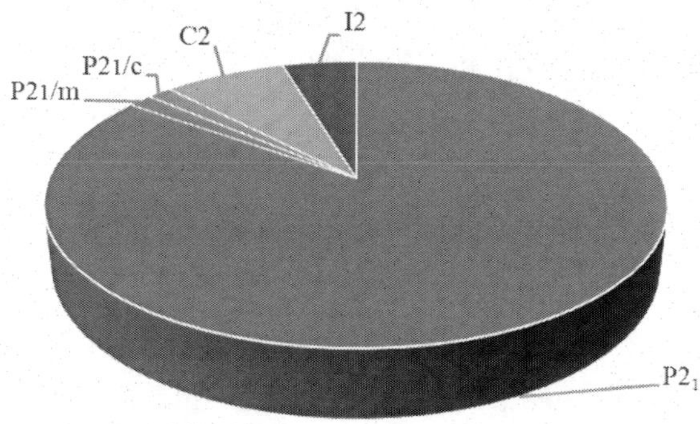

Figure 3. Distribution of space groups in monocrine crystals of steroid hormones.

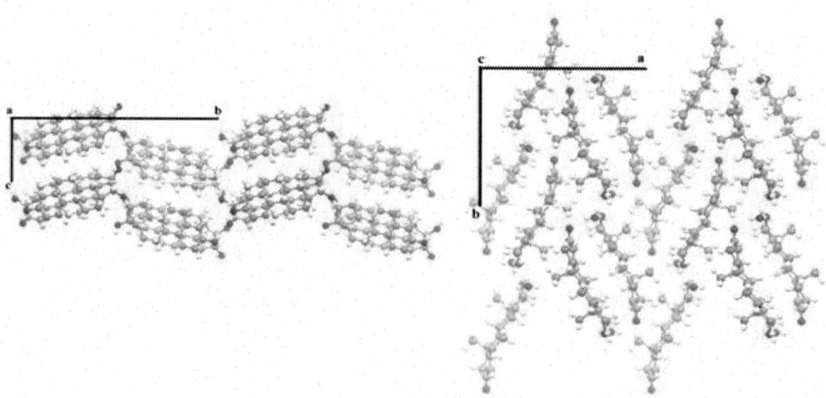

Figure 4. Typical structures of steroid hormones. Estriol sp.group $P2_1$ (left) and testosterone monohydrate sp.group $P2_12_12_1$ (right).

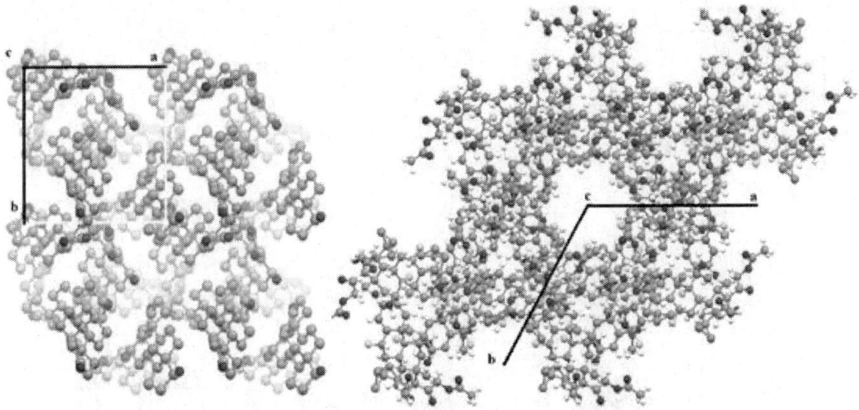

Figure 5. Highly symmetric structures of steroid hormones. Nandrolone monohydrate sp.group $P4_32_12$ (left) and α-difluprednate sp.group $P6_5$ (right).

As we see in Figure 4, typical structures of steroid hormones are presented. The screw axes of the second order are quite clearly visible. The packaging of molecules resembles quasilayers and herringbone-type structures. However, several structures were found on which axes of a higher (fourth and sixth) order are also found (Figure 5). I would also like to draw attention to the fact that steroid hormones are characterized by rich polymorphism, however, in the pharmaceutical industry, as a rule, only one modification of a substance is used. For example, quite a lot of polymorphic modifications are found on hormones such as Androsterone, Cortisone acetate, Cortisol, Diflorasone diacetate, Difluprednate, Estrone, Hydroxyprogesterone caproate, Progesterone, Testosterone, Tibolone (Table 1). Various modifications of steroid hormones are obtained by replacing the solvent in the synthesis of crystals. In the scientific literature, the term *solvent-induced phase transition* has increasingly begun to appear. We believe that in the near future a significant amount of polymophic modifications of steroid hormones, as well as crystalline hydrates and solvates based on them will be obtained.

Table 1. Crystallographic characteristics of steroid hormones

Hormone	Syngony	Modification	Space group	a, Å/ α, °	b, Å/ β, °	c, Å/ γ, °	Ref
21-hydroxy-3,20-dioxopregn-4-en-17-yl propionate monohydrate	monoclinic	-	P21	7.826 (6)/90	7.819 (5)/98.020 (10)	18.545 (7)/90	[1]
21-hydroxy-3,20-dioxopregn-4-en-17-yl propionate	orthorhombic	-	P212121	8.7794 (5)/90	8.9506 (5)/90	55.675 (3)/90	[1]
11-deoxycorticosterone-21-hemisuccinate	monoclinic	-	C2	21.896 (2)/90.00 (3)	7.596 (3)/105.944 (15)	14.291 (3)/90.00 (2)	[2]
11-ketoprogesterone	orthorhombic	-	P212121	6.2525 (4)/90	12.7576 (3)/90	22.4174 (5)/90	[3]
16α-hydroxyestrone	orthorhombic	-	P212121	10.176 (1)/90	18.461/90	7.8255 (4)/90	[4]
17α-estradiol hemihydrate	monoclinic	-	C2	18.935 (4)/90	7.0470 (14)/123.92 (3)	13.304 (3)/90	[5]
17α-hydroxyprogesterone	orthorhombic	-	P212121	9.831/90	23.468/90	7.839/90	[6]
	orthorhombic	-	P212121	9.836 (10)/90	23.466 (10)/90	7.819 (10)/90	[7]
17α-methyltestosterone hemihydrate	orthorhombic	-	P212121	6.374 (2)/90	12.807 (2)/90	43.398 (5)/90	[8]
19-norandrostenediol monohydrate	orthorhombic	-	P212121	10.330/90	6.360/90	24.780/90	[9]
19-nortestosterone	monoclinic	-	P21	10.045 (5)/90	10.761 (5)/89.36 (10)	14.497 (6)/90	[10]
	monoclinic	-	P21	9.755 (2)/90	11.467 (3)/101.07 (2)	14.196 (3)/90	[11]
21-deoxycortisone	trigonal	-	P31	7.297 (2)/90	7.297 (2)/90	30.304 (3)/120	[12]
2-hydroxyestradiol	orthorhombic	-	P212121	12.086 (2)/90	20.140 (4)/90	6.179 (1)/90	[13]
2-hydroxyestrone	orthorhombic	-	P212121	9.564 (1)/90	8.042 (1)/90	19.344 (2)/90	[14]
4-hydroxyestradiol	monoclinic	-	P21	7.0226 (8)/90	19.2395 (20)/107.806 (2)	5.7348 (6)/90	[15]
7α-methyltestosterone	monoclinic	-	P21	11.016 (5)/90	6.156 (4)/90	12.967 (4)/90	[16]
Adrenosterone	orthorhombic	-	P212121	9.263/90	26.531/90	6.477/90	[17]

Hormone	Syngony	Modification	Space group	a, Å/ α, °	b, Å/ β, °	c, Å/ γ, °	Ref
Aldosterone	orthorhombic	-	P212121	25.6/90	5.8/90	24.1/90	[18]
Amcinonide	monoclinic	-	P21	15.7400 (4)/90	15.1252 (4)/94.104 (1)	10.9849 (3)/90	[19]
Androstanolone benzoate	orthorhombic	-	P212121	10.860/90	16.241/90	6.236/90	[20]
Androstenedione	orthorhombic	-	P212121	12.978 (5)/90	16.949 (5)/90	7.365 (5)/90	[21]
Androsterone	monoclinic	-	P21	9.560 (20)/90	7.900 (30)/111.37 (3)	11.780 (20)/90	[22]
Androsterone	monoclinic	-	P21	7.30/90	10.40/100.0	22.20/90	[23]
	monoclinic	-	P21/m	7.20/90	10.40/100.0	22.00/90	[24]
	monoclinic	-	P21	11.95/90	7.70/111	9.45/90	[25]
	monoclinic	-	P21	9.445 (2)/90	7.954 (2)/111.251(3)	11.715 (3)/90	[26]
	monoclinic	-	P21	9.4045 (9)/90	7.9586 (8)/111.274(2)	11.6746 (12)/90	[27]
Androsterone hemihydrate	monoclinic	-	P21	7.4288 (8)/90	10.2892 (11)/97.553 (2)	22.174 (2)/90	[27]
Beclometasone	monoclinic	-	P21	12.0999 (3)/90	13.8673 (3)/90.1060 (10)	16.7971 (4)/90	[28]
Betamethasone dipropionate	orthorhombic	-	P212121	11.883 (4)/90	13.833 (4)/90	15.737 (5)/90	[29]
Bolandione	orthorhombic	-	P212121	7.2921 (15)/90	8.0475 (16)/90	26.378 (5)/90	[30]
	orthorhombic	-	P212121	8.0475/90	26.3777/90	7.2921/90	[31]
Boldenone	orthorhombic	-	P212121	12.302/90	18.644/90	7.065/90	[17]
Budesonide	orthorhombic	A	P212121	8.550 (1)/90	9.406 (1)/90	28.401 (3)/90	[32]
	orthorhombic	B	P212121	8.419 (1)/90	9.087 (1)/90	29.523 (4)/90	[32]
Chlormadinone acetate	orthorhombic	-	P212121	10.925 (6)/90	12.068 (6)/90	16.179 (9)/90	[33]
Ciclesonide	orthorhombic	-	P212121	6.6380 (10)/90	14.079 (3)/90	32.870 (7)/90	[34]
Clobetasol propionate	monoclinic	-	P21	7.6961 (3)/90	14.6036 (5)/95.739 (2)	10.4355 (5)/90	[35]
Clobetasone butyrate	orthorhombic	-	P212121	14.2345 (8)/90	18.8570 (10)/90	27.7142 (18)/90	[36]
Clostebol acetate	orthorhombic	-	P212121	7.7400 (10)/90	12.631 (2)/90	19.275 (2)/90	[37]
Corticosterone	monoclinic	-	P21	8.992 (2)/90	12.293 (2)/97,25	8.322 (2)/90	[38]
Cortisol	orthorhombic	I	P212121	10.0715 (6)/90	12.4240 (7)/90	30.441 (2)/90	[39]
	orthorhombic	I	P212121	12.441/90	30.496/90	10.139/90	[40]

Table 1. (Continued)

Hormone	Syngony	Modification	Space group	a, Å/ α, °	b, Å/ β, °	c, Å/ γ, °	Ref
	monoclinic	II	P21	12.4695 (14)/90	12.4629 (9)/103.040 (14)	12.5701 (16)/90	[37]
	orthorhombic	III	P212121	6.2232 (3)/90	15.7750 (10)/90	18.8776 (11)/90	[37]
	orthorhombic	III	P212121	6.435 (3)/90	15.626 (6)/90	18.912 (8)/90	[41]
Cortisone	orthorhombic	-	P212121	10.040/90	23.649/90	7.784/90	[42]
Cortisone acetate	orthorhombic	-	P212121	11.048 (2)/90	7.102 (2)/90	27.095 (5)/90	[43]
	monoclinic	-	P21	26.58/90	7,52/90	15.68/90	[44]
Cortisone acetate	orthorhombic	-	P212121	12.50/90	20.95/90	7.96/90	[44]
	orthorhombic	-	P212121	8.018 (4)/90	12.579 (7)/90	20.963 (1)/90	[45]
Cortivazol	monoclinic	-	C2	30.625 (5)/90	6.229 (2)/93.86 (2)	15.289 (2)/90	[46]
Cyproterone acetate	orthorhombic	-	P212121	10.757 (2)/90	12.266 (3)/90	16.623 (3)/90	[47]
Danazol	orthorhombic	-	P212121	6.595 (2)/90	10.519 (2)/90	26.166 (2)/90	[48]
	orthorhombic	-	P212121	6.601 (4)/90	10.530 (4)/90	26.177 (9)/90	[49]
Deflazacort	orthorhombic	-	P212121	11.2300 (5)/90	12.8161 (8)/90	16.1710 (10)/90	[50]
Desogestrel	monoclinic	-	P21	7.444 (5)/90	10.091 (5)/90	12.074 (5)/93.84 (5)	[51]
Desoximetasone	orthorhombic	-	P212121	11.319 (2)/90	22.803 (4)/90	7.381 (2)/90	[52]
Dexamethasone	orthorhombic	-	P212121	10.44 (1)/90	23.11 (2)/90	16.31 (1)/90	[53]
	orthorhombic	-	P212121	16.278 (3)/90	23.252 (5)/90	10.447 (4)/90	[54]
	orthorhombic	-	P212121	10.3640 (10)/90	16.1570 (10)/90	23.2060 (10)/90	[55]
Dexamethasone acetate monohydrate	orthorhombic	-	P212121	6.60090 (10)/90	15.7827 (3)/90	22.8607 (5)/90	[56]
Diflorasone diacetate	orthorhombic	DD1	P212121	29.386 (1)/90	10.4310 (9)/90	8.1422 (7)/90	[57]
	orthorhombic	DD2	P212121	15.2639 (10)/90	11.7506 (7)/90	13.8931 (11)/90	[57]
	monoclinic	DD3	P21	11.5276 (10)/90	13.8135 (11)/103.053 (6)	7.8973 (7)/90	[57]

Hormone	Syngony	Modification	Space group	a, Å/ α, °	b, Å/ β, °	c, Å/ γ, °	Ref
Diflorasone diacetate monohydrate	orthorhombic	DDW	P212121	30.311 (2)/90	10.6150 (9)/90	7.9337 (7)/90	[57]
Difluprednate	hexagonal	α	P65	16.676 (3)/90	16.676 (3)/90	18.728 (7)/120	[58]
	orthorhombic	β	P212121	26.80856 (89)/90	14.78705 (75)/90	6.54194 (22)/90	[58]
	orthorhombic	γ	P212121	24.22196 (69)/90	13.49846 (45)/90	7.76911 (29)/90	[58]
Dihydrotestosterone acetate	monoclinic	-	P21	7.055/90	48.170/115.95	6.182/90	[59]
Drospirenone	orthorhombic	-	P212121	12.236 (2)/90	12.592 (2)/90	12.879 (2)/90	[60]
	orthorhombic	-	P212121	12.2379 (5)/90	12.6069 (8)/90	12.8921 (8)/90	[61]
Dydrogesterone	orthorhombic	-	P212121	23.84/90	7.80/90	9.36/90	[62]
Epiandrosterone	monoclinic	-	P21	6.3058 (10)/90	21.497 (4)/109.305 (5)	6.5945 (10)/90	[63]
	monoclinic	-	P21	6.596 (1)/90	21.521 (1)/109.39 (1)	6.313 (1)/90	[64]
Epitestosterone	orthorhombic	-	P212121	11.423 (1)/90	20.151 (1)/90	7.003 (1)/90	[65]
Equilenin	orthorhombic	-	P212121	7.480 (7)/90	25.528 (26)/90	7.279 (7)/90	[66]
	orthorhombic	-	P212121	7.27709 (7)/90	7.32686 (6)/90	25.5179 (2)/90	[67]
Equilin	orthorhombic	-	P212121	6.5429 (8)/90	9.0345 (9)/90	23.894 (4)/90	[68]
Estradiol	orthorhombic	-	P212121	12.054/90	19.282/90	6.634/90	[69]
Estradiol valerate	monoclinic	-	P21	7.367870 (34)/90	19.84900 (6)/90.6845 (5)	13.99395 (10)/90	[70]
Estramustine	orthorhombic	-	P212121	21.278 (5)/90	24.342 (5)/90	8.7215 (17)/90	[71]
Estriol	monoclinic	-	P21	9.261 (3)/90	23.025 (2)/90	7.561 (2)/111.000	[72]
Estriol triacetate	monoclinic	-	P21	18.704 (19)/90	8.678 (9)/91.30	13.746 (14)/90	[66]
Estromustine	orthorhombic	-	P212121	8.021 (1)/90	38.544 (5)/90	7.021 (1)/90	[71]
Estrone	orthorhombic	II	P212121	18.520 (5)/90	7.815 (3)/90	10.083 (3)/90	[73]
	orthorhombic	II	P212121	10.043/90	18.424/90	7.787/90	[53]
	orthorhombic	II	P212121	7.787/90	10.043/90	18.424/90	[74]
	orthorhombic	II	P212121	7.773/90	9.983/90	18.412/90	[75]
	orthorhombic	II	P212121	7.7358 (1)/90	9.9137 (1)/90	18.3572 (2)/90	[76]

Table 1. (Continued)

Hormone	Syngony	Modification	Space group	a, Å/ α, °	b, Å/ β, °	c, Å/ γ, °	Ref
	monoclinic	III	P21	9.297/90	23.317/90	7.662/112.270	[77]
	monoclinic	III	P21	9.271 (5)/90	22.285 (5)/111.45(2)	7.610 (5)/90	[74]
	orthorhombic	I	P212121	7.463/90	12.188/90	16.301/90	[74]
Ethinylestradiol	orthorhombic	-	P212121	6.8664 (3)/90	20.9003 (11)/90	22.0448 (8)/90	[78]
Ethisterone	monoclinic	-	P21	6.473/90	21,055/105.94	6.487/90	[79]
	monoclinic	-	P21	6.43 (2)/90	21.17 (4)/105.6 (3)	6.48 (2)/90	[80]
Etonogestrel	monoclinic	-	P21	6.653 (1)/90	18.413 (1)/107.49 (1)	7.9653 (8)/90	[81]
Exemestane	orthorhombic	-	P212121	10.3335 (12)/90	12.1239 (14)/90	12.9953 (14)/90	[82]
Fludrocortisone	orthorhombic	-	P212121	10.087/90	23.710/90	7.660/90	[83]
	orthorhombic	-	P212121	10.02 (1)/90	23.61 (2)/90	7.69 (1)/90	[53]
	orthorhombic	-	P212121	10.088 (2)/90	23.723 (3)/90	7.663 (2)/90	[84]
Fluocinolone acetonide	orthorhombic	-	P212121	14.10/90	15.33/90	11.04/90	[52]
Fluocinonide	orthorhombic	-	P212121	14.169 (2)/90	6.297 (1)/90	28.097 (4)/90	[85]
Fluorometholone	monoclinic	-	P21	6.410 (4)/90	13.431 (3)/92.81 (3)	10.996 (3)/90	[86]
Fluoxymesterone	orthorhombic	-	P212121	13.495 (1)/90	19.553 (1)/90	6.594/90	[87]
Flurogestone acetate	monoclinic	-	P21	7.119 (2)/90	14.734 (3)/95.780 (10)	10.3780 (10)/90	[88]
Fluticasone propionate	monoclinic	I	P21	7.6493 (6)/90	14.1375 (12)/99.339 (7)	10.9833 (9)/90	[89]
	orthorhombic	II	P212121	23.2434 (9)/90	13.9783 (5)/90	7.6510 (3)/90	[90]
Furazabol	monoclinic	-	P21	18.747 (3)/90	6.346 (5)/99.86 (2)	15.647 (4)/90	[97]
Gestodene	orthorhombic	I	P212121	6.620 (3)/90	12.66 (1)/90	20.70 (1)/90	[92]
	orthorhombic	II	P212121	7.6427(7)/90	11.345(2)/90	20.214(3)/90	[92]
Halometasone monohydrate	orthorhombic	-	P212121	7.982 (3)/90	12.915 (5)/90	19.952 (8)/90	[93]
Hydrocortisone acetate	monoclinic	-	P2, P21, C2	8.85/90	13.64/102.3	8.84/90	[94]

Hormone	Syngony	Modification	Space group	a, Å/ α, °	b, Å/ β, °	c, Å/ γ, °	Ref
Hydroxyprogesterone acetate	orthorhombic	-	P212121	10.86 (1)/90	11.95 (2)/90	15.65 (5)/90	[95]
Hydroxyprogesterone caproate	orthorhombic	-	P212121	14.155 (2)/90	23.517 (7)/90	7.466 (1)/90	[96]
	triclinic	III	P1	7.5958 (4)/93.8615 (18)	11.0914 (6)/104.489 (2)	14.7532 (8)/102.1789 (19)	[97]
	monoclinic	II	P21	7.4390 (2)/90	22.6449 (7)/93.6888 (14)	14.0722 (4)/90	[97]
	orthorhombic	I	P212121	7.4417 (2)/90	14.0792 (3)/90	22.8711 (6)/90	[97]
Levonorgestrel	orthorhombic	-	P212121	6.565 (2)/90	12.788 (7)/90	20.658 (3)/90	[98]
	orthorhombic	-	P212121	20.673 (6)/90	12.797 (4)/90	6.567 (1)/90	[99]
Levonorgestrel acetate	orthorhombic	-	P212121	11.328 (6)/90	12.057 (3)/90	15.338 (3)/90	[100]
Levonorgestrel cyclobutylcarboxylate	orthorhombic	-	P212121	11.211 (2)/90	12.300 (4)/90	16.404 (3)/90	[100]
Levonorgestrel cyclopropylcarboxylate	orthorhombic	-	P212121	11.353 (3)/90	12.324 (9)/90	15.694 (4)/90	[100]
Medrogestone	monoclinic	-	P21	10.386 (2)/90	6.5267 (13)/95.60 (3)	14.507 (3)/90	[101]
Medroxyprogesterone acetate	monoclinic	-	P21	11.0780 (5)/90	10.5010 (4)/103.228	9.2690 (3)/90	[102]
Megestrol acetate	orthorhombic	-	P212121	10.843 (1)/90	11.744 (1)/90	15.815 (2)/90	[103]
Meprednisone	monoclinic	-	P21	11.4372 (4)/90	7.8901 (3)/97.896 (2)	21.1268 (9)/90	[104]
Mestanolone	monoclinic	-	P21	6.4321 (6)/90	21.773 (2)/104.482 (8)	6.3848 (8)/90	[105]
Mesterolone	orthorhombic	-	P212121	16.992/90	18.520/90	11.143/90	[106]
	orthorhombic	-	P212121	11.123 (3)/90	16.938 (4)/90	18.561 (3)/90	[107]
Mestranol	monoclinic	-	P21	6.8622 (8)/90	39.71 (3)/117.6 (2)	6.997 (5)/90	[108]
Metandienone	monoclinic	-	C2	28.317 (2)/90	9.4539 (5)/111.017 (9)	13.7684 (10)/90	[109]
Methylprednisolone	monoclinic	-	C2	21.597/90	9.202/114.39	10.916/90	[110]
Mometasone furoate	orthorhombic	-	P212121	12.3414 (7)/90	13.5838 (7)/90	14.9376 (8)/90	[111]
Mometasone furoate monohydrate	triclinic	-	P1	7.3208 (7)/73.251 (2)	8.4767 (8)/85.006 (2)	11.8136 (11)/69.344 (2)	[111]
	triclinic	-	P1	7.2367 (6)/73.617 (5)	8.4193 (6)/85.522 (6)	11.7507 (8)/69.492 (5)	[112]

Table 1. (Continued)

Hormone	Syngony	Modification	Space group	a, Å/ α, °	b, Å/ β, °	c, Å/ γ, °	Ref
Nandrolone	monoclinic	-	P21	10.045 (5)/90	10.761 (5)/89.36 (10)	14.497 (6)/90	[113]
	monoclinic	-	P21	9.755 (2)/90	11.467 (3)/101.07 (2)	14.196 (3)/90	[113]
Nandrolone monohydrate	tetragonal	-	P43212	11.238 (3)/90	11.238 (3)/90	26.604 (3)/90	[113]
Norethindrone acetate	orthorhombic	-	P212121	6.8302 (8)/90	12.164 (1)/90	23.846 (4)/90	[114]
Norethisterone	orthorhombic	-	P212121	20.803/90	12.185/90	6.542/90	[109]
	orthorhombic	-	P212121	6.541/90	12.159/90	20.774/90	[79]
	orthorhombic	-	P212121	6.514 (2)/90	12.066 (3)/90	20.748 (6)/90	[26]
Norethynodrel	orthorhombic	-	P212121	21.593 (13)/90	22.075 (13)/90	7.015 (8)/90	[116]
Oxandrolone	monoclinic	-	P21	9.146 (3)/90	9.591 (3)/91.38 (4)	9.512 (4)/90	[117]
Pentolame monohydrate	triclinic	-	P1	6.225 (1)/92.45 (1)	7.241 (1)/98.85 (1)	12.036 (1)/101.58 (1)	[118]
Prasterone	monoclinic	-	P21	6.261/90	21.920/107.81	6.432/90	[119]
Prednisolone	monoclinic	I	P21	6.3052(5)/90	12.9392 (7)/91.167 (9)	10.9984 (9)/90	[120]
	orthorhombic	II	P212121	5.9774 (3)/90	11.7916 (7)/90	25.5228 (14)/90	[120]
Prednisolone acetate	monoclinic	-	P21	8.519 (5)/90	13.915 (8)/102.688 (8)	8.985 (5)/90	[26]
Prednisolone hydrate	orthorhombic	-	P212121	6.3945 (3)/90	24.5789 (10)/90	24.8376 (11)/90	[120]
Prednisolone succinate	monoclinic	-	I2	21.1388 (3)/90	9.1628 (1)/98.1462 (8)	24.5916 (4)/90	[121]
	monoclinic	-	I2	21.0262 (3)/90	9.1138 (1)/98.3396 (7)	24.3796 (3)/90	[121]
	monoclinic	-	I2	20.9107 (3)/90	9.0536 (2)/98.9412 (9)	24.2008 (4)/90	[121]
Prednisone	orthorhombic	-	P212121	23.071/90	9.999/90	7.847/90	[122]
	orthorhombic	-	P212121	10.004 (10)/90	23.042 (40)/90	7.842 (10)/90	[7]
	orthorhombic	-	P212121	7.8381 (16)/90	9.940 (2)/90	23.020 (5)/90	[123]

Hormone	Syngony	Modification	Space group	a, Å/ α, °	b, Å/ β, °	c, Å/ γ, °	Ref
Pregnenolone	monoclinic	-	P21	6.188 (1)/90	11.863 (2)/91.91 (1)	12.058 (2)/90	[124]
	monoclinic	-	P21	6.1544 (10)/90	11.7359 (19)/92.426 (3)	11.9809 (19)/90	[125]
	monoclinic		P21	6.1663 (13)/90	11.8103 (17)/91.915 (27)	12.022 (7)/90	[126]
Progesterone	monoclinic	-	P21/c	10.9219 (17)/90	7.4400 (12)/93.362(3)	21.034 (3)/90	[127]
	orthorhombic	II	P212121	6.252 (2)/90	12.592 (3)/90	22.488 (2)/90	[128]
	orthorhombic	I	P212121	12.568/90	13.819 (10)/90	10.363 (10)/90	[129]
Progesterone	orthorhombic	I	P212121	12.559 (2)/90	13.798 (2)/90	10.340 (2)/90	[40]
	orthorhombic	I	P212121	10.287 (3)/90	12.524 (4)/90	13.701 (4)/90	[75]
	orthorhombic	I	P212121	10.2496 (7)/90	12.4830 (9)/90	13.6406 (9)/90	[125]
	orthorhombic	II	P212121	6.2089 (6)/90	12.5804 (12)/90	22.188 (2)/90	[125]
	orthorhombic	II	P212121	6.2062 (18)/90	12.573 (5)/90	22.120 (10)/90	[130]
	orthorhombic	I	P212121	10.380 (2)/90	12.810 (3)/90	13.890 (3)/90	[80]
Prolame	orthorhombic	-	P212121	8.356 (2)/90	13.343 (4)/90	16.119 (4)/90	[131]
Rimexolone	monoclinic	-	P21	11.76 (1)/90	11.75 (1)/-98.85 (8)	7.596 (4)/90	[132]
Silandrone	orthorhombic	-	P212121	13.222 (8)/90	14.092 (6)/90	11.523 (3)/90	[133]
Stanozolol	orthorhombic	-	P212121	19.4540 (7)/90	18.7009 (8)/90	10.8348 (4)/90	[134]
Stanozolol monohydrate	orthorhombic	-	P212121	7.2098 (5)/90	10.4654 (7)/90	26.0313 (13)/90	[134]
Testosterone	orthorhombic	-	P212121	7.940/90	13.632/90	15.952/90	[135]
	monoclinic	-	P21	14.45/90	11.09/110.5	10.88/90	[136]
	monoclinic	-	P21	14.720 (3)/90	11.080 (2)/113.34 (1)	10.868 (2)/90	[136]
Testosterone acetate	orthorhombic	-	P212121	12.6/90	18.1/90	7.8/90	[137]
	orthorhombic	-	P212121	12.800/90	18.169/90	7.856/90	[20]
Testosterone buciclate	orthorhombic	-	P212121	6.2987 (13)/90	10.500 (2)/90	39.680 (8)/90	[138]
Testosterone butyrate	orthorhombic	-	P212121	12.3/90	16.3/90	10.3/90	[137]
Testosterone formate	orthorhombic	-	P212121	21.7/90	12.5/90	12.9/90	[137]

Table 1. (Continued)

Hormone	Syngony	Modification	Space group	a, Å/ α, °	b, Å/ β, °	c, Å/ γ, °	Ref
Testosterone monohydrate	monoclinic	-	P21	9.622/90	8.099 (5)/97.81 (15)	11.149/90	[139]
	orthorhombic	-	P212121	15.966 (6)/90	13.609 (6)/90	7.952 (4)/90	[140]
	orthorhombic	-	P212121	13.63/90	15.95/90	7.94/90	[141]
Testosterone propionate	orthorhombic	-	P212121	12.6/90	20.3/90	7.6/90	[137]
	orthorhombic	-	P212121	7.586 (3)/90	12.513 (5)/90	19.869 (7)/90	[142]
Testosterone valerate	orthorhombic	-	P212121	12.3/90	16.7/90	10.3/90	[137]
Tibolone	monoclinic	-	P21	6.542 (2)/90	41.213 (9)/101.64 (2)	6.678 (2)/90	[143]
	triclinic	-	P1	6.542 (2)/87.05 (1)	6.6773 (8)/80.09 (2)	10.287 (2)/79.17 (2)	[144]
Triamcinolone	monoclinic	-	P21	6.0440 (3)/90	24.0156 (15)/101.937 (6)	12.7206 (6)/90	[145]
Triamcinolone acetonide	tetragonal	-	P41212	9.1692 (4)/90	9.1692 (4)/90	49.946 (4)/90	[146]

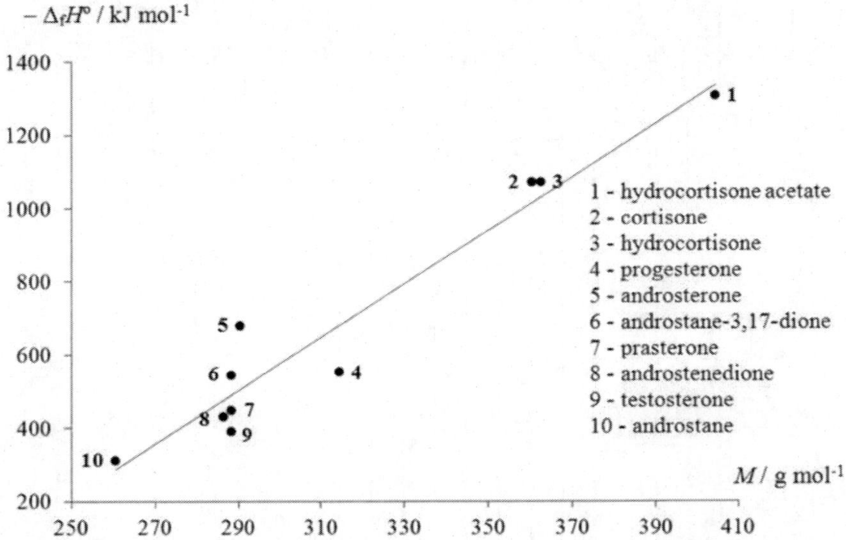

Figure 6. Dependence of the enthalpy of formation of steroids on the molecular weight.

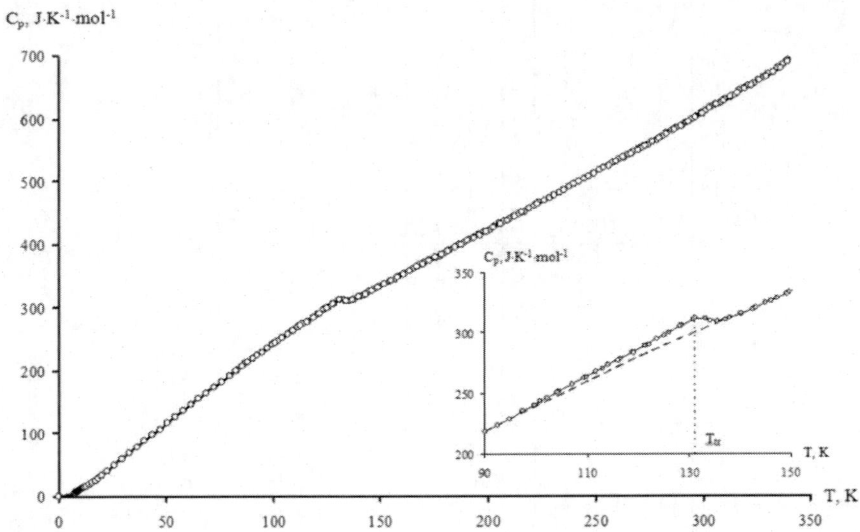

Figure 7. Temperature dependence of heat capacity of methylprednisolone aceponate.

Table 2. Thermodynamic characteristics of steroid hormones

Hormone	$-\Delta_c H^o(298)$ kJ/mol	$-\Delta_f H^o(298)$ kJ/mol	$S^o(T)$ J/(mol·K)	$-\Delta_f S^o(298)$ J/(mol·K)	$-\Delta_f G^o(298)$ kJ/mol	$\Delta_{fus}H$ kJ/mol	$\Delta_{fus}S$ J/(mol·K)
Androstenedione	10760 ± 30 [147]	431 [148]	-	-	-	-	-
Androsterone	11090 ± 30 [147]	678 [148]	-	-	-	-	-
Budesonide	-	-	-	-	-	34.7 [149]	-
Corticosterone	-	-	-	-	-	35.3 [150]	-
						33.32 [151]	
Cortisone	11200 ± 20 [147]	1070 [148]	-	-	-	36.8 [151]	-
Cortisone Acetate	-	-	-	-	-	38.43 [151]	-
Deoxycorticosterone	12030 ± 4 [147]	519 [148]	-	-	-	27.98 [151]	-
Dexamethasone	-	-	-	-	-	42.02 [151]	-
Dexamethasone acetate	-	-	-	-	-	37.72 [151]	-
Estrone acetate	-	-	-	-	-	15.0 [152]	-
Estrone propionate	-	-	-	-	-	23.0 [152]	-
Ethinylestradiol	-	-	-	-	-	27.57 [153]	-
Hydrocortisone	11480 ± 30 [147]	-	-	-	-	35.84 [151]	-
Hydrocortisone Acetate	12317.1 ± 11.6 [154]	1307.0 ± 11.6 [154]	-	-	-	53.64 [155]	-
Mestranol	-	-	-	-	-	-	34.55 [156]
Methylprednisolone	11898.9 ± 6.7 [157]	1045.8 ± 7.3 [157]	455.2	2141 ± 11 [157]	407 ± 8 [157 11]	-	-
Methylprednisolone aceponate	14304.4 ± 9.1 [158]	1465.3 ± 9.8 [158]	628.6	2592.9 ± 13.7 [158]	692 ± 10 [158]	-	-
Norethindrone	-	-	-	-	-	39.6 [159]	82.7 [159]
						39.6 [160]	
Norethindrone Acetate	-	-	-	-	-	27.3 [159]	56.9 [159]
Prasterone	11030 ± 20 [147]	448 [148]	-	-	-	-	-

Hormone	$-\Delta_c H^o(298)$ kJ/mol	$-\Delta_f H^o(298)$ kJ/mol	$S^o(T)$ J/(mol·K)	$-\Delta_f S^o(298)$ J/(mol·K)	$-\Delta_f G^o(298)$ kJ/mol	$\Delta_{fus}H$ kJ/mol	$\Delta_{fus}S$ J/(mol·K)
Prednisolone	11244.8 ± 7.6	1020.6	-	-	-	38.86 [151]	-
Prednisolone acetate	-	-	-	-	-	42.3 [155]	-
Progesterone	12000 ± 20 [147]	552 [148]	-	-	-	26.99 [151]	-
Stanolone (dihydrotestosterone)	10550 ±30 [147]	-	-	-	-	-	-
Testosteron	11080 ± 30 [147]	390 [148]	-	-	-	28.2 [161]	-
						29.45 [151]	
Testosterone acetate	-	-	-	-	-	27.88 [151]	-
Triamcinolone	-	-	-	-	-	42.56 [151]	-
Triamcinolone Acetonide	-	-	-	-	-	45.29 [150]	-
Triamcinolone diacetate	-	-	-	-	-	38.31 [150]	-

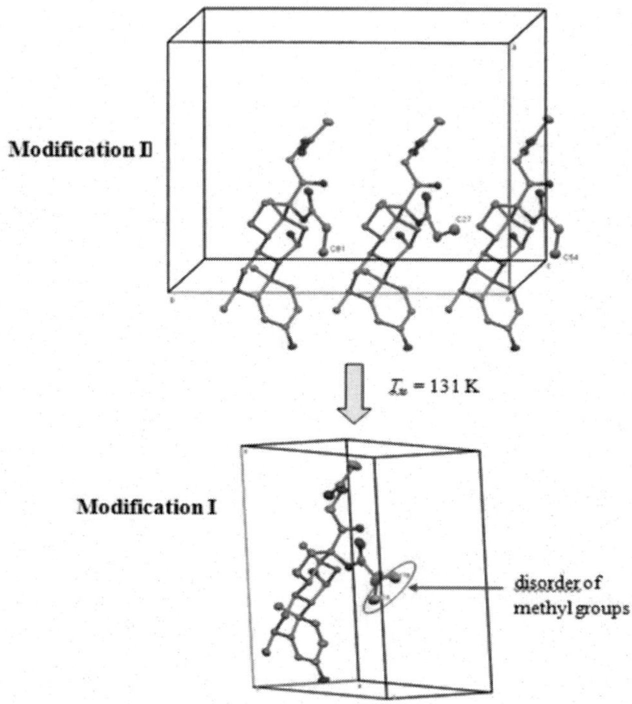

Figure 8. Structural changes at the phase transition in the methylprednisolone aceponate.

Information on the thermodynamic data in the literature is virtually absent in contrast to the structural data. Table 2 summarizes all the available experimental thermodynamic data (mainly standard enthalpies of formation) for steroid hormones. Correlation analysis showed that the dependence of the enthalpy of formation of steroids on the molecular weight is described by a linear equation (Figure 6). The only compound on which we discovered a reversible phase transition is methylprednisolone aceponate [158]. The transition temperature of the sample under study T_{tr} = 131.0 ± 0.1 K was estimated as the temperature of maximal heat capacity value within the temperature interval of the transition (Figure 7). The standard enthalpy of phase transition is $\Delta_{tr}H° = 220 \pm 2$ J·mol^{-1}. This is a very small value, which is associated with small structural changes. o clarify the nature of the phase transition, we grew a single crystal of this compound and performed X-ray diffraction studies. The transition to a high-temperature modification **I** (T = 150K) is accompanied by a transformation of the unit cell: $\mathbf{a}_{I}\leftrightarrow\mathbf{b}_{II}/3$, $\mathbf{b}_{I}\leftrightarrow\mathbf{a}_{II}$, $\mathbf{c}_{I}\leftrightarrow\mathbf{c}_{II}$, at the same time, the space group symmetry does not change. Methyl groups C27A and C27B are disordered with occupancies 0.637 (7) and 0.363 (7), respectively (Figure 8). A complete disorder of the methyl groups must conform to the ratio of the occupancies 0.5/0.5. The orientations of methyl groups are related as 1/3 and 2/3 in the modification **II**. Therefore, it can be expected that with increasing temperature the ratio of the occupancies will tend to 0.5/0.5. Temperature reduction increases the energy of intermolecular hydrogen bonds, as indicated by the increase in the angle D—H···A.

Conclusion

A systematic analysis of structural and thermodynamic data is carried out. It was shown that steroid hormones are characterized by rich isomorphism, but there are too few accurate physico-chemical data to quantitatively describe the areas of existence of a particular modification.

References

[1] P. Ferraboschi, M. C. Sala, R. Stradi, L. Ragonesi, C. Gagliardi, P. Lanzarotti, E. M. Ragg, M. Mori, F. Meneghetti, Steroids, 2017, 128, 95, doi: 10.1016/j.steroids.2017.09.003.

[2] R. Dey, S. Roychowdhury, P. Roychaudhury, *Journal of Chemical Crystallography,* 1999, 29, 1265, doi: 10.1023/A:1009504923424.

[3] V. K. Gupta, R. Goswami, K. N. Goswami, S. K. Mazumdar, K. K. Bhutani, Acta Crystallographica, *Section C: Crystal Structure Communications,* 1994, 50, 798, doi: 10.1107/S010827019300 65960.

[4] P. Strong, W. L. Duax, L. L. Engel, *Crystal Structure Communications,* 1976, 5, 815.

[5] E. A. Zhurova, V. V. Zhurov, D. Chopra, A. I. Stash, A. A. Pinkerton, *Journal of the American Chemical Society*, 2009, 131, 17260, doi: 10.1021/ja906057z.

[6] J. P. Declercq, G. Germain, M. van Meerssche, *Crystal Structure Communications,* 1972, 1, 9.

[7] G. van den Bossche, *Bulletin de la Societe Royale des Sciences de Liege* 1971, 40, 142.

[8] Z. Galdecki, P. Grochulski, Z. Wawrzak, *Journal of Crystallographic and Spectroscopic Research*, 1989, 19, 577, doi: 10.1007/BF01185393.

[9] D. J. Duchamp, J. A. Campbell, *Acta Crystallographica Section A: Cryst. Phys., Diffr. Theor. Crystallogr,* 1972, 28, S50.

[10] G. Precigoux, B. Busetta, C. Courseille, M. Hospital, Acta Crystallographica, *Section B: Struct. Crystallogr. Cryst. Chem*., 1975, 31, 1527, doi: 10.1107/S0567740875005638.

[11] M. M. Bhadbhade, K. Venkatesan, Acta Crystallographica, Section C: *Crystal Structure Communications,* 1984, 40, 1905, doi: 10.1107/S0108270184010003.

[12] R. Dey, V. Langer, P. Roychowdhury, S Roychowdhury, M. G. B. Drew, Acta Crystallographica, *Section C: Crystal Structure Communications,* 2005, 61, 201.

[13] D. C. Swenson, W. L. Duax, P. D. Strong, J. Weisz, M. S. Hershey, *Acta Crystallographica Section A: Cryst. Phys., Diffr., Theor. Crystallogr.,* 1981, 37, C56, doi: 10.1107/S0108767381097894.

[14] K. D. Onan, S. K. Panda, D. Cecchini, S. C. Chattoraj, T. F. Brennan, *Acta Crystallographica Section A: Cryst. Phys., Diffr., Theor. Crystallogr.,* 1981, 37, C73a.

[15] Z. Wawrzak, W. L. Duax, P. D. Strong, J. Weisz, *Journal of Steroid Biochemistry,* 1988, 29, 387.

[16] P. J. Cox, G. A. Sim, *Acta Crystallographica, Section B: Struct. Crystallogr. Cryst. Chem*., 1982, 38, 1360, doi: 10.1107/S0567740882005810.

[17] J. Ohrt, B. A. Haner, D. A. Norton, *Acta Crystallographica*, 1965, 19, 479, doi: 10.1107/S0365110X65003663.

[18] W. L. Duax, H. Hauptman, C. M. Weeks, D. A. Norton, *Journal of the Chemical Society D*, 1971, 1055, doi: 10.1039/c29710001055.

[19] S. Pagola, P. W. Stephens, *Cryst. Eng. Comm.*, 2012, 14, 5349, doi: 10.1039/c2ce25274k.

[20] J. Ohrt, B. A. Haner, D. A. Norton, *Acta Crystallographica,* 1965, 19, 479, doi: 10.1107/S0365110X65003663.

[21] B. Busetta, G. Comberton, C. Courseille, M. Hospital, C*rystal Structure Communications,* 1972, 1, 129.

[22] D. F. High, J. Kraut, *Acta Crystallographica,* 1966, 21, 88, doi: 10.1107/S0365110X66002378.

[23] R. Bouche, M. Draguet-Brughmans, *Microchimica Acta*, 1976, 403, doi: 10.1007/BF01220287.

[24] Banerjee, Zeitschrift fur Kristallographie, Kristallgeomie, Krystallphysik, *Kristallchemie,* 1971, 134, 152.

[25] J. D. Bernal, D. Crowfoot, *Zeitschrift fur Kristallographie, Kristallgeomie, Krystallphysik, Kristallchemie*, 1936, 93, 464.

[26] K. Shikii, H. Seki, S. Sakamoto, Y. Sei, H. Utsumie, K. Yamaguchi, *Chemical and Pharmaceutical Bulletin,* 2005, 53, 792, doi: 10.1248/cpb.53.792.

[27] A. T. Hulme, R. W. Lancaster, H. F. Cannon, *Cryst. Eng. Comm.*, 2006, 8, 309, doi: 10.1039/B602909D.

[28] K. A. Ketuly, A. H. A. Hadi, S. W. Ng, *Acta Crystallographica Section E: Structure Reports Online,* 2009, 65, o1822, doi: 10.1107/S1600536809025987.

[29] S. Wang, Y. Zheng, W. Zhang, *Acta Crystallographica Section E: Structure Reports Online,* 2004, 60, o1063, doi: 10.1107/S160053680401205X.

[30] K. J. Wood, M. Fronckowiak, W. L. Duax, M. B. Hursthouse, S. J. Lamond, L. Surplice, University of Southampton, *Crystal Structure Report Archive,* 1987, 1231, doi: 10.5258/ecrystals/1231.

[31] W. L. Duax, P. Strong, M. Glowka, M. B. Hursthouse, *CSD Communication*, 2014.

[32] J. Albertsson, A. Oskarsson, C. Svensson, *Acta Crystallographica, Section B: Struct. Crystallogr. Cryst. Chem.*, 1978, 34,3027, doi: 10.1107/S0567740878009966.

[33] R. J. Chandross, J. Bordner, *Acta Crystallographica, Section B: Struct. Crystallogr. Cryst. Chem.,* 1975, 31, 928, doi: 10.1107/S0567740875004165.

[34] M. P. Feth, J. Volz, U. Hess, E. Sturm, R.-P. Hummel, *Journal of Pharmaceutical Sciences*, 2008, 97, 3765, doi: 10.1002/jps.21223.

[35] M. Haramura, A. Tanaka, T. Akimoto, N. Hirayama, *Analytical Sciences: X-Ray Structure Analysis Online*, 2003, 19, x37, doi: 10.2116/analscix.19.x37.

[36] C. Nather, Inke Jeb, *Collection of Czechoslovak Chemical Communications,* 2011, 76(5), 553, doi: 10.1135/cccc2011027.

[37] E. Maccaroni, A. Mele, R. Del Rosso, L. Malpezzi, *Acta Crystallographica Section E: Structure Reports Online,* 2011, 67, o1952, doi: 10.1107/S1600536811026560.

[38] H. Campsteyn, L. Dupont, O. Dideberg, N. Mandel, *Acta Crystallographica, Section B: Struct. Crystallogr. Cryst. Chem.,* 1973,29, 1726, doi: 10.1107/S0567740873005364.

[39] V. Suitchmezian, I. Jess, C. Nather, *Journal of Pharmaceutical Sciences*, 2008, 97, 4516, doi: 10.1002/jps.21334.

[40] B. A. Haner, D. A. Norton, *Acta Crystallographica,* 1964, 17, 1610, doi:: 10.1107/S0365110X64003954.

[41] E. E. Castellano, P. Main, E. Westbrook, *Acta Crystallographica, Section B: Struct. Crystallogr. Cryst. Chem.*, 1980, 36, 3063, doi:: 10.1107/S0567740880010801.

[42] J. P. Declercq, G. Germain, M. van Meerssche, *Crystal Structure Communications,* 1972, 1, 13.

[43] J. P. Declercq, G. Germain, M. van Meerssche, *Crystal Structure Communications,* 1972, 1, 59.

[44] K. Callow, *Journal of Pharmacy and Pharmacology*, 1961, 13, 723.

[45] J. A. Kanters, A. de Koster, V. J. van Geerestein, L. A. van Dijck, *Acta Crystallographica, Section C: Crystal Structure Communications,* 1985, 41, 760, doi: 10.1107/S0108270185005406.

[46] E. W. Czerwinski, *Acta Crystallographica, Section C: Crystal Structure Communications*, 1991, 47, 2598, doi: 10.1107/S0108270191006200.

[47] R. J. Chandross, J. Bordner, *Acta Crystallographica, Section B: Struct. Crystallogr. Cryst. Chem.*, 1974, 30, 1581, doi: 10.1107/S0567740874005309.

[48] M. A.Viswamitra, R. Radhakrishnan, J. Bandekar, G. R. Desiraju, *Journal of the American Chemical Society*, 1993, 115, 4868, doi: 10.1021/ja00064a055.

[49] R. Dey, T. Banerjee, P. R. Chowdhury, S. Chaudhuri, *Journal of Chemical Crystallography*, 2001, 31, 263, doi: 10.1023/A:1014322611370.

[50] S .L. Cuffini, J. F. Ellena, Y. P. Mascarenhas, A. P. Ayala, H. W. Sielser, J. M. Filho, G. A. Monti, V. Aiassa, N. R. Sperandeo, *Steroids,* 2007, 72, 261, doi: 10.1016/j.steroids.2006.11.020.

[51] T. C. van Soest, L. A. van Dijck, F.J. Zeelen, Recueil des Travaux Chimiques des Pays-Bas, 1980, 99, 323.

[52] L. Dupont, O. Dideberg, H. Campsteyn, *Acta Crystallographica, Section B: Struct. Crystallogr. Cryst. Chem.*, 1974, 30, 514, doi: 10.1107/S0567740874003086.

[53] G. van den Bossche, *Bulletin de la Societe Royale des Sciences de Liege,* 1971, 40, 614.

[54] D. C. Rohrer, W. L. Duax, *Crystal Structure Communications,* 1977, 6, 123.

[55] J. W. Raynor, W. Minor, M. Chruszcz, *Acta Crystallographica Section E: Structure Reports Online*, 2007, 63, o2791, doi: 10.1107/S1600536807020806.

[56] R. P. da Silva, M. F. S. Ambrósio, L. A. Piovesan, M. C. R. Freitas, D. L. M. de Aguiar, B. A. C. Horta, E. Epprecht, R. A. da Silva San Gil, L. do Canto Visentin, *Journal of Pharmaceutical Sciences*, 2018, 107, 672, doi: 10.1016/j.xphs.2017.10.001.

[57] E. Maccaroni, G. B. Giovenzana, G. Palmisano, D. Botta, P. Volante, N. Masciocchi, *Steroids,* 2009, 74, 102, doi: 10.1016/j.steroids.2008.09.014.

[58] C. Giacobbe, G. Palmisano, G. B. Giovenzana, L. Giovannelli, R. Negri, N. Masciocchi, *Journal of Pharmaceutical and Biomedical Analysis,* 2015, 102, 305, doi: 10.1016/j.jpba.2014.09.028.

[59] D. A. Norton, C. T. Lu, A. E. Campbell, *Acta Crystallographica*, 1962, 15, 1189, doi: 10.1107/S0365110X6200314X.

[60] M. Parisi, E. Freire, M. Rusjan, J. M. Moreno, H. Bonadeo, D. Vega, *Journal of Molecular Structure*, 2013, 1040, 83, doi: 10.1016/j.molstruc.2013.01.040.

[61] Q. Wang, Q. M. Sun, S. S. Li, L. Q. Guo, Hui Li, *Powder Diffraction,* 2016, 31, 63, doi: 10.1017/S0885715615000822.

[62] P. B. Braun, J. Hornstra, J. I. Leenhouts, *Philips Research Reports*, 1969, 24, 427.

[63] Y. Zhang, P. Coppens, *CSD Communication*, 2002.

[64] C. M. Weeks, A. Cooper, D. A. Norton, H. Hauptman, J. Fisher, Acta Crystallographica, *Section B: Struct. Crystallogr. Cryst. Chem*., 1971, 27, 1562, doi: 10.1107/S0567740871004382.

[65] N. W. Isaacs, W. D. S. Motherwell, J. C. Coppola, O. Kennard, *Journal of the Chemical Society Perkin Transactions* 2, 1972, 2335, doi: 10.1039/p29720002335.

[66] J. M. Ohrt, B. A. Haner, D. A. Norton, *Acta Crystallographica*, 1967, 23, 1100, doi: 10.1107/S0365110X67004347.

[67] C. S. Frampton, D. D. MacNicol, *Acta Crystallographica Section E: Crystallographic Communications*, 2017, 73, 1223, doi: 10.1107/ S2056989017010532.

[68] M. W. Sawicki, N. Li, D. Ghosh, *Acta Crystallographica, Section C: Crystal Structure Communications,* 1999, 55, 425, doi: 10.1107/ S0108270198013250.

[69] D. A. Norton, G. Kartha, C. T. Lu, *Acta Crystallographica,* 1964, 17, 77, doi: 10.1107/S0365110X64000299.

[70] A. M. Wheatley, J. A. Kaduk, A. M. Gindhart, T. N. Blanton, *Powder Diffraction,* 2018, 33, 70, doi: 10.1017/S0885715618000039.

[71] J. S. Punzi, W. L. Duax, P. Strong, J. F. Griffin, M. M. Flocco, D. E. Zacharias, H. L. Carrell, K. D. Tew, J. P. Glusker, *Molecular Pharmacology*, 1992, 41, 569.

[72] Cooper, D. A. Norton, H. Hauptman, *Acta Crystallographica, Section B: Struct. Crystallogr. Cryst. Chem.,* 1969, 25, 814, doi: 10.1107/S0567740869003098.

[73] T. D. J. Debaerdemaeker, *Crystal Structure Communications,* 1972, 1, 39.

[74] B. Busetta, C. Courseille, M. Hospital, *Acta Crystallographica, Section B: Struct. Crystallogr. Cryst. Chem.*, 1973, 29, 298, doi: 10.1107/S0567740873002384.

[75] K. Shikii, S. Sakamoto, H. Seki, H. Utsumi, K. Yamaguchi, *Tetrahedron,* 2004, 60, 3487, doi: 10.1016/j.tet.2004.02.030.

[76] E. A. Zhurova, C. F. Matta, N. Wu, V. V. Zhurov, A. A. Pinkerton, *Journal of the American Chemical Society*, 2006, 128, 8849, doi: 10.1021/ja061080v.

[77] J. M. Ohrt, B. A. Haner, D. A. Norton, *Acta Crystallographica*, 1964, 17, 1611, doi: 10.1107/S0365110X64003966.

[78] J. Karpinska, A. Erxleben, P. McArdle, *Crystal Growth and Design*, 2013, 13, 1122, doi: 10.1021/cg301479c.

[79] J. Reisch, J. Zappel, G. Henkel, N. Ekiz-Gucer, *Monatshefte fuer Chemie,* 1993, 124, 1169, doi: 10.1007/BF00810025.

[80] C. G. Jones, M. W. Martynowycz, J. Hattne, T. J. Fulton, B. M. Stoltz, J. A. Rodríguez, H. M. Nelson, T. Gonen, *ACS Central Science,* 2018, 4, 1587, doi: 10.1021/acscentsci.8b00760.

[81] V. J. van Geerestein, J. A. Kanters, J. Kroon, *Acta Crystallographica, Section C: Crystal Structure Communications,* 1987, 43, 2398, doi: 10.1107/S0108270187087626.

[82] K. Gorlitzer, C. Bonnekessel, P. G. Jones, A. Palusczak, R. W. Hartmann, *Pharmazie,* 2006, 61, 575.

[83] C. M. Weeks, W. L. Duax, *American Crystallographic Association, Abstracts Papers* (Winter), 1972, 73.

[84] L. Dupont, O. Dideberg, H. Campsteyn, *Acta Crystallographica, Section B: Struct. Crystallogr. Cryst. Chem.,* 1972, 28, 3023, doi: 10.1107/S0567740872007381.

[85] Sung-Il Cho, B. M. Craven, *Acta Crystallographica, Section C: Crystal Structure Communications,* 1987, 43, 917, doi: 10.1107/S0108270187093582.

[86] Y. J. Park, M. Y. Lee, S. I. Cho, *Journal of the Korean Chemical Society,* 1992, 36, 812.

[87] W. L. Duax, P. D. Strong, M. E. Wolff, *Crystal Structure Communications,* 1979, 8, 985.

[88] P. Puig, J. Guillon, S. Massip, C. Jarry, J.-M. Leger, P. Hivorel, *Analytical Sciences: X-Ray Structure Analysis Online*, 2006, 22, x257, doi: 10.2116/analscix.22.x257.

[89] J. Cejka, B. Kratochvil, A. Jegorov, *Zeitschrift fur Kristallographie - New Crystal Structures,* 2005, 220, 143.

[90] B. M. Kariuki, K. Psallidas, K. D. M. Harris, R. L. Johnstone, R. W. Lancaster, S. E. Staniforth, S. M. Cooper, *Chemical Communications,* 1999, 1677, doi: 10.1039/a904702f.

[91] D. R. Lisgarten, R. A. Palmer, *Journal of Chemical Crystallography*, 1998, 28, 379, doi: 10.1023/A:1022416309763.

[92] V. J. van Geerestein, A. J. M. Duisenberg, M. H. G. Duits, J. A. Kanters, J. Kroon, *Acta Crystallographica, Section C: Crystal Structure Communications,* 1987, 43, 2402, doi: 10.1107/S0108270187087614.

[93] J. Reisch, G. Henkel, N. Ekiz-Gucer, G. Nolte, Liebigs Annalen, 1992, 63, doi: 10.1002/jlac.199219920112.

[94] J. W. Shell, Analytical Chemistry, 1955, 27, 1665, doi: 10.1021/ac60106a055.

[95] Rajnikant, Dinesh, Anshu, Mousmi, *Crystal Research and Technology,* 2004, 39, 353, doi: 10.1002/crat.200310194.

[96] Krstanovic, N. Cvetkovic, R. Oberti, Lj. Karanovic, *Acta Crystallographica, Section C: Crystal Structure Communications,* 1989, 45, 478, doi: 10.1107/S0108270188011874.

[97] J. Caplette, T. Frigo, M. Jozwiakowski, H. Shea, M. Mirmehrabi, P. Muller, *International Journal of Pharmaceutics,* 2017, 527, 42, doi: 10.1016/j.ijpharm.2017.05.031.

[98] P. Crabbe, E. O. Schlemper, *Bulletin des Societes Chimiques Belges* 1983, 92, 275, doi: 10.1002/bscb.19830920312.

[99] N. J. DeAngelis, T. H. Doyne, R. L. Grob, *Acta Crystallographica, Section B: Struct. Crystallogr. Cryst. Chem.,* 1975, 31, 2040, doi: 10.1107/S0567740875006802.

[100] P. Crabbe, E. O. Schlemper, *Bulletin des Societes Chimiques Belges*, 1983, 92, 275, doi: 10.1002/bscb.19830920312.

[101] P. Ferraboschi, P. Ciuffreda, S. Ciceri, P. Grisenti, C. Castellano, F. Meneghetti, *Steroids,* 2015, 104, 137, doi: 10.1016/j.steroids. 2015.09.007.

[102] W. L. Duax, V. Cody, J. Griffin, J. Hazel, C. M. Weeks, *Journal of Steroid Biochemistry*, 1978, 9, 901.

[103] M. Soriano-Garcia, A. Zambrano, M. Cabeza, E. Bratoeff, *Journal of Chemical Crystallography*, 2010, 40, 1115, doi: 10.1007/s10870-010-9805-8.

[104] D. Fernandez, D. Vega, J. A. Ellena, *Acta Crystallographica, Section C: Crystal Structure Communications*, 2003, 59, o187, doi: 10.1107/S0108270103003184.

[105] N. Takata, K. Shiraki, R. Takano, Y. Hayashi, K. Terada, *Crystal Growth and Design,* 2008, 8, 3032, doi: 10.1021/cg800156k.

[106] S. Danaci, E. Kendi, *Hacettepe Bulletin of Natural Sciences and Engineering,* 1983, 12, 171.

[107] S. Danaci, E. Kendi, F. G. Moers, H. Behm, P. T. Buerskens, *Acta Crystallographica, Section C: Crystal Structure Communications,* 1988, 44, 1677, doi: 10.1107/S0108270188006274.

[108] T. Steiner, B. Lutz, J. van der Maas, N. Veldman, A. M. M. Schreurs, J. Kroon, J. A. Kanters, *Chemical Communications,* 1997, 191, doi: 10.1039/a606758a.

[109] J. Karpinska, A. Erxleben, P. McArdle, *Acta Crystallographica Section E: Structure Reports Online,* 2013, 69, o60, doi: 10.1107/ S1600536812049860.

[110] J. P. Declercq, G. Germain, M. van Meerssche, *Crystal Structure Communications,* 1972, 1, 5.
[111] X. Chen, M. Carillo, R. C. Haltiwanger, P. Bradley, *Journal of Pharmaceutical Sciences,* 2005, 94, 2496, doi: 10.1002/jps.20470.
[112] P. Müller, *Acta Crystallographica Section C: Structural Chemistry*, 2015, 71, 1080, doi: 10.1107/S2053229615021476.
[113] X. Zeng, X. Xiong, H. Yang, B. Tang, Q. Du, Q. Hou, Z. Suo, H. Li, *Journal of Pharmaceutical Sciences,* 2018, 107, 1928, doi: 10.1016/j.xphs.2018.01.011.
[114] T. Steiner, *Acta Crystallographica, Section C: Crystal Structure Communications,* 1996, 52, 2261, doi: 10.1107/ S0108270 196002417.
[115] J. P. Mornon, G. Lepicard, J. Delettre, *Comptes Rendus Seances de l'Academie des Sciences, Series C (Chimiques),* 1976, 282, 387.
[116] C. Lemini, R. Jaimez, R. A. Toscano, *Revista de la Sociedad Quimica de Mexico,* 2004, 48, 256.
[117] D. F. Rendle, J. Trotter, *Journal of the Chemical Society Perkin Transactions* 2, 1975, 1361, doi: 10.1039/p29750001361.
[118] A. T. McPhail, P. A. Luhan, P. S. W. Tschang, K. D. Onan, *Journal of the Chemical Society Perkin Transactions* 2, 1977, 379, doi: 10.1039/p29770000379.
[119] G. P. Stahly, S. Bates, M. C. Andres, B. A. Cowans, *Crystal Growth and Design*, 2006, 6, 925, doi: 10.1021/cg050479c.
[120] V. Suitchmezian, I. Jess, J. Sehnert, L. Seyfarth, J. Senker, C. Nather, *Crystal Growth and Design,* 2008, 8, 98, doi: 10.1021/ cg700866f.
[121] E. Nishibori, T. Ogura, S. Aoyagi, M. Sakata, *Journal of Applied Crystallography,* 2008, 41, 292, doi: 10.1107/S0021889808001520.
[122] V. M. Tseikinskii, V. I. Simonov, V. B. Ribakov, N. N. Petropavlov, *Bioorganicheskaia Khimiia*, 1979, 5, 1677.
[123] V. Suitchmezian, I. Jess, C. Nather, *CSD Communication*, 2008.
[124] J. Bordner, G. L. A. Hennessee, R. J. Chandross, *Crystal Structure Communications,* 1978, 7, 513.

[125] R. W. Lancaster, P. G. Karamertzanis, A. T. Hulme, D. A. Tocher, T. C. Lewis, S. L. Price, *Journal of Pharmaceutical Sciences*, 2007, 96, 3419, doi: 10.1002/jps.20983.

[126] T. Friscic, R. W. Lancaster, L. Fabian, P. G. Karamertzanis, *Proceedings of the National Academy of Sciences of the United States of America,* 2010, 107, 13216, doi: 10.1073/pnas.0915142107.

[127] R. W. Lancaster, P. G. Karamertzanis, A. T. Hulme, D. A. Tocher, D. F. Covey, S. L. Price, *Chemical Communications*, 2006, 4921, doi: 10.1039/B611599C.

[128] H. Campsteyn, L. Dupont, O. Dideberg, *Acta Crystallographica, Section B: Struct. Crystallogr. Cryst. Chem.*, 1972, 28, 3032, doi: 10.1107/S0567740872007393.

[129] E. F. Serantoni, A. Krajewski, R. Mongiorgi, L. Riva di Sanseverino, R. Cameroni, *Crystal Structure Communications,* 1975, 4, 189.

[130] Sarkar, D. Ragab, S. Rohani, *Crystal Growth and Design*, 2014, 14, 4574, doi: 10.1021/cg5006727.

[131] J. M. Fernandez-G, M. F. Rubio-Arroyo, M. Soriano-Garcia, R. A. Toscano, M. del Carmen Perez-Cesar, C. Rubio-Poo, J. J. Mandoki, A. de la Pena, C. Lemini, N. Mendoza-Patino, F. Cruz, *Steroids,* 1985, 45, 151, doi: 10.1016/0039-128X(85)90044-3.

[132] J. C. Barnes, J. D. Paton, R. T. Logan, J. Redpath, *Acta Crystallographica, Section C: Crystal Structure Communications*, 1985, 41, 763, doi: 10.1107/S0108270185005418.

[133] C. M. Weeks, H. Hauptman, D. A. Norton, Crystal Structure Communications, 1972, 1, 79.

[134] J. Karpinska, A. Erxleben, P. McArdle, *Crystal Growth and Design*, 2011, 11, 2829, doi: 10.1021/cg101651p.

[135] B. N. Das, Zeitschrift fur Kristallographie, Kristallgeomie, Krystallphysik, *Kristallchemie,* 1971, 134, 154.

[136] P. J. Roberts, R. C. Pettersen, G. M. Sheldrick, N. W. Isaacs, O. Kennard, *Journal of the Chemical Society Perkin Transactions* 2, 1973, 1978, doi: 10.1039/p29730001978.

[137] P. J. F. Griffiths, K. C. James, M. Rees, *Acta Crystallographica*, 1965, 19, 149a.

[138] N. W. Alcock, K. J. Sanders, A. Rodger, *Acta Crystallographica Section E: Structure Reports Online*, 2004, 60, o348, doi: 10.1107/S1600536804000820.

[139] B. Busetta, C. Courseille, F. Leroy, M. Hospital, *Acta Crystallographica, Section B: Struct. Crystallogr. Cryst. Chem.*, 1972, 28, 3293, doi: 10.1107/S0567740872007861.

[140] G. Precigoux, M. Hospital, G. van den Bosche, *Crystal Structure Communications,* 1973, 2, 435.

[141] A. L. Thakkar, N. D. Jones, H. A. Rose, L. G. Tensmeyer, N. A. Hall, *Acta Crystallographica, Section B: Struct. Crystallogr. Cryst. Chem.*, 1970, 26, 1184, doi: 10.1107/S0567740870003801.

[142] J. Reisch, N. Ekiz-Gucer, M. Takacs, G. Henkel, Liebigs Annalen, 1989, 595, doi: 10.1002/jlac.1989198901105.

[143] J. P. Declercq, M. van Meerssche, F. J. Zeelen, Recueil des Travaux Chimiques des Pays-Bas, 1984, 103, 145.

[144] Schouten, J. A. Kanters, *Acta Crystallographica, Section C: Crystal Structure Communications,* 1991, 47, 1754, doi: 10.1107/S0108270190014196.

[145] V. Suitchmezian, I. Jess, C. Nather, *Crystal Growth and Design,* 2007, 7, 69, doi: 10.1021/cg060434r.

[146] C. Nather, I. Jess, Angewandte Chemie, *International Edition*, 2006, 45, 6381, doi: 10.1002/anie.200601468.

[147] D. Paoli, J. C. Garrigues, H. Patin, Combustion microcalorimetry: application to steroids, *C. R. Acad. Sci. Paris*, 1969, 268, 780-783.

[148] NIST.

[149] F. L. Mota, A. P Carneiro, A. J. Queimada, S. P. Pinho, E. A. Macedo, Temperature and solvent effects in the solubility of some pharmaceutical compounds: Measurements and modeling, *European Journal of Pharmaceutical Sciences,* 2009, 37, 3-4, 499-507.

[150] C. M. Wassvik, A. G. Holmén, R. Draheim, P. Artursson, C. A. S. Bergström, Molecular Characteristics for Solid-State Limited Solubility, *J. Med. Chem.,* 2008, 51, 10, 3035-3039.

[151] Regosz, A.; Chmielewska, A.; Pelplinska, T.; Kowalski, P., Prediction of the solubility of steroid hormones, *Pharmazie*, 1994, 49, 5, 371.

[152] S. Yano, M. Eiraku, M. Tomiyama, T. Sugihara, Mesogenicity of testosterone and estrone derivatives, *Liquid Crystals*, 1990, 7, 4, 537-544. doi:10.1080/02678299008033830.

[153] J. A. H. van Laarhoven, M. A. B. Kruft,; H. Vromans, In vitro release properties of etonogestrel and ethinyl estradiol from a contraceptive vaginal ring, *International Journal of Pharmaceutics*, 2002, 232, 1-2, 163-173.

[154] A. V. Knyazev, N. N. Smirnova, A. S. Shipilova, V. N. Larina, E. V. Gusarova, S. S. Knyazeva, *Journal of Thermal Analysis and Calorimetry,* 123, 3, 2016, 2201-2206.

[155] Z. Cenuse, V. Meltzer, DSC characterization of drugs containing cyclopentanophenanthrene nucleus, An. Univ. Bucuresti. *Chimie,* 1997, 6, 101.

[156] G. De Maury, A. Chauvet, J. Masse, Etude thermoanalytique de quelques steroides. III. Estrogenes, *Thermochimica Acta*, 1985, 87, 189-202.

[157] A. V. Knyazev, V. N. Emel'yanenko, N. N. Smirnova, D. H. Zaitsau, O. V. Stepanova, A. S. Shipilova, A. V. Markin, E. V. Gusarova, S. S. Knyazeva, S. P. Verevkin, *Journal of Chemical Thermodynamics* 107, 2017, 37-41.

[158] A. V. Knyazev, V. N. Emel'yanenko, N. N. Smirnova, O. V. Stepanova, A. S. Shipilova, A. V. Markin, Y.S. Samosudova, E. V. Gusarova, S. S. Knyazeva, S. P. Verevkin, *Journal of Chemical Thermodynamics* 103, 2016, 244-248.

[159] Lewis, G. A.; Enever, R. P., Solution thermodynamics of some potentially long-acting norethindrone derivatives. I. Variation of enthalpies and entropies of fusion with temperature, *Int. J. Pharm.*, 1979, 22, 203-214.

[160] E. S. Domalski,; E. D. Hearing, *J. Phys. Chem. Ref*, 1996, III, 25, 1,1.

[161] C. M. Wassvik, A. G. Holmén, C. A. S. Bergström, I. Zamora, P. Artursson, *European Journal of Pharmaceutical Sciences*, 2006, 29, 3-4, 294-305.

In: A Closer Look at Hormones
Editor: Alexander V. Knyazev
ISBN: 978-1-53617-254-6

Chapter 5

"VITAMIN" D AND PROGESTERONE: FRATERNAL TWINS?

Giovanni Monastra[1], Beatrice Orrù[2], Gianpiero Forte[2] and Vittorio Unfer[3,*]

[1]Department of Experimental Medicine, Sapienza University of Rome, Rome, Italy

[2]Department of Research and Development, Lo.Li. Pharma, Rome, Italy

[3]Faculty of Medicine and Psychology, Department of Developmental and Social Psychology, Sapienza University of Rome, Rome, Italy

ABSTRACT

This chapter focuses on the activities of "vitamin" D on the female reproductive system. The initial overview outlines the main structural and metabolic features of this group of compounds, emphasizing that the classification as vitamin is outdated by now and fails to describe its multifaceted metabolic activities. These compounds should be rather

---

* Corresponding Author's Email: vunfer@gmail.com.

regarded as steroid hormones (secosteroids, specifically), and the nomenclature reconsidered accordingly. The following sections of the chapter aim to unravel the role of these hormones in the menstrual cycle, and their interplay with progesterone during the early stages of pregnancy and the physiologic development to delivery. In these processes, hormones D and progesterone exert several analogous activities, regulating complementary pathways. Preclinical and clinical studies allow to compare such activities both in-vivo and in-vitro. In case of pregnancy, the levels of progesterone and hormones D increase in a similar manner, inducing endometrial receptivity and supporting the development of implantation. Such combined action is crucial from the luteal phase onwards, and supplementation with hormones D proved beneficial in increasing fertility and preventing negative pregnancy outcomes. For this reason, the time for the starting of the therapy must be very carefully defined.

Keywords: "vitamin" D, progesterone, female reproductive system

ABBREVIATIONS

$1,25(OH)_2D_3$	calcitriol
$25(OH)D_3$	calcidiol
CRH (F)	corticotrophin-releasing hormone (factor)
CYP	cytochrome P450
DBP	vitamin D binding protein
ICSI	intracytoplasmic sperm injection
IFN	interferon
IL	interleukin
IVF	in-vitro fertilization
NK	natural killer
PTB	preterm birth
Th	T helper
Treg	T regulatory
VDR	vitamin D receptor

1. D-Compounds: An Overview

Rickets and the Discovery of "Vitamin" D

For long time "vitamin" D was only related to calcium metabolism [1], and its medical applications were exclusively limited to curing the impairment of calcium homeostasis. Such perspective was remnant of the circumstances surrounding its discovery, in the quest for curing rickets, and persisted until very recently [2].

Early reports of bony deformities in infants, likely ascribable to rickets, date back to the second century AD [3]. The Greek physicians Galen (129 – circa 200/216 AD) and Soranus of Ephesus (1st/2nd century AD) described this condition in children living in the poor and dark neighborhoods of the imperial Rome.

But rickets is truly a modern age pathology [4], mainly connected to pollution and novel life habits, which drastically reduced people's exposure to sunlight. The disease became indeed endemic in the South-Western England at the beginning of the industrial revolution [5]. In the mid-17th century Whistler and Glisson independently described an unknown ailment, identifying a relationship between age and onset of symptoms. In 1634 the term rickets appeared for the first time as death cause in the London Bills of Mortality [6]. Glisson recognized the disease as non-congenital, nor inherited or contagious, and the first investigations on its etiology suggested a connection between incidence and dietary habits.

Remedies to rickets were found two centuries later, when the disease became rampant in polluted cities in Europe and Northern United States, and consisted of sunbaths and dietary animal fat. Analysis of geographical incidence, and empirical observations of anti-rachitic properties of cod-liver oil were crucial factors in curing rickets. McCollum later figured out that a fat-soluble component of animal food, different from vitamin A, restores physiological calcium metabolism and bone growth. Three vitamins had already been discovered back then, and named after the first letters of the alphabet. With the same logic, McCollum called such new component vitamin D.

At the beginning of 1900s, experiments on rats demonstrated the influence of irradiated food intake on bone growth [7], suggesting a connection between light exposure and calcium metabolism [8]. But the rationale behind the benefits of sunbaths in treating rickets was elucidated much later [9], when it was discovered that humans endogenously synthesize "vitamin" D from a cholesterol precursor [10].

Structure and Pathways of "Vitamin" D Biosynthesis: Analogies with Steroid Hormones

The general name vitamin D [11] includes a family of similar structures, Cholecalciferol (D3) and Ergocalciferol (D2) being the most biologically relevant analogues. They are of animal and vegetable origin, respectively, and differ for the structure of the aliphatic side chain. In humans, under normal conditions, the skin produces at least 80% of the daily required cholecalciferol from 7-dehydrocholesterol, in a two-step process that require an initial photochemical ring-opening reaction and a subsequent a thermal isomerization. Human skin can thus be considered an endocrine gland which secretes a pro-hormone that undergoes two enzymatic hydroxylation reactions, yielding first calcidiol (or calcifediol), 25(OH)D, and eventually calcitriol, 1,25(OH)2D (Figure 1). The former is the most long-lived, abundant form of "vitamin" D (more appropriately, hormone D) in the bloodstream, whereas the latter is the short-lived, most active form.

"Vitamin" D travels in blood bound to a protein carrier, a feature typical of the steroid hormones: about 85% is bound to the vitamin D binding protein (DBP), which is a serum α2-globulin (MW: 52-59 kDa) synthesized in the liver and discovered in 1975 [12-14]; the remainder 15% binds to albumin and less than 0.5% exists in the unbound, free form. Ergocalciferol have a slightly lower affinity for DBP than its counterpart of animal origin, with higher serum concentration of its metabolites in the free form [15].

All known actions of "vitamin" D in cells are mediated by the Vitamin D Receptor (VDR), a DNA-binding protein (MW = 50÷60 kDa) with biochemical properties resembling those of receptors for steroid and thyroid hormones [16]. VDR is ubiquitously expressed in human tissues in multiple isoforms [17, 18], likely to tailor the response to compounds D. Like all steroid hormones, "vitamin" D interacts with its receptor located both on the cell membrane (associated with caveolae) and in the nucleus. The receptor-hormone complex in some cases can bind to DNA, activating or repressing the transcription of one or more genes, and analogous pathway can be followed by "vitamin" D.

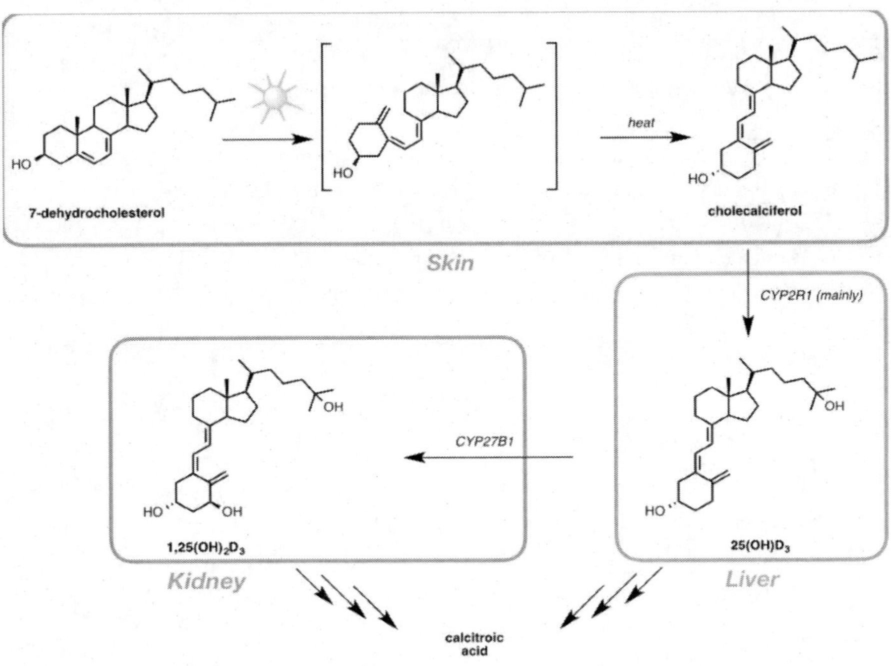

Figure 1. Main metabolic pathway of compounds D_3. Sunlight promotes the synthesis of cholecalciferol from 7-dehydrocholesterol in the skin. Cholecalciferol is then converted into $25(OH)D_3$ by CYP2R1 enzyme in the liver, and eventually into $1,25(OH)_2D_3$ by CYP27B1 enzyme in the kidneys. Both $25(OH)D_3$ and $1,25(OH)_2D_3$ can be further oxidized by CYP24A1 enzyme to yield different catabolic metabolites, whose major component is calcitroic acid.

Considering all these characteristics (Table 1), the label vitamin has consistently started to appear simplistic and misleading. Many authors nowadays agree that these species are fully-fledged hormones, and the nomenclature should be reconsidered in this perspective.

Throughout this chapter, hormones D will refer to calcidiol and calcitriol, while vitamin D will exclusively indicate cholecalciferol (and/or ergocalciferol) and exogenous supplements administered to patients. Collectively, all these species are referred to as compounds D.

Table 1. Steroid hormones and "vitamin" D: Common features

Steroid Hormones	"Vitamin" D
Endogenously produced by glands.	≥ 80% produced by the skin upon exposure to sunlight.
Mostly bound to plasma transport proteins. Only their free form reaches the target cells.	Bound to Vitamin D binding protein (DBP), produced in the liver.
Receptors typically located in the cytoplasm and in nucleus of target cells.	Vitamin D receptor (VDR) localized within: i) caveolae of the plasma membrane; ii) nucleus.
Receptor-hormone complex binds to DNA, activating or repressing gene transcription.	VDR-Vitamin D complex binds to DNA, modifying gene expression.

Although systemic levels of hormones D are maintained by enzymes in the liver and kidneys (and testes in males) [19, 20], all target organs express the enzymes to produce local concentration of hormones D. Just alike, they also express the enzyme responsible for the degradation of both calcidiol and calcitriol. Local production of hormones D and the discovery of several metabolites with VDR affinity greatly complicate outlining the whole regulation of compounds D metabolism. It comprises an elaborate network of feedback systems that tunes specific and diverse activities in different tissues, under various conditions. Some cytochrome P450 (CYP) enzymes make up the machinery that controls biosynthesis and activity of hormones D [21]. Microsomal CYP2R1 is the enzyme that produces most of the circulating 25(OH)D [22]. Mitochondrial CYP27A1, originally thought to be the primary 25OHase in this process, later proved to be one of several enzymes that provide a backup system to produce calcidiol in case of impaired CYP2R1 pathway [23]. CYP27B1 is the only known human 1α-OHase that converts 25(OH)D into the more active 1,25(OH)2D [24]. A single hydroxylase, CYP24A1 [25], catalyzes all the reactions to degrade both forms of hormones D to a major catabolic metabolite, calcitroic acid – a hydrophilic, low VDR-affinity compound, which is eventually excreted [26].

In an oversimplified view – functional to the purpose of this chapter – homeostasis of hormones D, especially the concentration of calcitriol, is tightly regulated through controlling expression and activity of CYP27B1 and 24A1. In response to low calcium concentration in plasma, parathormone (PTH) induces the expression of CYP27B1 and, hence, the biosynthesis of 1,25(OH)2D. Calcitriol triggers a negative feedback that inhibits the activity of CYP27B1, halting its own synthesis. At the same time, increased concentrations of 1,25(OH)2D upregulates the expression of CYP24A1, promoting its catabolism and the degradation of 25(OH)D. Calcitriol also induces the expression of the Fibroblast Growth Factor FGF-23, which contributes to control the concentration of 1,25(OH)2D down-regulating PTH and CYP27B1, and up-regulating CYP24A1 [27].

Origin and Evolution of Compounds D and VDR – Ancestral Roles

"Vitamin" D formed in the evolution of marine organisms to aerobic life [28], when cholesterol became a major constituent of pulmonary surfactant in lungfishes, the first air breathers. Cholesterol biosynthesis is in fact an extremely high oxygen demanding process, which could have preserved life from the detrimental effects of an oxidizing environment [29]. Compounds D evolved as metabolites of cholesterol, to keep its concentration under tight regulation and prevent health hazards [30]. At the same time, shifting habitat – from oceans (calcium concentration ~ 10 mM) to freshwater (calcium concentration 0.5÷1 mM), and eventually to land – organisms created a strategy to regulate intracellular calcium concentrations based on compounds D [31]. Indeed, calcium is involved in numerous cellular signaling pathways, and maintaining a gradient between different cell compartments is crucial.

Interestingly, compounds D and VDR evolved separately, to eventually converge at a certain point in the evolution timeline. VDR was an adaptive receptor that neutralized a wide array of xenobiotics, marine biotoxins with structures partially similar to calcitriol, which later became VDR agonist of choice. The ancient protective role of VDR is still detectable in humans, where it promotes CYP3A4 expression in the colon to eliminate the toxic lithocholic acid [30]. Similarly, the ancient protective role of compounds D evolved in the modulation of the innate immune response. Toll-like receptors (located on monocytes, macrophages, dendritic cells, mast cells, neutrophils, B and T cells, fibroblasts, endothelial and epithelial cells) up-regulate CYP27B1 expression, and calcitriol stimulates the production of Cathelicidin Anti-Microbial Peptide (CAMP) [32]. CAMP, in association with human defensins released upon hormone D stimulation, acts against pathogens like mycobacterium tuberculosis, by disrupting the bacterial biofilm or stimulating autophagy. Such antimicrobial activity is exerted in the epithelium in case of injuries, and in the placenta, to provide an antibacterial environment and prevent fetal infections.

Intestinal Absorption of Compounds D

In healthy subjects, compounds D are absorbed at the intestinal level, with an average efficiency of 78% (ranging between 55% and 99%). The main absorption site in human gut is unknown, but uptake in rats occurs both in the jejunum and in the ileum. Non-hydroxylated forms (ergocalciferol and cholecalciferol) cross the intestinal barrier by passive diffusion and/or by active transport, mediated by specific proteins: SR-B1 (Scavenger Receptor class B type 1), CD36 (Cluster Determinant 36) and NPC1L1 (Niemann-Pick C1-Like 1). Such proteins are also involved in the uptake of cholesterol, vitamin E [33, 34], and carotenoids [35-37]. Vitamin D uptake is likely protein-mediated at low concentrations (e.g., dietary intake), whereas passive absorption becomes prevalent at high concentrations (e.g., pharmacological administration). Through the enterocytes, vitamin D enters the endoplasmic reticulum, then the chylomicrons and finally reaches the lymphatic pathway. On the other hand, hydroxylated forms (calcidiol and calcitriol) are absorbed three times more efficiently and reach the portal vein pathway [38].

Compounds D are mainly stored in muscle, liver and adipose tissues due to their lipophilic nature [39]. Such characteristic is also responsible for their slow turnover in the body (half-life ~2 months). 25(OH)D3 has serum concentration in the range 25–200 nmol/L and a half-life of ~15 d, whereas cholecalciferol and 1,25(OH)2D3 have a much shorter half-life (Table 2) [40]. Due to its stability and high concentration, calcidiol is the most reliable marker for evaluating the levels of hormones D in the body.

Table 2. Steroid Half-life of compounds D [40-42]

Compound D	Half-life
cholecalciferol	24 hours
$25(OH)D_3$	15-20 days
$1,25(OH)_2D_3$	4-15 hours

Studies on vitamin D levels in plasma highlighted that serum peak of cholecalciferol appears the day after administration, but 25(OH)D3 rises slowly, peaking on 7 days after administration. Such delay is likely ascribable to metabolic conversion in the liver [43].

Factors That May Influence Vitamin D Absorption

Food – Food matrices have apparently little to negligible effect, but the area has scarcely been investigated and further in-depth studies are desirable [38].

Drugs – Many common drugs may interfere with vitamin D absorption and/or subsequent metabolism, reducing the circulating levels of hormones D. They include: bile acid sequestrants (colestipol and cholestyramine), lipase inhibitors (orlistat), statins, antimicrobials (rifampin, isoniazid, hydroxychloroquine), antiepileptic drugs, corticosteroids, immunosuppressive agents (cyclosporine, tacrolimus), chemotherapeutic agents (etoposide, epipodophyllotoxin, cyclosphosphamide, ifosfamide, vincristine, vinblastine, paclitaxel, docetaxel, irinotecan, tamoxifen and imatinib), highly active antiretroviral agents, histamine H2-receptor antagonists [44].

Smoking – Smoking may lead to depleted levels of circulating hormones D. Decreased circulating 25(OH)D, up to 10%, was reported for perimenopausal smokers [45], with a positive correlation with smoking pack-years [46]. Analogous effect (10-15% decrease) was observed in pregnant smokers. Unlike 1,25(OH)2D, 25(OH)D can cross the placenta [47] and depleted maternal calcidiol levels may result in a low fetal hormone D status.

2. From the Bones to Other Therapeutic Areas

In the mid-1930s a group of researchers reported on hormonal properties of a group of molecules with cholesterol-like structure, a seventeen-carbon cyclopentanoperhydrophenanthrene scaffold. They observed that injection of each of these compounds, dissolved in sesame oil, induced estrus in ovariectomized rats and that the classic steroid skeleton was responsible for such effect. In particular:

> "marked activity being shown by ergosterol and its irradiation product calciferol (crystalline vitamin D), and still greater activity by neo-ergosterol" [48].

Further investigations highlighted that ergosterol and 7-dehydrocholesterol act synergistically with progesterone in sustaining pregnancy [49] or obtaining implantation in ovariectomized rabbits [50], when administered at non-estrogenic nor progestogenic doses.

Forgotten for quite some time, these results were reconsidered only between the end of the sixties and the beginning of the seventies, when new studies shed light on activities of calcitriol that go beyond bone growth and rickets treatment. Around that time, the structure of calcitriol was elucidated, and its nuclear receptor (VDR) identified [51, 52]. For the first time, "vitamin" D was defined as a steroid hormone [53, 54] – namely a secosteroid:

> "It is possible that 1,25- dihydroxycholecalciferol should be reclassified as a steroid hormone. Certainly, its secretion by the kidney followed by its selective accumulation in the target intestinal mucosa where it exerts its characteristic physiological effect on calcium metabolism satisfies the minimal definitions of a hormone" [55].

Almost three decades later, this concept became quite widely accepted within the scientific community [56], and well established lately [57]. Nonetheless a minority of scholars disagrees with this view, leaving the classification of "vitamin" D and its metabolites still a debated topic [58].

3. Compounds D and Steroid Hormones

Observing that human skin produces cholecalciferol was an essential step in reclassifying "vitamin" D and its metabolites. Upon exposure to UV-B radiation, the C9-C10 bond of 7-dehydrocholesterol undergoes photolytic cleavage with the opening of the B-ring – hence the name secosteroid. Sharing a common ancestor, the structure of compounds D is closely related to corticoids and sex hormones such as testosterone, progesterone and estrogens. In sunny areas, endogenous production of cholecalciferol meets the daily requirements in healthy young subjects, but dark skin decreases sunlight absorption. Blacks and Hispanics in the US, for example, generally lack adequate levels of endogenous cholecalciferol, outlining an ethnic predisposition for low "vitamin" D status. Other physical conditions represent significant and independent risk factors for the reduced concentration of circulating hormones D. These include obesity, hypertension and low levels of high-density lipoprotein [59].

As mentioned, steroid hormones bind to transport proteins in plasma and, according to the free hormone hypothesis [60], only the small unbound fraction reaches its receptors in target cells. Likewise, hormones D largely bind to DBP and albumin. Some authors suggest that the free hormone hypothesis applies also in this case, but the question has remained debated to this date. Compounds D bound to DBP, in fact, can cross the cell membranes of specific organs and tissues (namely placenta, parathyroid and renal proximal convoluted tubules) through endocytosis by the cubilin/megalin complex [61], locally increasing the concentration of available hormone.

Hormones are chemical messengers that bind to membrane receptors [62-64]. Signal transmission occurs through activation of the second messenger pathway that induces rapid cellular responses, ranging from seconds (e.g., opening of ion channels) to 10-60 minutes (e.g., activation of phosphatidylinositol-3'-kinase or endothelial nitric oxide synthase). Unlike growth factors and peptide hormones, steroid hormones can cross the cell membrane and interact with nuclear receptors that bind to DNA, triggering or repressing gene transcription [65]. This genomic response commonly takes hours to days and may be blocked by transcription or translation inhibitors. Hormones D as well exert either genomic and nongenomic actions [63]: interaction with VDR on cell membrane induces the formation of second messengers (cAMP, diacylglycerol, inositol triphosphate, arachidonic acid) or protein phosphorylation; interaction with nuclear VDR may initiate a genetic response [66]. In the nucleus, liganded VDR forms a heterodimer with the Retinoid-X-Receptor (RXR). VDR/RXR subsequently binds to DNA regions known as Vitamin D Responsive Elements (VDREs) and modifies the expression of several genes by recruiting numerous coactivators or corepressors [67].

The characteristics of the receptor's activity, along with the discovery of many enzymes responsible for the metabolism of this group of molecules, strongly suggest implications of hormones D in several metabolic and regulation processes [68]. In addition, VDR is almost ubiquitously distributed in human tissues and organs, including ovarian, myometrial and endometrial cells, and the placenta (Table 3) [69-73].

Such considerable expression of VDR in the female reproductive system, along with the timing in calcitriol regulation, clarifies the involvement of low hormone D status in fertility and in the onset of pregnancy complications.

Table 3. Human tissues expressing VDR

Adipose	Kidney	Pituitary
Adrenal	Liver (fetal)	Placenta
Bone	Lung	Prostate
Bone marrow	Lymphocytes (B & T)	Retina
Brain	Muscle, embryonic	Skin
Breast	Muscle, smooth	Stomach
Cartilage	Osteoblast	Testis
Colon	Ovary	Thymus
Epididymis	Pancreas β cell	Thyroid
Hair follicle	Parathyroid	Uterus (myometrium and endometrium)
Intestine	Parotid	Muscle, cardiac

Modified from: Norman AW. From vitamin D to hormone D: fundamentals of the vitamin D endocrine system essential for good health, Am J Clin Nutr. 2008 Aug;88(2):491S-499S.

4. Progesterone Release: The Luteal Phase

The regulation of the female reproductive system involves the interplay of many actors, whose concentration must reach appropriate levels in specific timeframes. Progesterone and calcitriol, in particular, are essential for successful gestation, and their metabolism is tightly controlled throughout the menstrual cycle.

Menstrual cycle alternates two main stages: 1) follicular or proliferative phase (1st to 14th day of the menstrual cycle); 2) luteal or secretive phase (15th to 28th day of the menstrual cycle, normally). In the first, the follicles inside the ovary develop to prepare the ovulation. In the second, the corpus luteum (CL), a transient endocrine gland that forms at this stage, secretes progesterone (hence secretory phase). CL undergoes rapid neovascularization under the influence of Vascular endothelial growth factor (VEGF) and Fibroblast growth factor (FGF), both upregulated in the luteinized granulosa cells and critical for embryo implantation [74, 75]. In response to low stimulation of the luteinizing hormone and appropriate levels of cholesterol, luteal cells synthesize progesterone in a pulsatile way, resulting in a 7-fold concentration change within a few hours. In the luteal phase, progesterone inhibits proliferation of endometrium and induces a receptive state for blastocyst implantation that maintains early pregnancy [76].

In this scenario, hormones D modulate the innate and adaptive immune system, control cell proliferation and regulate metabolic processes, affecting embryo implant, cytokine production and immune response to infections [77-80].

5. Role of Hormones D and Progesterone in Early Pregnancy [81]

Hormones D

Calcitriol metabolism heavily changes during menstrual cycle and pregnancy, with pregnant women exhibiting higher values of circulating hormones D, independently of calcium homeostasis [82-84]. As mentioned, the receptor VDR is present in the ovaries, uterus and placenta; it is expressed in endometrial stromal cells throughout the entire menstrual cycle and in early pregnant decidual cells [85], and its number constantly increases in the first and second trimester of pregnancy. Correspondingly, the levels of CYP27B1 mRNA remain constant in endometrial stromal cells through the menstrual cycle, but they significantly increase in decidual cells [85]. It is remarkable that hormones D can be detected also in the seminal fluid [86], and support a successful onset of pregnancy [87, 88]. This widespread co-expression of VDR and CYP27B1 [89] provides the basis for an autocrine/paracrine action of hormones D that modulates both acquired and innate immunity at the fetal-maternal interface, specifically the inflammatory response that occurs in the initial stage of pregnancy [90]. In the implantation process, blastocysts produce pro-inflammatory cytokine IL-1β, which stimulates endometrial dendritic cells and macrophages to express CYP27B1, and consequently increases calcitriol concentration [85]. TNF, the other main pro-inflammatory cytokine, is much less effective in this process. In vitro experiments proved that calcitriol inhibits the proliferation of T helper 1 (Th1) cells, decreasing the production of their cytokines (IFN-γ, IL-2, TNF-α); at the same time, it stimulates the secretion of T helper 2 (Th2) cytokines, including IL-10 (Figure 2) [91]. Furthermore, calcitriol significantly decreases the cytotoxicity of natural killer (NK) cells in a dose-dependent manner and suppresses the secretion of pro-inflammatory cytokines (IFN-γ and TNF-α), while increasing the release of IL-10, CSF2, IL-1 and IL-6 [91], modulating the inflammatory response that physiologically develops at the initial stage of pregnancy [90]. In patients with recurrent miscarriages, supplementation with vitamin D significantly lowers the frequency of T helper 17 (Th17), reducing the Th17/Treg ratio in peripheral blood and the inflammatory state [92]. Analogous effect was observed in the induction and upregulation of the suppressive activity of T regulatory (Treg) cells on dermal inflammation in patients with skin-draining lymph nodes [93].

On the endometrium, binding of hormones D to VDR upregulates the expression of Calbindin and Osteopontin [85], essential proteins for embryo implantation and placenta development and ultimately for successful pregnancy [85, 94-96]. Calbindin, whose expression was believed to be independent of hormones D, is a calcium binding protein that modulates intracellular calcium levels and influences the endometrial receptivity for implantation [97, 98]. Endometrial expression of calbindin mRNA is indeed highly regulated during this phase. Osteopontin, also known as Secreted Phosphoprotein 1 (SPP1), is involved in implantation and decidualization processes [99], acting as signal transductor at the uterine-placental interface and regulating immune cell behavior and cytokine production. Between the luteal phase and the 12th week of pregnancy, hormones D induce the synthesis of molecules with beneficial effects (activin A, VEGF, CRH – previously indicated as CRF) [100-102] and inhibits the production of those with detrimental effects (estrogens, TNF-α and other inflammatory cytokines) [102-105].

Hormones D directly modulate also the biosynthesis of progesterone. In human cumulus granulosa cells, they significantly enhance the expression of 3β-HSD, which converts pregnenolone into progesterone and increases the local concentration of the latter [106]. Analogous behavior can be observed in cultures of human ovarian cells [107].

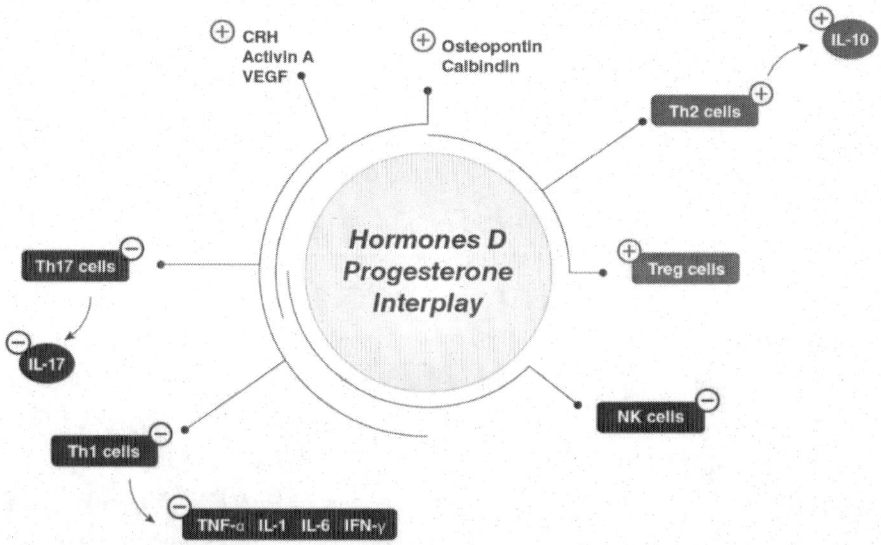

Figure 2. Common immunomodulatory activities exerted by hormones D and progesterone in early pregnancy. Hormones D and progesterone induce anti-inflammatory pathways (Th2 and Treg cells) and stimulate the release of corticotropin-releasing hormone (CRH), activin A, vascular endothelial growth factor (VEGF), osteopontin and calbindin. Concurrently, they inhibit pro-inflammatory pathways (NK, Th1 and Th17 cells), supporting pregnancy since its inception.

Progesterone

The receptor of progesterone is expressed in the glandular epithelium and stroma of human endometrium during the menstrual cycle [108].

Progesterone and calcitriol mutually affect each other's metabolism. The former induces the expression of VDR gene in human T cells (Treg, Th1 and Th17, in various cytokine conditions; spleen CD4+) and in the uterus of pregnant or progestin-injected mice [109]. Experiments using mifepristone, RU486, an antagonist of progesterone receptor, demonstrated that this effect is highly specific.

Progesterone-induced overexpression of VDR increases the sensitivity of T cells to calcitriol in different species, modulating the immune response that may prevent embryo implantation. Calcitriol regulation of T cells, in fact, leads to the increase of Tregs and their associated molecules (FoxP3, CD8, LAP-TGF-β1), and to the suppression of effector T cells, which exert a pro-inflammatory activity. If siRNA is used to knock down VDR gene, progesterone and calcitriol significantly lose the ability to inhibit T cell differentiation into effector T cells [109].

Progesterone also affects the response of the immune system, with mechanisms similar to hormones D (Figure 2): promotes the formation of Th2 cells and the release of their anti-inflammatory cytokines [110]; regulates the differentiation of fetal T cells into Treg cells in human naive cord blood [111]; suppresses the genes related to Th1 and Th17, concurrently enhancing Th2-related genes in lymphocytes from pregnant cows [112], where it also inhibits the release of IFN-γ and IL-17 in a dose-dependent manner [112]; prevents degranulation and release of perforin from of NK cells, neutralizing their cytotoxicity [113]; induces the release of osteopontin in human decidua [114] and calbindin in the uterus of Egyptian buffalo [95]; between the luteal phase and the 12th week of pregnancy, stimulates the synthesis of molecules with beneficial effects (activin A, CRH, VEGF), while suppressing the release of those with detrimental effects (estrogens or TNF) [113, 115-118]. In mononuclear cells from human placental blood, progesterone antagonizes the effects of lipopolysaccharide (LPS) and significantly reduces the release of TNF-α,

IL-1β, IL-6, IL-8, and MIP-1α. In-vitro experiments showed increased levels of IL-10 after treatment with LPS and progesterone [119].

These evidences indicate that progesterone and hormones D act in a combined manner, mutually potentiating their anti-inflammatory effects to sustain early pregnancy.

Role in Advanced Pregnancy: The "Time Factor"

The role of hormones D and progesterone drastically changes through the course of pregnancy. While they increase the levels of corticotrophin-releasing hormone (CRH) in the initial phase, they inhibit its biosynthesis in advanced stages of gestation [120, 121], preventing preterm birth (PTB). CRH interaction with its receptor induces the transition from uterine quiescence to myometrial contractions and stimulates the fetal pituitary gland to produce adrenocorticotropic hormone (ACTH). ACTH increases the production of dehydroepiandrosterone sulphate (DHEA-S) and cortisol from the fetal adrenal glands. In the placenta, conversion of DHEA-S to estrogens causes myometrial contractions, while cortisol stimulates in fetal lungs the production of surfactant protein A (SP-A), inducing uterine contractility and facilitating delivery.

6. Hormones D in Pregnancy and in Assisted Reproductive Technology (ART)

Low maternal 25(OH)D status represents a risk factor for PTB [122], and PTB risk seems to be also related to both fetal and maternal VDR polymorphisms [123]. Physiologically, labor is a self-regulated inflammatory process, conditioned by the balance between innate and adaptive immune cells [124]. Depleted circulating calcidiol is a marker of pro-inflammatory immune response, which increases the risk for PTB occurrence [125, 126].

Low hormones D levels in pregnancy are also related to reduced fetal growth and adverse neonatal outcomes, in particular shortened gestational age (SGA) and diminished head circumference (HC):

> "Maternal vitamin D deficiency (<30 nmol/L) has an effect on offspring anthropometric parameters, being associated with lower birthweight and HC in their offspring (compared to mothers with concentrations ≥30 nmol/L); both vitamin D deficiency and insufficiency (<50 nmol/L) are related to a higher risk of SGA and PTB. Maternal 25(OH)D concentrations ≥75 nmol/L were not observed to be associated neither with birthweight, nor SGA nor PTB" [127].

Such outcomes are independent of ethnicity, but seem to be recurrent in pregnancies whose first and second trimester are in winter or spring [128]. PTB risk, assessed against vitamin D supplementation, is significantly reduced in pregnant women with restored physiological values of 25(OH)D. In particular, subjects that reach 25(OH)D values ≥ 40 ng/ml prior to delivery have a 60% lower PTB occurrence than those with serum values < 40 ng/ml [129].

Hormones D are fundamental for successful pregnancy progression, and low levels are associated with risk of miscarriage and recurrent pregnancy loss. A study on 1072 women in childbearing age demonstrated a relationship between low hormone D status and infertility [130]. Estimations of 25(OH)D levels before treatment protocol for ICSI revealed a positive significant correlation with number of oocytes, fertilized oocytes and endometrial thickness. In particular, high levels of calcidiol are necessary to develop the proper endometrial thickness for embryo implantation after ICSI. The correlation between 25(OH)D concentrations and clinical pregnancy remains a controversial topic. While some authors state that clinical pregnancy rates increase with higher calcidiol levels, others report opposite effects [131-134]. Improved fertility and fetal conditions [135], as well as successful ART results [136, 137], were achieved restoring optimal hormone D status through supplementation [138]. Concerning the safety in pregnant women, vitamin D supplementation in doses up to 4,000 IU/day for 24-28 weeks proves to be completely safe [139, 140], without evidences of impact on calcium homeostasis or onset of bone pathologies.

CONCLUSION

Hormones D play essential roles in female reproductive system, providing support from the luteal phase of the menstrual cycle onward. Acting in concert with progesterone, they modulate the immune response associated with the inflammatory state in early pregnancy and prepare the endometrium receptivity, favoring embryo implantation. Hormones D and progesterone regulate each other's metabolism and exert a combined action [141], providing the basis for a successful pregnancy. In cases of low serum levels of hormones D, and in assisted reproductive procedures, supplementation proved beneficial in restoring a physiological state, and is recommended to reduce the risk of miscarriages and negative pregnancy outcomes.

REFERENCES

[1] Norman AW: Vitamin D metabolism and calcium absorption. *The American Journal of Medicine* 1979, 67:989-998.

[2] Norman AW: The History of the Discovery of Vitamin D and Its Daughter Steroid Hormone. *Annals of Nutrition and Metabolism* 2012, 61:199-206.

[3] Rajakumar K: Vitamin D, Cod-Liver Oil, Sunlight, and Rickets: A Historical Perspective. *Pediatrics* 2003, 112:e132-e135.

[4] Wharton B, Bishop N: Rickets. The Lancet 2003, 362:1389-1400.

[5] Dunn PM: Francis Glisson (1597–1677) and the "discovery" of rickets. *Archives of Disease in Childhood - Fetal and Neonatal Edition* 1998, 78:F154-F155.

[6] O'Riordan JL, Bijvoet OL: Rickets before the discovery of vitamin D. *Bonekey Rep* 2014, 3:478.

[7] Steenbock H: The induction of growth promoting and calcifying properties in a ration by exposure to light. *Science* 1924, 60:224-225.

[8] Carpenter KJ, Zhao L: Forgotten Mysteries in the Early History of Vitamin D. *The Journal of Nutrition* 1999, 129:923-927.

[9] Webb AR, Holick MF: The Role of Sunlight in the Cutaneous Production of Vitamin D3. *Annual Review of Nutrition* 1988, 8:375-399.

[10] Wolf G: The discovery of vitamin D: the contribution of Adolf Windaus. *J Nutr* 2004, 134:1299-1302.

[11] DeLuca HF: Overview of general physiologic features and functions of vitamin D. *Am J Clin Nutr* 2004, 80:1689s-1696s.

[12] Speeckaert MM, Speeckaert R, van Geel N, Delanghe JR: Vitamin D binding protein: a multifunctional protein of clinical importance. *Adv Clin Chem* 2014, 63:1-57.

[13] Delanghe JR, Speeckaert R, Speeckaert MM: Behind the scenes of vitamin D binding protein: More than vitamin D binding. *Best Practice & Research Clinical Endocrinology & Metabolism* 2015, 29:773-786.

[14] Cooke NE, Haddad JG: Vitamin D binding protein (Gc-globulin). *Endocr Rev* 1989, 10:294-307.

[15] Tsuprykov O, Chen X, Hocher C-F, Skoblo R, Lianghong Y, Hocher B: Why should we measure free 25(OH) vitamin D? *The Journal of Steroid Biochemistry and Molecular Biology* 2018, 180:87-104.

[16] Haussler MR: Vitamin D receptors: nature and function. *Annu Rev Nutr* 1986, 6:527-562.

[17] Poon AH, Gong L, Brasch-Andersen C, Litonjua AA, Raby BA, Hamid Q, Laprise C, Weiss ST, Altman RB, Klein TE: Very important pharmacogene summary for VDR. *Pharmacogenetics and genomics* 2012, 22:758-763.

[18] Wang Y, Zhu J, DeLuca HF: Where is the vitamin D receptor? *Arch Biochem Biophys* 2012, 523:123-133.

[19] Foresta C, Selice R, Di Mambro A, Strapazzon G: Testiculopathy and vitamin D insufficiency. *Lancet* 2010, 376:1301.

[20] Choudhary D, Jansson I, Stoilov I, Sarfarazi M, Schenkman JB: Expression patterns of mouse and human CYP orthologs (families 1–4) during development and in different adult tissues. *Archives of Biochemistry and Biophysics* 2005, 436:50-61.

[21] Prosser DE, Jones G: Enzymes involved in the activation and inactivation of vitamin D. *Trends Biochem Sci* 2004, 29:664-673.

[22] Zhu JG, Ochalek JT, Kaufmann M, Jones G, Deluca HF: CYP2R1 is a major, but not exclusive, contributor to 25-hydroxyvitamin D production in vivo. *Proceedings of the National Academy of Sciences of the United States of America* 2013, 110:15650-15655.

[23] Thacher TD, Levine MA: CYP2R1 mutations causing vitamin D-deficiency rickets. *J Steroid Biochem Mol Biol* 2017, 173:333-336.

[24] Takeyama K-i, Kitanaka S, Sato T, Kobori M, Yanagisawa J, Kato S: 25-Hydroxyvitamin D₃ 1α-Hydroxylase and Vitamin D Synthesis. *Science* 1997, 277:1827-1830.

[25] Jones G, Prosser DE, Kaufmann M: 25-Hydroxyvitamin D-24-hydroxylase (CYP24A1): its important role in the degradation of vitamin D. *Arch Biochem Biophys* 2012, 523:9-18.

[26] Yu OB, Arnold LA: Calcitroic Acid-A Review. *ACS Chem Biol* 2016, 11:2665-2672.

[27] Petkovich M, Helvig C, Epps T: Chapter 80 - CYP24A1 Regulation in Health and Disease. In *Vitamin D* (Third Edition). Edited by Feldman D, Pike JW, Adams JS. San Diego: Academic Press; 2011: 1525-1554.

[28] Galea AM, Brown AJ: Special relationship between sterols and oxygen: were sterols an adaptation to aerobic life? *Free Radic Biol Med* 2009, 47:880-889.

[29] Brown AJ, Galea AM: Cholesterol as an evolutionary response to living with oxygen. *Evolution* 2010, 64:2179-2183.

[30] Newmark H, Dantoft W, Ghazal P: Evolutionary Origin of the Interferon-Immune Metabolic Axis: The Sterol-Vitamin D Link. *Front Immunol* 2017, 8:62.

[31] Bikle DD: Vitamin D: an ancient hormone. *Experimental Dermatology* 2011, 20:7-13.

[32] Shuler FD, Hendrix J, Hodroge S, Short A: Antibiotic-like actions of vitamin D. *W V Med J* 2013, 109:22-25.

[33] Reboul E, Klein A, Bietrix F, Gleize B, Malezet-Desmoulins C, Schneider M, Margotat A, Lagrost L, Collet X, Borel P: Scavenger receptor class B type I (SR-BI) is involved in vitamin E transport across the enterocyte. *The Journal of Biological Chemistry* 2006, 281:4739-4745.

[34] Narushima K, Takada T, Yamanashi Y, Suzuki H: Niemann-pick C1-like 1 mediates alpha-tocopherol transport. *Mol Pharmacol* 2008, 74:42-49.

[35] Reboul E, Abou L, Mikail C, Ghiringhelli O, Andre M, Portugal H, Jourdheuil-Rahmani D, Amiot MJ, Lairon D, Borel P: Lutein transport by Caco-2 TC-7 cells occurs partly by a facilitated process involving the scavenger receptor class B type I (SR-BI). *Biochem J* 2005, 387:455-461.

[36] van Bennekum A, Werder M, Thuahnai ST, Han CH, Duong P, Williams DL, Wettstein P, Schulthess G, Phillips MC, Hauser H: Class B scavenger receptor-mediated intestinal absorption of dietary beta-carotene and cholesterol. *Biochemistry* 2005, 44:4517-4525.

[37] Moussa M, Landrier JF, Reboul E, Ghiringhelli O, Comera C, Collet X, Frohlich K, Bohm V, Borel P: Lycopene absorption in human intestinal cells and in mice involves scavenger receptor class B type I but not Niemann-Pick C1-like 1. *J Nutr* 2008, 138:1432-1436.

[38] Borel P, Caillaud D, Cano NJ: Vitamin D bioavailability: state of the art. *Crit Rev Food Sci Nutr* 2015, 55:1193-1205.

[39] Mawer EB, Backhouse J, Holman CA, Lumb GA, Stanbury SW: The distribution and storage of vitamin D and its metabolites in human tissues. *Clin. Sci.* 1972, 43:413-431.

[40] Jones G: Pharmacokinetics of vitamin D toxicity. *Am. J. Clin. Nutr.* 2008, 88:582s-586s.

[41] Haddad JG, Jr., Rojanasathit S: Acute administration of 25-hydroxycholecalciferol in man. *J. Clin. Endocrinol. Metab.* 1976, 42:284-290.

[42] Levine BS, Singer FR, Bryce GF, Mallon JP, Miller ON, Coburn JW: Pharmacokinetics and biologic effects of calcitriol in normal humans. *J. Lab. Clin. Med.* 1985, 105:239-246.

[43] Heaney RP, Armas LA, Shary JR, Bell NH, Binkley N, Hollis BW: 25-Hydroxylation of vitamin D3: relation to circulating vitamin D3 under various input conditions. *Am. J. Clin. Nutr.* 2008, 87:1738-1742.

[44] Robien K, Oppeneer SJ, Kelly JA, Hamilton-Reeves JM: Drug-vitamin D interactions: a systematic review of the literature. *Nutr. Clin. Pract.* 2013, 28:194-208.

[45] Brot C, Jensen LB, Sorensen OH: Bone mass and risk factors for bone loss in perimenopausal Danish women. *J Intern. Med.* 1997, 242:505-511.

[46] Piazzolla G, Castrovilli A, Liotino V, Vulpi MR, Fanelli M, Mazzocca A, Candigliota M, Berardi E, Resta O, Sabbà C, Tortorella C: Metabolic syndrome and Chronic Obstructive Pulmonary Disease (COPD): The interplay among smoking, insulin resistance and vitamin D. *PLOS ONE* 2017, 12:e0186708.

[47] Kovacs CS: Chapter 39 - Fetus, Neonate, and Infant. In *Vitamin D* (Fourth Edition). Edited by Feldman D: Academic Press; 2018: 679-708.

[48] Cook JW, Hewett CL, Lawson W: The Oestrogenic Activity of some Condensed-Ring Compounds in Relation to their other Biological Activities. *Proceedings of the Royal Society of London Series B* 1934, 114:272-286.

[49] Feyel-Cabanes T: Maintien de la grossesse chez le lapine gestante castrée par une association hormone-stérol. *C R Seances Soc Biol Fil* 1949, 143:1454-1458.

[50] Chambon Y: [Synergic action of vitamin D and of progesterone in obtaining ovum implantation in the castrated pregnant rabbit]. *C R Seances Soc Biol* Fil 1951, 145:955-959.

[51] Haussler MR NA: Chromosomal receptor for a vitamin D metabolite. *Proc Natl Acad Sci* USA 1969, 62:155–162.

[52] Tsai HC NA: Studies on the mode of action of calciferol. VI. Effect of 1,25-dihydroxy-vitamin D3 on RNA synthesis in the intestinal mucosa. *Biochem Biophys Res Commun* 1973, 54 622–627.

[53] Haussler MR MJ, Norman AW: The association of a metabolite of vitamin D3 with intestinal mucosa chromatin in vivo. *J Biol Chem* 1968, 243:4055–4064.

[54] Myrtle JF HM, Norman AW. : Evidence for the biologically active form of cholecalciferol in the intestine. *J Biol Chem* 1970, 245:1190–1196.

[55] Norman AW MJ, Midgett RJ, Nowicki HG, Williams V, Popjak G.: 1,25-dihydroxycholecalciferol: identification of the proposed active form of vitamin D3 in the intestine. *Science* 1971, 17351-54:51-54.

[56] Intakes SCotSEoDR: *Dietary reference intakes: calcium, phosphorus, magnesium, Vitamin D, and fluoride.* Washington (DC): National Academy Press; 1997.

[57] Norman AW: From vitamin D to hormone D: fundamentals of the vitamin D endocrine system essential for good health. *Am J Clin Nutr* 2008, 88:491S-499S.

[58] Vieth R: Why "Vitamin D" is not a hormone, and not a synonym for 1,25-dihydroxy-vitamin D, its analogs or deltanoids. *J Steroid Biochem Mol Biol* 2004, 89-90:571-573.

[59] Forrest KY, Stuhldreher WL: Prevalence and correlates of vitamin D deficiency in US adults. *Nutr Res* 2011, 31:48-54.

[60] Siiteri PK, Murai JT, Raymoure WJ, Kuhn RW, Hammond GL, Nisker JA: The Serum Transport of Steroid Hormones11This investigation was supported in part by USPHS, NIH Grant Awards HD 10328, CA 27702, and a grant from the Andrew W. Mellon Foundation. In *Proceedings of the 1981 Laurentian Hormone*

Conference. Volume 38. Edited by Greep RO. Boston: Academic Press; 1982: 457-510.

[61] Bikle DD, Malmstroem S, Schwartz J: Current Controversies: Are Free Vitamin Metabolite Levels a More Accurate Assessment of Vitamin D Status than Total Levels? *Endocrinology and Metabolism Clinics of North America* 2017, 46:901-918.

[62] Norman AW, Mizwicki MT, Norman DP: Steroid-hormone rapid actions, membrane receptors and a conformational ensemble model. *Nat Rev Drug Discov* 2004, 3:27-41.

[63] Norman AW: Minireview: vitamin D receptor: new assignments for an already busy receptor. *Endocrinology* 2006, 147:5542-5548.

[64] Levin ER: Extranuclear steroid receptors are essential for steroid hormone actions. *Annu Rev Med* 2015, 66:271-280.

[65] O'Malley BW: Mechanisms of action of steroid hormones. *N Engl J Med* 1971, 284:370-377.

[66] Haussler MR, Jurutka PW, Mizwicki M, Norman AW: Vitamin D receptor (VDR)-mediated actions of 1alpha,25(OH)(2)vitamin D(3): genomic and non-genomic mechanisms. *Best Pract Res Clin Endocrinol Metab* 2011, 25:543-559.

[67] Nezbedova P, Brtko J: *1α,25-dihydroxyvitamin D3 inducible transcription factor and its role in the vitamin D action.* 2004.

[68] Mizwicki MT, Menegaz D, Yaghmaei S, Henry HL, Norman AW: A molecular description of ligand binding to the two overlapping binding pockets of the nuclear vitamin D receptor (VDR): structure-function implications. *The Journal of steroid biochemistry and molecular biology* 2010, 121:98-105.

[69] Valdivielso JM: The physiology of vitamin D receptor activation. *Contrib Nephrol* 2009, 163:206-212.

[70] Christakos S, Hewison M, Gardner DG, Wagner CL, Sergeev IN, Rutten E, Pittas AG, Boland R, Ferrucci L, Bikle DD: Vitamin D: beyond bone. *Ann N Y Acad Sci* 2013, 1287:45-58.

[71] Acker GM, Garabedian M, Guillozo H, Pecquinot MA, Balsan S: In vitro formation of 25-hydroxyvitamin D3 metabolites in

endometrium: dependence on the hormonal status of the rat. *Endocrinology* 1982, 111:2103-2109.

[72] Delvin EE, Arabian A, Glorieux FH, Mamer OA: In vitro metabolism of 25-hydroxycholecalciferol by isolated cells from human decidua. *J Clin Endocrinol Metab* 1985, 60:880-885.

[73] Kachkache M, Rebut-Bonneton C, Demignon J, Cynober E, Garabedian M: Uterine cells other than stromal decidual cells are required for 1,25-dihydroxyvitamin D3 production during early human pregnancy. *FEBS Lett* 1993, 333:83-88.

[74] Chakraborty I, Das SK, Dey SK: Differential expression of vascular endothelial growth factor and its receptor mRNAs in the mouse uterus around the time of implantation. *J Endocrinol* 1995, 147:339-352.

[75] Sak ME, Gul T, Evsen MS, Soydinc HE, Sak S, Ozler A, Alabalik U: Fibroblast growth factor-1 expression in the endometrium of patients with repeated implantation failure after in vitro fertilization. *Eur Rev Med Pharmacol Sci* 2013, 17:398-402.

[76] Young SL: Oestrogen and progesterone action on endometrium: a translational approach to understanding endometrial receptivity. *Reprod Biomed Online* 2013, 27:497-505.

[77] Niswender GD, Juengel JL, Silva PJ, Rollyson MK, McIntush EW: Mechanisms controlling the function and life span of the corpus luteum. *Physiol Rev* 2000, 80:1-29.

[78] Mesen TB, Young SL: Progesterone and the luteal phase: a requisite to reproduction. *Obstet Gynecol Clin North Am* 2015, 42:135-151.

[79] Herr D, Bekes I, Wulff C: Regulation of Endothelial Permeability in the Corpus Luteum: A Review of the Literature. *Geburtshilfe Frauenheilkd* 2013, 73:1107-1111.

[80] Davis JS, Rueda BR: The corpus luteum: an ovarian structure with maternal instincts and suicidal tendencies. *Front Biosci* 2002, 7:d1949-1978.

[81] Monastra G, De Grazia S, De Luca L, Vittorio S, Unfer V: Vitamin D: a steroid hormone with progesterone-like activity. *Eur Rev Med Pharmacol Sci* 2018, 22:2502-2512.

[82] Kumar R: The metabolism of 1,25-dihydroxyvitamin D3. *Endocr Rev* 1980, 1:258-267.

[83] Hartwell D, Riis BJ, Christiansen C: Changes in vitamin D metabolism during natural and medical menopause. *J Clin Endocrinol Metab* 1990, 71:127-132.

[84] Salle BL, Delvin EE, Lapillonne A, Bishop NJ, Glorieux FH: Perinatal metabolism of vitamin D. *Am J Clin Nutr* 2000, 71:1317S-1324S.

[85] Vigano P, Lattuada D, Mangioni S, Ermellino L, Vignali M, Caporizzo E, Panina-Bordignon P, Besozzi M, Di Blasio AM: Cycling and early pregnant endometrium as a site of regulated expression of the vitamin D system. *J Mol Endocrinol* 2006, 36:415-424.

[86] Blomberg Jensen M: Vitamin D and male reproduction. *Nat Rev Endocrinol* 2014, 10:175-186.

[87] Boisen IM, Bøllehuus Hansen L, Mortensen LJ, Lanske B, Juul A, Blomberg Jensen M: Possible influence of vitamin D on male reproduction. *The Journal of Steroid Biochemistry and Molecular Biology* 2017, 173:215-222.

[88] Lorenzen M, Boisen IM, Mortensen LJ, Lanske B, Juul A, Blomberg Jensen M: Reproductive endocrinology of vitamin D. *Molecular and Cellular Endocrinology* 2017, 453:103-112.

[89] Hewison M, Zehnder D, Bland R, Stewart PM: 1alpha-Hydroxylase and the action of vitamin D. *J Mol Endocrinol* 2000, 25:141-148.

[90] Evans KN, Nguyen L, Chan J, Innes BA, Bulmer JN, Kilby MD, Hewison M: Effects of 25-hydroxyvitamin D3 and 1,25-dihydroxyvitamin D3 on cytokine production by human decidual cells. *Biol Reprod* 2006, 75:816-822.

[91] Ota K, Dambaeva S, Kim MW, Han AR, Fukui A, Gilman-Sachs A, Beaman K, Kwak-Kim J: 1,25-Dihydroxy-vitamin D3 regulates NK-cell cytotoxicity, cytokine secretion, and degranulation in women with recurrent pregnancy losses. *Eur J Immunol* 2015, 45:3188-3199.

[92] Rafiee M, Gharagozloo M, Ghahiri A, Mehrabian F, Maracy MR, Kouhpayeh S, Pieper IL, Rezaei A: Altered Th17/Treg Ratio in Recurrent Miscarriage after Treatment with Paternal Lymphocytes and Vitamin D3: a Double-Blind Placebo-Controlled Study. *Iran J Immunol* 2015, 12:252-262.

[93] Gorman S, Geldenhuys S, Judge M, Weeden CE, Waithman J, Hart PH: Dietary Vitamin D Increases Percentages and Function of Regulatory T Cells in the Skin-Draining Lymph Nodes and Suppresses Dermal Inflammation. *J Immunol Res* 2016, 2016: 1426503.

[94] Choi KC, Leung PC, Jeung EB: Biology and physiology of Calbindin-D9k in female reproductive tissues: involvement of steroids and endocrine disruptors. *Reprod Biol Endocrinol* 2005, 3: 66.

[95] Emam MA, Abouelroos ME, Gad FA: Expression of calbindin-D9k and vitamin D receptor in the uterus of Egyptian buffalo during follicular and luteal phases. *Acta Histochem* 2016, 118:471-477.

[96] Labudzynskyi D, Shymanskyy I, Veliky M: Role of vitamin D3 in regulation of interleukin-6 and osteopontin expression in liver of diabetic mice. *Eur Rev Med Pharmacol Sci* 2016, 20:2916-2919.

[97] Luu KC, Nie GY, Salamonsen LA: Endometrial calbindins are critical for embryo implantation: evidence from in vivo use of morpholino antisense oligonucleotides. *Proc Natl Acad Sci* U S A 2004, 101:8028-8033.

[98] Minoura H, Findlay JK, Salamonsen LA, Jian Wang YL, Nie G-Y: Complex Regulation of Calcium-Binding Protein D9k (Calbindin-D9k) in the Mouse Uterus During Early Pregnancy and at the Site of Embryo Implantation1. *Biology of Reproduction* 2000, 62:27-36.

[99] Johnson GA, Burghardt RC, Bazer FW, Spencer TE: Osteopontin: roles in implantation and placentation. *Biol Reprod* 2003, 69:1458-1471.

[100] Woeckel VJ, van der Eerden BC, Schreuders-Koedam M, Eijken M, Van Leeuwen JP: 1alpha,25-dihydroxyvitamin D3 stimulates activin

A production to fine-tune osteoblast-induced mineralization. *J Cell Physiol* 2013, 228:2167-2174.
[101] Wierzbicka JM, Zmijewski MA, Piotrowska A, Nedoszytko B, Lange M, Tuckey RC, Slominski AT: Bioactive forms of vitamin D selectively stimulate the skin analog of the hypothalamus-pituitary-adrenal axis in human epidermal keratinocytes. *Mol Cell Endocrinol* 2016, 437:312-322.
[102] Song J, Li Y, An R: Vitamin D restores angiogenic balance and decreases tumor necrosis factor-alpha in a rat model of pre-eclampsia. *J Obstet Gynaecol Res* 2017, 43:42-49.
[103] Halder SK GS, Al-Hendy A: Vitamin D exhibits antiestrogenic effects in human uterine leiomyoma cells. *Fertil Steril* 2010, 94:S219–S220.
[104] Krishnan AV, Swami S, Peng L, Wang J, Moreno J, Feldman D: Tissue-selective regulation of aromatase expression by calcitriol: implications for breast cancer therapy. *Endocrinology* 2010, 151:32-42.
[105] Barrera D, Diaz L, Noyola-Martinez N, Halhali A: Vitamin D and Inflammatory Cytokines in Healthy and Preeclamptic Pregnancies. *Nutrients* 2015, 7:6465-6490.
[106] Merhi Z, Doswell A, Krebs K, Cipolla M: Vitamin D alters genes involved in follicular development and steroidogenesis in human cumulus granulosa cells. *J Clin Endocrinol Metab* 2014, 99:E1137-1145.
[107] Parikh G, Varadinova M, Suwandhi P, Araki T, Rosenwaks Z, Poretsky L, Seto-Young D: Vitamin D regulates steroidogenesis and insulin-like growth factor binding protein-1 (IGFBP-1) production in human ovarian cells. *Horm Metab Res* 2010, 42:754-757.
[108] Mote PA, Balleine RL, McGowan EM, Clarke CL: Colocalization of progesterone receptors A and B by dual immunofluorescent histochemistry in human endometrium during the menstrual cycle. *J Clin Endocrinol Metab* 1999, 84:2963-2971.
[109] Thangamani S, Kim M, Son Y, Huang X, Kim H, Lee JH, Cho J, Ulrich B, Broxmeyer HE, Kim CH: Cutting edge: progesterone

directly upregulates vitamin d receptor gene expression for efficient regulation of T cells by calcitriol. *J Immunol* 2015, 194:883-886.

[110] Piccinni MP, Giudizi MG, Biagiotti R, Beloni L, Giannarini L, Sampognaro S, Parronchi P, Manetti R, Annunziato F, Livi C, et al.: Progesterone favors the development of human T helper cells producing Th2-type cytokines and promotes both IL-4 production and membrane CD30 expression in established Th1 cell clones. *J Immunol* 1995, 155:128-133.

[111] Lee JH, Ulrich B, Cho J, Park J, Kim CH: Progesterone promotes differentiation of human cord blood fetal T cells into T regulatory cells but suppresses their differentiation into Th17 cells. *J Immunol* 2011, 187:1778-1787.

[112] Maeda Y, Ohtsuka H, Tomioka M, Oikawa M: Effect of progesterone on Th1/Th2/Th17 and regulatory T cell-related genes in peripheral blood mononuclear cells during pregnancy in cows. *Vet Res Commun* 2013, 37:43-49.

[113] Druckmann R, Druckmann MA: Progesterone and the immunology of pregnancy. *J Steroid Biochem Mol Biol* 2005, 97:389-396.

[114] Qu X, Yang M, Zhang W, Liang L, Yang Y, Zhang Y, Deng B, Gao W, Liu J, Yang Q, et al.: Osteopontin expression in human decidua is associated with decidual natural killer cells recruitment and regulated by progesterone. *In Vivo* 2008, 22:55-61.

[115] Florio P, Rossi M, Sigurdardottir M, Ciarmela P, Luisi S, Vigano P, Grasso D, Fiore G, Cobellis L, Di Blasio AM, Petraglia F: Paracrine regulation of endometrial function: interaction between progesterone and corticotropin-releasing factor (CRF) and activin A. *Steroids* 2003, 68:801-807.

[116] Makrigiannakis A, Margioris AN, Chatzaki E, Zoumakis E, Chrousos GP, Gravanis A: The decidualizing effect of progesterone may involve direct transcriptional activation of corticotrophin-releasing hormone from human endometrial stromal cells. *Mol Hum Reprod* 1999, 5:789-796.

[117] Kim M, Park HJ, Seol JW, Jang JY, Cho YS, Kim KR, Choi Y, Lydon JP, Demayo FJ, Shibuya M, et al.: VEGF-A regulated by

progesterone governs uterine angiogenesis and vascular remodelling during pregnancy. *EMBO Mol Med* 2013, 5:1415-1430.

[118] Kumar P, Magon N: Hormones in pregnancy. *Niger Med J* 2012, 53:179-183.

[119] Preciado-Martinez E, Garcia-Ruiz G, Flores-Espinosa P, Bermejo-Martinez L, Espejel-Nunez A, Estrada-Gutierrez G, Razo-Aguilera G, Granados-Cepeda M, Helguera-Repetto AC, Irles C, Zaga-Clavellina V: Progesterone suppresses the lipopolysaccharide-induced pro-inflammatory response in primary mononuclear cells isolated from human placental blood. *Immunol Invest* 2018, 47:181-195.

[120] Vrachnis N, Malamas FM, Sifakis S, Tsikouras P, Iliodromiti Z: Immune aspects and myometrial actions of progesterone and CRH in labor. *Clin Dev Immunol* 2012, 2012:937618.

[121] Wang B, Cruz Ithier M, Parobchak N, Yadava SM, Schulkin J, Rosen T: Vitamin D stimulates multiple microRNAs to inhibit CRH and other pro-labor genes in human placenta. *Endocr Connect* 2018.

[122] Qin LL, Lu FG, Yang SH, Xu HL, Luo BA: Does Maternal Vitamin D Deficiency Increase the Risk of Preterm Birth: A Meta-Analysis of Observational Studies. *Nutrients* 2016, 8.

[123] Tolppanen A-M, Sayers A, Fraser WD, Lawlor DA: Association of serum 25-hydroxyvitamin D₃ and D₂ with academic performance in childhood: findings from a prospective birth cohort. Journal of Epidemiology and *Community Health* 2012, 66:1137-1142.

[124] Bollapragada S, Youssef R, Jordan F, Greer I, Norman J, Nelson S: Term labor is associated with a core inflammatory response in human fetal membranes, myometrium, and cervix. *Am J Obstet Gynecol 2009*, 200:104.e101-111.

[125] Müller K, Diamant M, Bendtzen K: Inhibition of production and function of interleukin-6 by 1,25-dihydroxyvitamin D3. *Immunology Letters* 1991, 28:115-120.

[126] Muller K, Odum N, Bendtzen K: 1,25-dihydroxyvitamin D3 selectively reduces interleukin-2 levels and proliferation of human T cell lines in vitro. *Immunol Lett* 1993, 35:177-182.

[127] Tous M, Villalobos M, Iglesias L, Fernandez-Barres S, Arija V: Vitamin D status during pregnancy and offspring outcomes: a systematic review and meta-analysis of observational studies. *Eur J Clin Nutr* 2019.

[128] Staples J, Ponsonby A-L, Lim L: Low maternal exposure to ultraviolet radiation in pregnancy, month of birth, and risk of multiple sclerosis in offspring: longitudinal analysis. *BMJ* 2010, 340:c1640.

[129] McDonnell SL, Baggerly KA, Baggerly CA, Aliano JL, French CB, Baggerly LL, Ebeling MD, Rittenberg CS, Goodier CG, Mateus Nino JF, et al.: Maternal 25(OH)D concentrations >/=40 ng/mL associated with 60% lower preterm birth risk among general obstetrical patients at an urban medical center. *PLoS One* 2017, 12:e0180483.

[130] Pagliardini L, Vigano P, Molgora M, Persico P, Salonia A, Vailati SH, Paffoni A, Somigliana E, Papaleo E, Candiani M: High Prevalence of Vitamin D Deficiency in Infertile Women Referring for Assisted Reproduction. *Nutrients* 2015, 7:9972-9984.

[131] Ozkan S, Jindal S, Greenseid K, Shu J, Zeitlian G, Hickmon C, Pal L: Replete vitamin D stores predict reproductive success following in vitro fertilization. *Fertil Steril* 2010, 94:1314-1319.

[132] Anifandis GM, Dafopoulos K, Messini CI, Chalvatzas N, Liakos N, Pournaras S, Messinis IE: Prognostic value of follicular fluid 25-OH vitamin D and glucose levels in the IVF outcome. *Reproductive Biology and Endocrinology* 2010, 8:91.

[133] Firouzabadi RD, Rahmani E, Rahsepar M, Firouzabadi MM: Value of follicular fluid vitamin D in predicting the pregnancy rate in an IVF program. *Arch Gynecol Obstet* 2014, 289:201-206.

[134] Ciepiela P, Duleba AJ, Kowaleczko E, Chelstowski K, Kurzawa R: Vitamin D as a follicular marker of human oocyte quality and a

serum marker of in vitro fertilization outcome. *J Assist Reprod Genet* 2018, 35:1265-1276.

[135] Kwak-Kim J, Skariah A, Wu L, Salazar D, Sung N, Ota K: Humoral and cellular autoimmunity in women with recurrent pregnancy losses and repeated implantation failures: A possible role of vitamin D. *Autoimmun Rev* 2016, 15:943-947.

[136] Abdullah UH, Lalani S, Syed F, Arif S, Rehman R: Association of Vitamin D with outcome after intra cytoplasmic sperm injection. *J Matern Fetal Neonatal Med* 2017, 30:117-120.

[137] Chu J, Gallos I, Tobias A, Tan B, Eapen A, Coomarasamy A: Vitamin D and assisted reproductive treatment outcome: a systematic review and meta-analysis. *Hum Reprod* 2018, 33:65-80.

[138] Hollis BW, Wagner CL: Vitamin D supplementation during pregnancy: Improvements in birth outcomes and complications through direct genomic alteration. *Mol Cell Endocrinol* 2017, 453:113-130.

[139] Hollis BW, Johnson D, Hulsey TC, Ebeling M, Wagner CL: Vitamin D supplementation during pregnancy: double-blind, randomized clinical trial of safety and effectiveness. *J Bone Miner Res* 2011, 26:2341-2357.

[140] Dawodu A, Saadi HF, Bekdache G, Javed Y, Altaye M, Hollis BW: Randomized controlled trial (RCT) of vitamin D supplementation in pregnancy in a population with endemic vitamin D deficiency. *J Clin Endocrinol Metab* 2013, 98:2337-2346.

[141] Kim CH: A functional relay from progesterone to vitamin D in the immune system. *DNA Cell Biol* 2015, 34:379-382.

In: A Closer Look at Hormones
Editor: Alexander V. Knyazev
ISBN: 978-1-53617-254-6

Chapter 6

THE EVOLUTION OF THE GROWTH HORMONE *LOCUS* IN PRIMATES: A HISTORY OF GENETIC INNOVATION

Antonio Alí Pérez-Maya[1,*], Jesús Antonio Rea-Carrillo[1] and Hugo Alberto Barrera-Saldaña[2]

[1]Department of Biochemistry and Molecular Medicine of Medicine School of Autonomous University of Nuevo León, Monterrey, Nuevo León, México

[2]School of Medicine and Health Sciences, Monterrey Institute of Technology, Monterrey, Nuevo León, México

ABSTRACT

Pituitary growth hormone (GH, somatotropin) is a polypeptide hormone synthesized and secreted by the somatotroph cells of the anterior pituitary gland in all vertebrates. It regulates body growth and has a

---

* Corresponding Author's Email: bioquimicomty@gmail.com.

variety of metabolic actions. In most mammals, including strepsirrhines, the GH *locus* comprises a single gene. In primates, after the divergence of tarsiers and anthropoids, but before the separation of new world monkeys and catarrhines, the GH *locus* experienced a dramatic evolution compared to the relatively constant evolutionary rate exhibited by the remaining orders of mammals except, but to a lesser degree, by the artiodactyls. At least two independent *locus* expansions, one in platyrrhines and another in catarrhines accompanied this rapid evolution. The first characterized by a series of tandem gene duplications and the second by events of divergence, gene conversion, and rearrangements that led to the expansion of intergenic regions (IGRs). The rapid evolution of the genes in the GH-cluster during primate evolution thus involved not only gene duplication and divergence, but also changes in regulatory mechanisms so that these genes are expressed in either the pituitary or the placenta, although with low expression in several other tissues. In our laboratory, we study the evolutionary path followed by this gene family in primates by analyzing the sequences of their genomic *locus* in the genome of different species, using as a reference the human GH *locus*, whose anatomy (Chen *et al.*, 1989) and expression (Canizales, 1996) our laboratory helped to reveal years ago. After this solving of the organization and composition of our species *locus*, we continued to do so with the *loci* of lemur (*Lemur catta*), chimpanzee (*Pan troglodytes*), gorilla (*Gorilla gorilla*), rhesus monkey (*Macaca mulatta*), and vervet (*Cercopithecus aethiops*). From their analyses we have gained insights into the structure of representative *loci* of primates along the evolutionary scale. The comparative genomics of their genes and intergenic regions allowed us to begin understanding the dramatic evolution of this fascinating *locus* of the genome of primates, and the one of our own genome.

Keywords: growth hormone, primates, somatotropin

1. Structure of the Human GH *Locus*

The GH *locus* in humans is located in band q24.2 on the long arm of chromosome 17 (Baumann, 2009; Harper *et al.*, 1982; George *et al.*, 1981; Timasheva *et al.*, 2013). It covers an area of approximately 50,000 bases pairs (bp) and contains five structurally conserved genes that are oriented in the same transcriptional direction (Baumann, 2009; Chen *et al.*, 1989). The *locus* is flanked on its 5′ end by the *CD79B* gene, which is specific of

B lymphocytes (Cajiao *et al.*, 2004; Yoo *et al.*, 2006) and on its 3´ end by the testis-specific *TCAM1P* pseudogene (Kurihara and Kimura, 2015; Kimura *et al.*, 2004).

At its 5'-end, all the GH *loci* initiates with the *GH-1* gene (referred to as normal or *GH-N*), whose expression is specific of the somatotrophs of the anterior pituitary gland. In the human *locus*, the remaining four genes are expressed in the syncytiotrophoblast of the placenta. The functionality of the second gene (*hCSH-L*) has been discuss for years, but remains unclear; it was originally considered a pseudogene and in fact, alternative splicing suggests that the majority of the transcripts would be unable to express a secreted protein, although evidence of its expression in placenta has been reported (Barrera-Saldaña *et al.*, 1983; Reséndez-Pérez *et al.*, 1990; Baumann, 2009; Timasheva *et al.*, 2013). The third and fifth genes (*hCSH-A* and *hCSH-B* also known as *CSH1* and *CSH2* respectively) codifying for an identical mature hormone (Barrera-Saldaña *et al.*, 1982). The fourth gene is the *hGH-V* (also known as *GH2*), which codifies for a variant of GH produced by the placenta and that replaces the hypophyseal GH in its function during pregnancy (Baumann, 2009; Chen *et al.*, 1989).

Of the ~ 50 kb genomic region that spans the GH *locus*, transcriptional units of its five genes do not occupy more than 2 kb each, being separated by four intergenic regions (~ 38 kb): two of 13 kb, one of 6.5 kb, and one of 6 kb (Chen *et al.*, 1989). In these regions, both repetitive sequences, as well as elements involved in the regulation of the gene's expression, have been identified (Figure 1). In the human *locus*, there are 48 copies of repetitive sequences of the *Alu* family and one of the *Kpn* family. *Alu* insertions occur in both directions and three are truncated. It should be noted that over 20% of the nucleotide sequences of the hGH *locus* correspond to these repetitive elements, whose distribution pattern suggests their possible role in the evolutionary pathway that shaped it (Chen *et al.*, 1989).

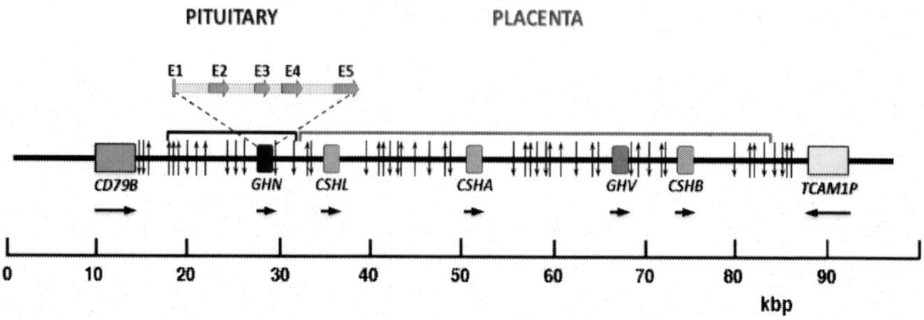

Figure 1. Genes, intergenics, and *Alu* elements of the human GH *locus*. Canonical anatomy of the *GH/CSH* genes, and tissues where are expressed. Arrangement and orientation (horizontal arrows) of genes that are separated by intergenic regions whose distribution of repetitive elements of the *Alu* family (vertical arrows) are indicated. Distances can be deduced from the scale shown.

2. THE GENES

Despite their differences in tissue specificity, expression levels, and functions of codified products, the *GH/CSH* genes display a very high similarity in their nucleotide sequences (over 95%) and identical anatomy: consisting of five exons separated by four introns of similar sizes. The structural analysis of the globular proteins that the genes encode for has revealed that they possess four alpha helixes and contain two disulfide bonds. Each newly synthesized pre-hormone has a 26 amino acid (aa) signal peptide that directs its transport across the rough endoplasmic reticulum membrane to the Golgi apparatus, where they are secreted in their mature form into the bloodstream (Handwerger and Freemark, 2000).

The *hGH-N* gene generates as its main product, a 191 aa (hGH 22 kDa) single-chain protein with two disulfide bridges. During maturation of the transcript, another two isoforms can be produced: a 20 kDa (176 aa) protein that represents ~ 10% of the GH put out by the pituitary gland, which lacks residues 32-46 (from the use of an alternative splice site in exon 3) and a third isoform that has been described as a product of this gene, which has a molecular weight of 17.5 kDa and lacks residues 32-71 derived from the omission of exon 3. This isoform is present only in pathological conditions when the *GH-N* gene has mutated (genetic GH deficiency type II) and it does not appear to be significantly expressed in normal conditions (Baumann, 2009; Ranke and Wit, 2018; Reséndez-Pérez *et al.*, 1990).

The most abundant product of the *GH-V* gene is a single polypeptide of 191 aa (22 kDa) that replaces GH-N during pregnancy. This hormone has two disulfide bonds, similar in structure to GH-N (approximately 93% identical), but with differences in 13 residues, which are responsible for making this protein more basic. Four isoforms have been observed as resulting from the *hGH-V* gene in the placenta in late pregnancy (Baumann, 2009; Frankenne *et al.*, 1987-A; Wallis, 2014). One results from the retention of intron D of the gene's primary transcript, whose translation gives rise to a protein with a molecular weight of 26 kDa (230 aa). This isoform, which is believed to be a membrane protein, exhibits a notable divergence in its terminal carboxyl with respect to the main product of this gene (Reséndez-Pérez *et al.*, 1990). Another isoform that has been detected in the placenta differs from the major isoform, in that it is glycosylated at the asparagine residue in position 140 which increases its weight to 25 kDa (Reséndez-Pérez *et al.*, 1990). A fourth isoform (24 kDa, 210 aa) results from a deletion of four bp (GTGG) in the 5' of intron D of its pre-mRNA which results in an alternative second donor splice site; besides being found in the placenta, this isoform has also been found in the testicles (Reséndez-Pérez *et al.*, 1990). A fifth isoform results from the deletion of 45 bp in exon 3 of its mRNA, which generates a 20 kDa protein (similar to the hGH-N of equal molecular weight) (Reséndez-Pérez *et al.*, 1990).

3. Gene Products: The Hormones of the *Locus* hGH in Physiology

The main function of hGH-N is to regulate linear postnatal growth and various metabolic activities. It is involved in processes such as glucose homeostasis, protein synthesis in muscle tissue, lipolysis in adipose tissue, tissue repair, and bone regeneration (Gómez and Villabona, 2014). It has also been implicated in learning and memory, neurogenesis, and neuroprotection (Sonntag *et al.*, 2005).

The hGH-V synthesis is restricted to the syncytiotrophoblast of the placenta from 5 to 8 weeks of pregnancy (Cooke *et al.*, 1988-A; Cooke *et al.*, 1988-B; Timasheva *et al.*, 2013). Its levels are associated with intrauterine growth rate, birth weight, and fetal levels of insulin-like growth factor I (IGF-I) (Chellakooty *et al.*, 2004; Rygaard *et al.*, 1998; Velegrakis *et al.*, 2017; Verhaeghe *et al.*, 2002). This hormone has somatogenic and lactogenic activity and, as mentioned above, replaces hGH-N during the second half of pregnancy. In addition, it has been implicated in the increase of IGF-I levels in maternal serum (Frankenne *et al.*, 1987-A; Frankenne *et al.*, 1987-B; Hennen *et al.*, 1990; Macleod *et al.*, 1991; Nickel *et al.*, 1990; Velegrakis *et al.*, 2017).

A similar role has been ascribed to human CSH (hCSH), whose levels have been associated with good placental function and fetal levels of IGF-1 (Handwerger and Freemark, 2000; Karabulut *et al.*, 2001; Velegrakis *et al.*, 2017). This 22 kDa hormone is thought to play an important role as an insulin antagonist in the regulation of maternal metabolism during pregnancy, causing nitrogen retention and glucose intolerance. Opposing effects to insulin of hCSH include inhibition of glycogen synthesis, glycolysis, lipogenesis, and of glucose entry into cells; also, the activation of glycogenolysis, proteolysis in muscle, and gluconeogenesis (Hartog, 2008; Josimovich and Atwood, 1964; Velegrakis *et al.*, 2017). The purpose of these actions is to increase the plasma concentration of glucose, preserving it for consumption by organs such as the maternal brain, and providing adequate amounts of glucose and nutrients to the fetus. It also exerts direct somatotrophic activity on fetal tissues (Felber *et al.*, 1972; Williams and Coltart, 1978). It can also accelerate lipolysis, which is why it is thought to release fatty acids during pregnancy to ensure an energy supply to the fetus (Kim and Felig, 1971; Walker *et al.*, 1991), an increase in the synthesis of this hormone, which represents 10% of the total production of placental proteins, occurs during the third trimester of pregnancy (Jacquemin *et al.*, 1994).

4. Gene Regulation

There is a strict regulation of GH at both the metabolic and genetic levels. At the metabolic level, the hGH is regulated by a large number of mechanisms such as the presence of GH-releasing hormone, somatostatin, or ghrelin, followed by secretion systems, and a complex mechanism of signal transduction. These processes have been widely reported in a large number of studies (Ranke & Wit, 2018). These start a response at the gene, mRNA, and protein levels, through feedback loops from the hypothalamus, the liver, and the adrenal and thyroid glands (Reséndez-Pérez *et al.*, 1990). Major regulators of the *GH-N* gene include the pituitary-specific factor (Pit-1), glucocorticoids, thyroid hormone, specificity protein 1 (Sp1), and cAMP response element binding protein (CREBP) (Reséndez-Pérez *et al.*, 1990). Gene expression is tightly controlled by mechanisms operating at supra-gene levels of nuclear organization, chromosome structure, and DNA modifications (Kimura *et al.*, 2004).

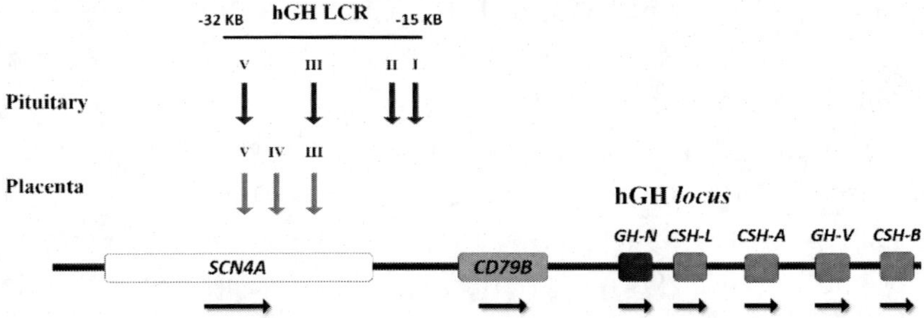

Figure 2. hGH *locus* control region. The hypophyseal *hGH-N* gene (blue box) and four placental expression genes (orange squares) that make up the hGH *locus* are shown. White and green rectangles represent unrelated genes, *SCN4A* and *CD79B*. Horizontal arrows show the transcriptional orientation of the genes. The position of each hypersensitive site in the *locus* control region (LCR) of the hGH *locus* is marked with vertical arrows. Modified from (Shewchuk *et al.*, 2006).

The human *locus* is under the effect of a LCR that contributes to the genes tissue-specific pattern of expression. This type of element has the ability to activate the expression of target genes and establish, in chromatin, independent domains of transcriptional regulation. It is also characterized by the presence of sites that are hypersensitive to DNase I enzyme (Hypersensitive Sites, HS), identified as HSI to HSV and located between 15 and 32 kb before the *locus* (Figure 2) (Hunsaker *et al.*, 2012; Kimura *et al.*, 2004). This hypersensitivity is associated with changes in chromatin structure, which in turn indicates transcriptional activity events of target genes (Elefant *et al.*, 2001; Kimura *et al.*, 2007; Shewchuk *et al.*, 2002).

Pit-1 is essential for the coordinated transcriptional control of the *hGH-1* gene cluster (Reséndez-Pérez *et al.*, 1990). Pit-1 binding in the HS I and HS II sites is necessary and sufficient to direct regulated tissue-specific expression, more so than binding at the promoter of the *hGH-N* gene (Reséndez-Pérez *et al.*, 1990). Control of the specified set of genes via the LCR is coordinated through the formation of domains of histone modifications (hyperacetylation or methylation) and chromatin looping (Reséndez-Pérez *et al.*, 1990). There is also evidence that ongoing non-coding transcription of the *CD79B* gene is essential for regulation by the LCR. The continuous non-coding transcription of the nearby gene, *CD79B*, allows for the constant presence of RNA polymerase II (RNA pol II) and an open chromatin conformation. The regulated transcription of the *GH* gene cluster begins with Pit-1 binding at HS-I and in the promoter of the *hGH-N* gene, which then triggers an interaction between the two regions accomplished by looping. Looping of chromatin allows for the juxtaposition of the *hGH-N* promoter into a region of high transcriptional activity, the LCR/*CD79B* transcriptional domain that is dense with RNA pol II (Reséndez-Pérez *et al.*, 1990).

Differential expression of members of the hGH *locus* gene cluster in the placenta and pituitary depends largely on the control exerted by the various subgroups of the LCR (Elefant *et al.*, 2001; Jones *et al.*, 1995; Kimura *et al.*, 2004). Histones at the internal regions of the multigene *locus* are selectively acetylated in placental chromatin and one of these chromatin regions includes precisely the P element of the placental genes located in front of these and which is involved in the repression of their expression in the hypophysis (Kimura *et al.*, 2004). HSI and HSII sites are specific for expression in the pituitary, and acetylation of histones in a region centered on these sites is associated with expression of the *hGH-N* gene in this gland. The HSIV site is specific for placenta and the acetylation of histones associated with placental gene expression is limited to the region spanning HSIII-HSV. The HSIII and HSV sites, located at 28 kb and 32 kb upstream of the *hGH-N* gene, respectively, are found in both cell types. HSI and HSII sites flank the 5' end of *CD79B* gene. The three remaining sites (HSIII-HSV) are located on the second half of the gene encoding the protein that forms the specific sodium channel skeletal muscle (*SCN4A*). The region HS I/II contains three Pit-1 binding sites that act as an enhancer of the pituitary *hGH-N* promoter (A. P. Kimura *et al.*, 2007; Reséndez-Pérez *et al.*, 1990). Pit-1 also binds to proximal (-92 to -65) and distal (- 130 to - 105) sites on this promoter. The hyper-acetylation of H3 and H4 histones facilitate access of Pit-1 to all these sites. Both the promoter and HSI/II region are required for the efficient expression of *hGH-N* (Reséndez-Pérez *et al.*, 1990).

Genetic Elements in the Establishment of the Regulatory Network of GH *Locus*

Proximal Promoter

Several response elements are found in the *hGH-N* promoting region. The TATA box and the Sp-1 element are completely preserved in all genes. The initiator binding site (InrE), required for efficient promoter activity and maximum potentiator activity, is generally well conserved, as it is the cyclic AMP response element (CRE), contrasting this with the situation of non-primate mammals (Pérez-Maya *et al.*, 2016).

Distal and proximal sequences of Pit-1 elements are highly conserved within the gorilla genes *gCSH-A2*, *gCSH-A3*, and *gCSH-B*, the orangutan *oCSH-B* and the chimpanzee *cCSH-A1*, all being identical to those of the

GH-N gene. A striking exception to this conservation of the Pit-1 elements is seen in the proximal element for the *GH-V* genes, which shows five nucleotide replacements, shared in orangutan, gorilla, chimpanzee, and human. This is probably a reflection of the low expression levels of *GH-V* in the placenta, compared to the high level expression of *GH-N* in the anterior lobe of the pituitary gland (Pérez-Maya *et al.*, 2016).

The thyroid hormone response element (TRE) seems to be more important in the placenta than in the pituitary gland. The TRE that precede the *CSH* genes are quite variable, but for most of them the positions -116, -112, and -110 are G, G, and A, respectively; these sites were identified as critical for the performance of this element in *hCSH-A/B*. In the chimp case, gene conversion events may have influenced the variation of this region and the same might be true for the gorilla. The variability observed in TRE suggests that the pattern of response to thyroid hormones may differ considerably (Pérez-Maya *et al.*, 2016).

Transcription

The transcriptional control is a key point of gene expression regulation. In this category, the action of promoters (Figure 3), the presence of enhancer regulatory sequences, and the interaction between multiple activator proteins or inhibitors that act by binding to specific sequences of DNA recognition are included. Transcriptional control not only deals with "turning on" or "turning off" genes, but also with regulating the transcription intensity of "turned on" genes (Reséndez-Pérez *et al.*, 1990).

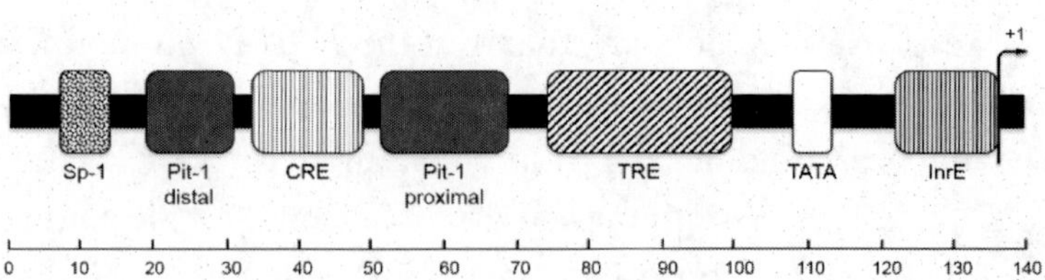

Figure 3. Distribution of regulatory elements in the proximal promoter of the hGH *locus* genes. The organization of regulatory elements of the proximal promoter that have been identified in *hGH-N* gene are shown. The arrow represents the starting site of the transcription.

Analysis of evolutionary conserved sequences identified a conserved element located between the proximal and distal Pit-1 binding sites. This element was shown to bind a zinc finger protein, Zn-15 (Reséndez-Pérez *et al.*, 1990). Mutation of the Zn-15 binding site decreased *hGH-N* expression more than 100-fold (Reséndez-Pérez *et al.*, 1990).

Expression of the *hGH-N* gene in the somatotrophs of the anterior pituitary is directed by the transcription factor Pit-1, also known as GHF1 (Karin *et al.*, 1991). Since Pit-1 is a factor that is specific for pituitary tissue (Ingraham *et al.*, 1988), it is apparently not involved in the induction of placental gene expression. This factor has two conserved domains: an important POU domain for transcriptional activation and a homeobox (HB) domain that intervenes in DNA binding; both are located in the carboxyl terminus of the protein (Meis *et al.*, 1997).

Pit-1 belongs to the POU family of transcription factors that includes Pit-1, Oct-1 and Oct-2, and Unc-86 (Reséndez-Pérez *et al.*, 1990). These highly conserved transcription factors each have a homeodomain that is linked to another domain known as the POU-specific domain. Most other family members are highly expressed in the nervous system. Pit-1 was first discovered as the transcription factor that is necessary for expression of the genes responsible for the synthesis of GH and prolactin (PRL), and this definition was later extended to include thyroid stimulating hormone β, or TSHβ (Reséndez-Pérez *et al.*, 1990). The 5´-proximal 320 bp of the human *GH* gene promoter is necessary and sufficient for selective expression of the *GH* gene in mouse pituitary. This region contains multiple Pit-1 sites and a thyroid hormone response element. Although Pit-1 is required for normal somatotroph differentiation, it is not the final differentiating factor because it is produced before GH expression and in two other distinct hormone-secreting cell types (Reséndez-Pérez *et al.*, 1990). Pit-1 contains a bipartite DNA binding domain that resembles the classic helix-turn-helix motif that is necessary to bind in the double helix major groove. Further, both the homeodomain and POU specific domain are required for high affinity binding to DNA sequences. The highly conserved Pit-1 sites in the human *GH-N* gene are - 127 /- 107 (AGCTTCTAAATTATCCATTA) and - 87 /- 72 (CCATGCATAAATGTA) (Reséndez-Pérez *et al.*, 1990).

Interestingly, the spacing between the two sites, consensus sequence, and AT-rich, determines the outcome of expression for the corresponding gene. Depending on the binding site, Pit-1 will induce expression of the *GH-N* gene and suppression of the one of prolactin in somatotrophs, while the opposite is observed in lactotrophs (Reséndez-Pérez *et al.*, 1990).

Glucocorticoid Response Element

A glucocorticoid response element (GRE) has been reported in the first intron of *hGH-N* and *hGH-V*. This element is completely conserved in chimpanzee's *GH-N* (Pérez-Maya *et al.*, 2012), but not in the orthologue genes of gorilla or orangutan, nor in any of the *CSH* genes. Interestingly, it is fully conserved in the *GH-V* gene of the four hominids, suggesting that it may play a role in the regulation of this placental gene (Pérez-Maya *et al.*, 2016).

Repressor

Although the promoters of placental genes also present Pit-1 binding sites, these are subject to transcriptional repression mediated by a conserved region of 263 nucleotides known as P element or pituitary inhibitor, located approximately 2 kb in front of the transcription start site of the human genes that are expressed in the placenta. It is believed that this element inhibits the expression of these genes in the pituitary and/or activates their expression in the placenta. (Norquay *et al.*, 2001; Pérez-Maya *et al.*, 2016). The corresponding sequences are found in gorilla and orangutan and can play the same role. However, it should be noted that a "P element" similar sequence is found upstream to the *GH-N* gene in many other mammals, including the marmoset and dog, making its function as a pituitary suppressor less clear (Pérez-Maya *et al.*, 2016).

The P element contains two domains: PSE-A and PSE-B (Figure 4). The first interacts with members of the Regulatory Factor X (RFX) family, while the second interacts with Nuclear Factor-1 (NF-1) (Norquay *et al.*, 2001; Dodd *et al.*, 2003; Nachtigal *et al.*, 1993). A group of common *trans* factors (Sp1, USF/MLTF, and NF-I) is involved in establishing the baseline transcription level of the *hGH-N* gene in the hypophysis (Jacquemin *et al.*, 1994).

Although many questions about the transcriptional regulation in the placenta of the *GH/CSH* genes remain, progress has been made in this regard. For *hCSHs* genes, another regulatory element referred to as an initiator or InrE, which is located between the TATA box and the transcription initiation site has been identified. Mutations in this sequence reduce expression of the corresponding promoter by 60% (Jiang *et al.*, 1995).

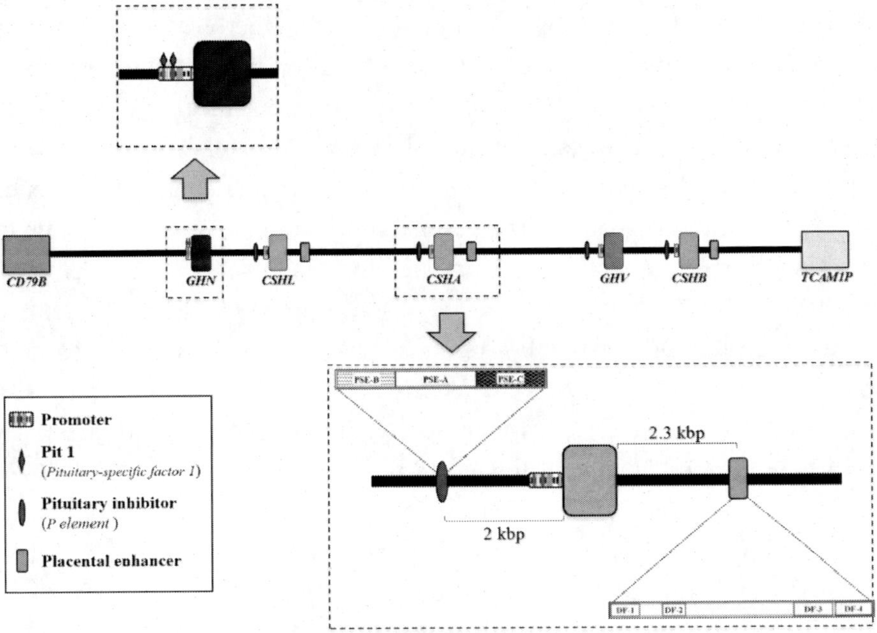

Figure 4. Details of the relevant sites for gene regulation at the hGH *locus*. Both Pit-1 sites present in the *hGH-N* promoter are outlined as green diamonds before the gene. The site of the pituitary inhibitor or P element is represented as red ovals before each placental gene (their PSE-A, PSE-B, and PSE-C domains are amplified). The placental enhancer is represented as a rectangle after the *CSHs* genes [the four sites for binding placental nuclear proteins (DF-1, DF-2, DF-3 and DF-4) are amplified].

Enhancer

Approximately 2.3 kb after the *hCSH* genes, there is an enhancer which increases the expression of these in the placenta, (Barrera-Saldaña, 1998; Jacquemin *et al.*, 1994; Rogers *et al.*, 1986). Protection assays against digestion with DNase I have revealed, in this enhancer, the presence of four binding sites, called DF-1, DF-2, DF-3, and DF-4, for placenta nuclear proteins (Jacquemin *et al.*, 1994); it has been shown that the activity of the enhancer in the placenta is mediated by a synergism between the DF-3 and DF-4 sites (Jacquemin *et al.*, 1994). It has also been reported that the function of this enhancer is independent of its orientation and relative position (Rogers *et al.*, 1986); its activity also requires cooperative binding to multiple sites located within this element, enhancer factor transcription-1 (TEF-1) and a specific enhancer factor called placenta CSEF-1. Enhancers of the three *hCSHs* genes show a high similarity in their sequence; however, copies in *CSHA* and *CSHL* genes show a weak transcriptional stimulation activity compared to *CSHB*, due to mutations in their respective binding sites to DF-3 proteins (Jacquemin *et al.*, 1996).

Although the DF-3 and DF-4 regions appear to be the most important, it has been shown that an AGAA insertion, upstream to the DF-2 domain of the enhancer, decreases its functionality. Interestingly this 4 base pairs insertion is not present in the enhancers that follow the *hCSH-B*, *cCSH-B*, *gCSHA3*, and *gCSH-B* genes. The remaining *CSH* gene enhancers show this insertion, suggesting reduced functionality (Pérez-Maya *et al.*, 2016). In addition to this insertion, there are some substitutions that may affect the enhancer of these genes (Jacquemin *et al.*, 1996). A substitution at position 208 (G by A) in the DF-3 domain is found in all *CSH* genes of orangutan, gorilla, and chimpanzee, suggesting that the enhancer here is much less active than in human *CSH-B*. Another substitution at position 228 (T for A) may decrease enhancer activity in all *CSH* genes except *hCSHB*, *cCSH-B*, *gCSH-B*, *oCSH-L*, and *oCSH-B* (Pérez-Maya *et al.*, 2016).

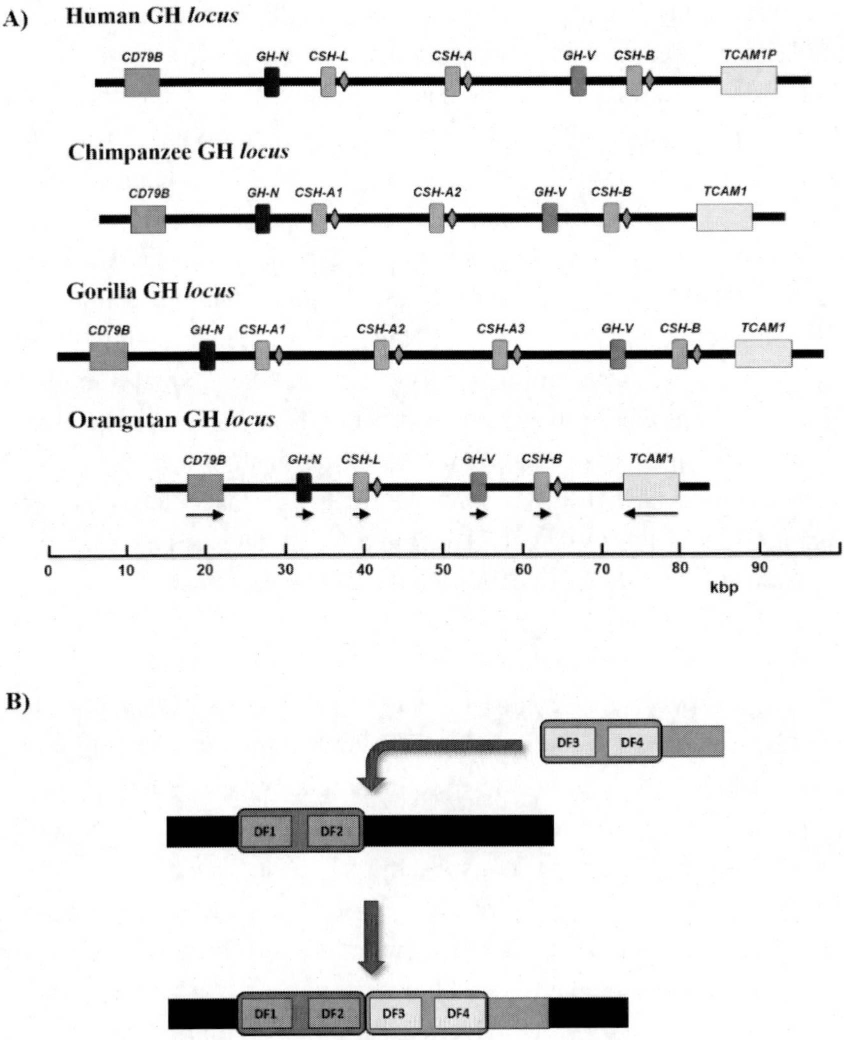

Figure 5. A) Comparative analysis of the distribution of hAT-Charlie elements. Blue boxes show the genes that form the GH *loci*, the solid lines represent the IGRs and the orange diamonds indicate the hAT-Charlie MER5B. The hGH *locus* is shown here in reverse orientation. Arrows indicate direction of transcription for orangutan; the same direction applies for other species. B) Contribution of the hAT-Charlie elements to the placental enhancer origin. The upper section represents the possible mechanism by which the insertion of a MER5B element (orange box) introduces DF-3 and DF-4 domains into a sequence containing DF-1 and DF-2 located behind some *GH* gene. The placental enhancer resulting from this insertion is represented in the lower part.

An interesting feature of the enhancer that had not been observed previously is that each of the enhancer elements that follow the three *CSH* genes overlaps a transposable element (TE) MER5B, a member of the super hat-Charlie family of DNA transposons (Figure 5A). As indicated in Figure 5B, the overlap includes domains DF3 and DF4, which are the main contributors to the enhancement activity (Jacquemin *et al.*, 1996). In prosimians and dogs where there is only one gene in the GH *locus*, there is also a MER5B element conserved. This suggests that the element was present long before the gene was duplicated and was subsequently selected (or eliminated) to play a role in regulating of CSH expression. It has previously been proposed that the extraction of TE to provide regulators of gene expression played an important role in the evolution of the maternal side of the placenta. MER5B's contribution to the placental enhancer of *CSH* suggests that the same can be applied to the fetal side, although the extent of this has not yet been determined (Pérez-Maya *et al.*, 2016).

Repetitive DNA Elements

During vertebrate evolution, the transposable elements have contributed repeatedly regulatory and coding sequences for the host, leading to the emergence of new gene regulators and lineage-specific functions. In all vertebrate lineages, there is evidence of genomic rearrangements mediated by transposable elements, such as insertions, deletions, inversions, and duplications potentially associated with or following speciation events. Taken together, these observations indicate that the transposable elements are the main causes of the genomic and biological diversity of vertebrates, with possible important roles in speciation and major evolutionary transitions (Böhne *et al.*, 2008).

Chromosomal rearrangements occur throughout the genome, however, they occur more frequently in regions rich in low copy number repeats or repetitive elements, such as LTR, LINE, *Alu*, MIR, and elements of DNA MER2. Several points of recurrent breaks in deletions and chromosomal duplications have been located in low copy number repeats and these points are usually large and highly homologous regions. These rearrangements are produced by homologous non-allelic recombination between the low copy number repeats copies. In contrast, few studies have examined the mechanisms responsible for non-recurrent rearrangements of breakpoints in non-homologous regions. Previous analyses have revealed deletions originated from *Alu* /*Alu* recombination events and others that result from a binding mechanism of ends non-homologous. The breakpoints have been found in low copy number repeats and within repetitive elements, including *Alu* and MER5B. These findings suggest that higher-order genomic architecture, such as low copy number repeats, and small repetitive sequences, such as *Alu* elements, can mediate chromosomal deletions through homologous and non-homologous mechanisms (Shaw and Lupski, 2005).

In the human *locus*, there are 48 copies of *Alu* elements. An *Alu* Sx repeat is found in the 3' flanking region of both the *hGH-N* and *hGH-V* genes. The deletion of this repetition in reporter constructs containing 3' flanking sequences of the *hGH-N* gene, increased the reporter activity in the transfected GC pituitary cells, suggesting that this region contains a repressor element. This region of the *hGH-N* gene contains the *Alu* Sx repeat between nucleotides 2228 and 2501 (where the start codon is at position +559 bp). A refined mapping of this silencing element revealed that it comprises four discrete entities that include a repressor domain core (CRD), an anti-silencing domain (ASE) which contains elements that mediate the orientation-independent silencing activity, and two domains that flank the CRD/ASE, which modulates the silencing activity in a CRD-dependent manner. The upstream modulator domain is also necessary for the orientation-independent silencer function (Burbano *et al.*, 2010).

An important repressor element is found between nucleotides 2,306 and 2,465. However, since a series of deletions between nucleotides 2,306 and 2,572 exhibit a gradual increase in activity, it has been suggested that multiple repressor elements may be distributed across this 266 bp sequence. This region includes most of the *Alu* element, indicating that a repressor is located in that element (Trujillo *et al.*, 2006).

The *Alu* repeats of the GH *locus* may have played a role not only in the generation of the *locus*, by involving in the duplication of genes, but also, just downstream of the *hGH-N* gene, *cis*-acting elements that control the transcriptional regulation of duplicate genes. This raises the question of whether other *Alu* repeats at the *locus* can affect the transcription of neighboring genes. At this point, we can only speculate based on the similarities of the *Alu* sequences flanking the *GH*s and *CSH*s genes. The *Alu* repeat located in the 3' region of the *hGH-N* and *hGH-V* genes belongs to the Sx subfamily, while the *Alu* sequences flanking the *CSH*s genes are of the Sz class in the 5' regions and of the Sq subfamily in the 3' regions. In fact, the 3' flanking region of the *hGH-N* and *hGH-V* genes are identical in about 92%. Therefore, it is very likely that the *Alu* element within the 3' flanking region of the *hGH-V* gene also has a silencing element (Trujillo *et al.*, 2006).

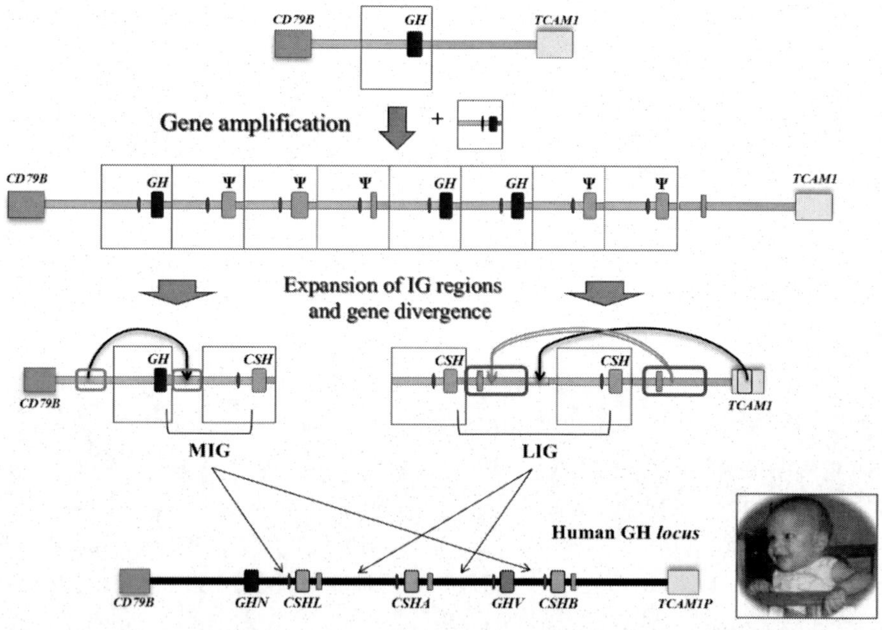

Figure 6. Evolution of the GH *locus* in primates. Likely mechanism that shaped the anatomy of the GH *locus* in its evolution from a unigenic *locus* in prosimians to a multigenic arrangement in humans is represented. The expansion of the GH *locus* occurred due to an increase in the number of genes and in the length of the intergenic regions that originated the medium intergenic (MIG) and long intergenic (LIG) regions accompanied by gene divergence events.

As in the human, in the chimpanzee genome there are also 48 copies of the *Alu* element, along with a series of additional elements related to SINE and LINE. The only clear difference in the content of repetitive elements between human and chimpanzee *loci* is an element related to a LINE-1 placed in an upstream position of *CSH-A2*/*CSH-A*, present in chimpanzee but not in human (Pérez-Maya *et al.*, 2012; Chen *et al.*, 1989).

5. Evolution of GH *Locus*

GH, PRL, prolactin 2 (PRL2), and somatolactin (SL) belong to the same hormone family (somatolactogens) and have a wide range of effects involving development, osmoregulation, metabolism, and growth stimulation. It has been proposed that hormone and receptor family were expanded by gene duplications in the early evolution of vertebrates and in agreement with this hypothesis the four members share a common exon organization. The proteins show a characteristic framework of conserved amino acids, with four cysteine residues forming two disulfide bridges that stabilize their 4-helix tertiary structure (Ocampo *et al.*, 2018; Wallis, 2014). The discovery of a single family member, a GH-like sequence in the jawless sea lamprey (*Petromyzon marinus*) was taken as evidence that genetic duplications in an ancestor of jawed vertebrates (*Gnathostomata*) led to the appearance of GH, PRL, and SL before their radiation occurred. Afterward, there was gain and loss of specific genes in each lineage, ray-finned fish (*Actinopterygii*) and lobe-finned fish (*Sarcopterygii*). This includes duplication and diversification of the GH branch in primates and of the PRL branch in rodents and ruminants. Recent studies identified a GH-like counterpart in the lancelet *Branchiostoma japonicum*. This was the first recognition of a hormone protein and a supposed gene of the GH family in specie of invertebrates. This led to the suggestion that GH is the ancestral member, with PRL and SL emerging up to the vertebrate lineage (Ocampo *et al.*, 2018).

The triplet of GH family genes (*GH*, an ancestral *PRL/PRL2*, and *SL*) appears to have been present even before vertebrate radiation. A scenario has been proposed in which local duplication of genes resulted in this early triplet. Subsequently, the classic *PRL* gene and its paralogue *PRL2* emerged through the first of the two rounds of vertebrate tetraploidization, increasing to four the number of family members. Afterward, the third round of tetraploidization in teleost fish doubled the *SL* gene, resulting in *SLa* and *SLb*. In addition to these events, several gains and losses of lineage-specific genes have occurred, particularly at the base of tetrapod and amphibian evolution. In passerine birds, the duplication of the *GH* gene was followed by rapid evolution. In mammals, gene replications gave rise to families of proteins expressed mainly in the placenta. In the case of superior primates, there is also a comparable group of genes expressed in the placenta, but these derive from multiple duplications of the *GH* gene (Ocampo *et al.*, 2018).

Reconstructing the History of the Human GH *Locus* Evolution

Primates emerged in the Tertiary Period of the Cenozoic Era and are classified into two suborders: strepsirrhines and haplorrhines. In the Oligocene, anthropoids evolved from a common ancestor and split into three groups: *Ceboidea* or New World Monkeys, *Cercopithecoidea* or Old World Monkeys, and *Hominoidea* or Apes, which includes man. The GH family in primates originated from an initially unigenic evolutionary branch that continued without large changes until the appearance of the platyrrhines (*Ceboidea*) (Jacquemin *et al.*, 1996). Starting with prosimians, GH is encoded by a single gene, whose sequence is very similar to the ancestral gene found in non-primate mammals. However, within higher primates occurred an accelerated GH *locus* evolution that resulted in a substantial difference (about 35%) between human GH and non-primate mammalian GH sequences (Figure 6) (Ocampo *et al.*, 2018; Wallis, 2014).

It has been postulated that in the beginning *locus* expansion events caused by duplications occurred due to *Alu* type repetitive sequences that surrounded the ancestral *locus*. The slow evolutionary change is interrupted by short episodes where the rate of evolution increased from 10 to 100 times. This evolution is observed in a changes by non-synonym mutations; however, this does not occur for synonym substitutions which do not affect protein structure. This and other evidence suggest that, at least in some cases, accelerated evolution was adaptive and associated with subtle changes in biological activity. One possible mechanism that explains the accelerated evolution is “function change”, in which a protein has two or more biological actions and the pressure for rapid change could have stopped after gene duplication, with one of the duplicated genes retaining one function of the hormone and the second gene adopting the other one (Jacquemin *et al.*, 1996; Wallis, 2014).

The multiple duplications of genes in the GH *locus* and the divergence of its members was followed by changes that led to differences in tissue expression, with at least two specificities: pituitary and placenta. In old-world monkeys and higher primates, sequences similar to the terminal end of the *TCAM1* gene are observed on the long intergenic sequences of the *locus*. These sequences were probably added during duplication events that have occurred in this site (Pérez-Maya *et al.*, 2016).

The structure of the GH *locus* differs considerably between primates. Rhesus monkey (*Macaca mulatta*) and baboon (*Papio hamadryas*) have six genes/pseudogenes in the cluster while gibbon (*Hylobates leucogenys*) has seven, marmoset (*Callithrix jacchus*) has eight genes/pseudogenes, and the white-fronted capuchin (*Cebus albifrons*) contain at least 40. Human and chimpanzee (*Pan troglodytes*) have five (Pérez-Maya *et al.*, 2012), while the gorilla (*Gorilla gorilla gorilla*) and orangutan (*Pongo abelii*) contain respectively, six and four genes (Figure 5A). The mature proteins (i.e., excluding the signal peptide) encoded by the gorilla and orangutan *GH-N* genes are identical to human GH-N. These differ from chimpanzee GH just in one residue (residue 126 is Glu in chimpanzee, Gly in human, gorilla, and orangutan) (Pérez-Maya *et al.*, 2016).

CONCLUSION

The comparison of genomes, genes, and proteins of different species is the most intuitive and straightforward approach to interpret "the instruction manual" of genomes. Comparative genomic analyses have revealed unexpected and complex patterns of changes during the evolution of primates. Phylogenetic analysis is given us the unprecedented opportunity to reconstruct the evolutionary history of the human genome. Comparative genomic analyses of the GH *locus* provide a framework for understanding the evolution of its structure at the DNA level and the corresponding impact on gene expression regulation and function of the encoded hormones. In our laboratory, we have been investigating the evolutionary path followed by this gene family in primates, studying the sequences of GH *locus* in representative species and referencing them to the human GH *locus* whose anatomy and expression, our working team also helped to reveal years ago.

REFERENCES

Barrera-Saldana, H. (1998). Growth hormone and placental lactogen*: Biology, medicine and biotechnology*. *Gene*, (Vol. *211*). https://doi.org/10.1016/S0378-1119(98)00092-4.

Barrera-Saldana, H., Robberson, D. & Saunders, G. F. (1982). Transcriptional products of the human placental lactogen gene. *The Journal of biological chemistry*, (Vol. *257*).

Barrera-Saldana, H., Seeburg, P. H. & Saunders, G. F. (1983). Two structurally different genes produce the same secreted human placental lactogen hormone. *Journal of Biological Chemistry*, *258*(6), 3787–3793.

Baumann, G. P. (2009). Growth hormone isoforms. *Growth Hormone and IGF Research*, *19*(4), 333–340. https://doi.org/10.1016/j.ghir.2009.04.011.

Böhne, A., Brunet, F., Galiana, D., Schultheis, C. & Volff, J. N. (2008). Transposable elements as drivers of genomic and biological diversity in vertebrates. *Chromosome Research*, *16*(1), 203–215. https://doi.org/10.1007/s10577-007-1202-6.

Burbano, H. A., Hodges, E., Green, R. E., Briggs, A. W., Krause, J., Meyer, M. & Pääbo, S. (2010). Targeted investigation of the Neandertal genome by array-based sequence capture. *Science (New York, N.Y.)*, *328*(5979), 723–725. https://doi.org/10.1126/science.1188046.

Cajiao, I., Zhang, A., Yoo, E. J., Cooke, N. E. & Liebhaber, S. A. (2004). Bystander gene activation by a *locus* control region. *The EMBO Journal*, *23*(19), 3854–3863. https://doi.org/10.1038/sj.emboj.7600365.

Canizales, M. E. (1996). Comparative study of gene promoters of the hGH-hPL complex that is expressed in placenta. *Master´s thesis*, Autonomous University of Nuevo León.

Chellakooty, M., Vangsgaard, K., Larsen, T., Scheike, T., Falck, J., Legarth, J. & Juul, A. (2004). A Longitudinal Study of Intrauterine Growth and the Placental Growth Hormone (GH)-Insulin-Like Growth Factor I Axis in Maternal Circulation: Association between Placental GH and Fetal Growth. *The Journal of clinical endocrinology and metabolism*, (Vol. *89*). https://doi.org/10.1210/jc.2003-030282.

Chen, E., Liao, C. Y., Smith, D., Barrera-Saldaña, H., Gelinas, R. & Seeburg, P. H. (1989). The human growth hormone *locus*: Nucleotide sequence, biology, and evolution. *Genomics*, (Vol. *4*). https://doi.org/10.1016/0888-7543(89)90271-1.

Cooke, N. E., Ray, J., Watson, M. A., Estes, P. A., Kuo, B. A. & Liebhaber, S. A. (1988). Human growth hormone gene and the highly homologous growth hormone variant gene display different splicing patterns. *The Journal of Clinical Investigation*, *82*(1), 270–275. https://doi.org/10.1172/JCI113582.

Cooke, N., Ray, J., Emery, J. & Liebhaber, S. (1988). Two distinct species of human growth hormone-variant mRNA in the human placenta predict the expression of novel growth hormone proteins. *The Journal of biological chemistry*, (Vol. 263).

Dodd, J. G., Norquay, L. D., Sheppard, P., Gregoire, S., Yang, X., Cattini, P. A. & Reith, W. (2003). RFX1 and NF-1 Associate with P Sequences of the Human Growth Hormone *Locus* in Pituitary Chromatin. *Molecular Endocrinology*, *17*(6), 1027–1038. https://doi.org/10.1210/me.2003-0025.

Elefant, F., Su, Y., Liebhaber, S. & Cooke, N. (2001). Patterns of histone acetylation suggest dual pathways for gene activation by a bifunctional *locus* control region. *The EMBO journal*, (Vol. *19*). https://doi.org/10.1093/emboj/19.24.6814.

Felber, J., Zaragoza, N., Benuzzi-Badoni, M. & Genazzani, A. (1972). The Double Effect of Human Chorionic Somatomammotropin (HCS) and Pregnancy on Lipogenesis and on Lipolysis in the Isolated Rat Epididymal Fat Pad and Fat Pad Cells. *Hormone and metabolic research = Hormon- und Stoffwechselforschung = Hormones et métabolisme*, (Vol. *4*). https://doi.org/10.1055/s-0028-1094070.

Frankenne, F., Pirens, G., Gomez, F. & Hennen, G. (1987). Discovery of a placental variant of human growth hormone: biochemistry, physiology and implication in the secretion of hypophyseal forms. *Reproduction, nutrition, development*, (Vol. *27*).

Frankenne, F., Rentier-Delrue, F., Scippo, M. L., Martial, J. & Hennen, G. (1987). Expression of the growth hormone variant gene in the placenta. *The Journal of clinical endocrinology and metabolism*, (Vol. *64*). https://doi.org/10.1210/jcem-64-3-635.

George, D. L., Phillips, J. A., Francke, U. & Seeburg, P. H. (1981). The genes for growth hormone and chorionic somatomammotropin are on the long arm of human chromosome 17 in region q21 to qter. *Human Genetics*, *57*(2), 138–141. https://doi.org/10.1007/BF00282009.

Gómez-Sáez, J. M. & Villabona, C. (2014). *Chapter 16 - Neurohypophysis diseases*. In P. Farreras Valentí & C. B. T.-F.-R. M. I. M. y N. E. (Décimo sétimo E. Rozman (Eds.) (pp. 178–180). Madrid: Elsevier. https://doi.org/https://doi.org/10.1016/B978-84-9022-595-0.00016-8.

Handwerger, S. & Freemark, M. (2000). The roles of placental growth hormone and placental lactogen in the regulation of human fetal growth and development. *Journal of Pediatric Endocrinology and Metabolism*. https://doi.org/10.1515/JPEM.2000.13.4.343.

Harper, M., Barrera-Saldaña, H. & Saunders, G. F. (1982). Chromosomal localization of the human placental lactogen-growth hormone gene cluster to 17q22-24. *American journal of human genetics*, (Vol. *34*).

Hartog, M. (2008). Human chorionic somatomammotropin and its clinical significance. *Clinical Endocrinology*, (Vol. *1*). https://doi.org/10.1111/j.1365-2265.1972.tb00392.x.

Hennen, G., Caufriez, A., Cantraine, F., Englert, Y., Frankenne, F., Golstein, J. & Copinschi, G. (1990). Placental growth hormone as a potential regulator of maternal IGF-I during human pregnancy. *American Journal of Physiology-Endocrinology and Metabolism*. Department of Obstetrics and Gynaecology, Free University of Brussels, Belgium. FAU - Frankenne, F. https://doi.org/10.1152/ajpendo.1990.258.6.e1014.

Hunsaker, T. L., Jefferson, H. S., Morrison, J. K., Franklin, A. J. & Shewchuk, B. M. (2012). POU1F1-mediated activation of *hGH-N* by deoxyribonuclease I hypersensitive site II of the human growth hormone *locus* control region. *Journal of Molecular Biology*, *415*(1), 29–45. https://doi.org/10.1016/j.jmb.2011.11.001.

Ingraham, H. A., Chen, R., Mangalam, H. J., Elsholtz, H. P., Flynn, S. E., Lin, C. R. & Rosenfeld, M. G. (1988). A tissue-specific transcription factor containing a homeodomain specifies a pituitary phenotype. *Cell*, 55(3), 519–529. https://doi.org/https://doi.org/10.1016/0092-8674(88)90038-4.

Jacquemin, P., Alsat, E., Oury, C., Belayew, A., Muller, M., Evain-Brion, D. & Martial, J. (1996). The Enhancers of the Human Placental Lactogen B, A, and L Genes: Progressive Activation During *In Vitro* Trophoblast Differentiation and Importance of the DF-3 Element in Determining Their Respective Activities. *DNA and cell biology*, (Vol. *15*). https://doi.org/10.1089/dna.1996.15.845.

Jacquemin, P., Oury, C., Peers, B., Morin, A., Belayew, A. & Martial, J. (1994). Characterization of a single strong tissue-specific enhancer downstream from the three human genes encoding placental lactogen. *Molecular and cellular biology*, (Vol. *14*). https://doi.org/10.1128/MCB.14.1.93.

Jiang, S. W., Shepard, A. & Eberhardt, N. (1995). An Initiator Element Is Required for Maximal Human Chorionic Somatomammotropin Gene Promoter and Enhancer Function. *The Journal of biological chemistry*, (Vol. *270*). https://doi.org/10.1074/jbc.270.8.3683.

Jones, B., Monks, B., Liebhaber, S. & Cooke, N. (1995). The human growth hormone gene is regulated by a multicomponent *locus* control region. *Molecular and cellular biology*, (Vol. *15*). https://doi.org/10.1128/MCB.15.12.7010.

Josimovich, J. B. & Atwood, B. L. (1964). Human placental lactogen (HPL), a trophoblastic hormone synergizing with chorionic gonadotropin and potentiating the anabolic effects of pituitary growth hormone. *American Journal of Obstetrics and Gynecology*, *88*(7), 867–879. https://doi.org/https://doi.org/10.1016/0002-9378(64)90734-3.

Karabulut, A., Layfield, R. & Pratten, M. (2001). Growth promoting effects of human placental lactogen during early organogenesis: A link to insulin-like growth factors. *Journal of anatomy*, (Vol. *198*). https://doi.org/10.1046/j.1469-7580.2001.19860651.x.

Karin, M., Theill, L., Castrillo, J., Mccormick, A. & Brady, H. (1991). Cell Type Specific Expression of the Growth Hormone Gene and its Control by GHF-1. *Nihon Naibunpi Gakkai zasshi*, (Vol. *66*). https://doi.org/10.1507/endocrine1927.66.12_1205.

Kim, Y. J. & Felig, P. (1971). Plasma chorionic somatomammotropin levels during starvation in mid-pregnancy. *Journal of Clinical Endocrinology and Metabolism*. https://doi.org/10.1210/jcem-32-6-864.

Kimura, A., Liebhaber, S. & Cooke, N. (2004). Epigenetic Modifications at the Human Growth Hormone *Locus* Predict Distinct Roles for Histone Acetylation and Methylation in Placental Gene Activation.

Molecular endocrinology (Baltimore, Md.), (Vol. *18*). https://doi.org/10.1210/me.2003-0468.

Kimura, A. P., Sizova, D., Handwerger, S., Cooke, N. E. & Liebhaber, S. A. (2007). Epigenetic Activation of the Human Growth Hormone Gene Cluster during Placental Cytotrophoblast Differentiation. *Molecular and Cellular Biology*, *27*(18), 6555–6568. https://doi.org/10.1128/mcb.00273-07.

Kurihara, M. & Kimura, A. P. (2015). Characterization of the human *TCAM1P* pseudogene and its activation by a potential dual promoter-enhancer: Comparison with a protein-coding mouse orthologue. *FEBS Letters*, *589*(4), 540–547. https://doi.org/10.1016/j.febslet.2015.01.023.

Macleod, J., Worsley, I., Ray, J., Friesen, H., Liebhaber, S. & Cooke, N. (1991). Human Growth Hormone-Variant Is a Biologically Active Somatogen and Lactogen*. *Endocrinology*, (Vol. *128*). https://doi.org/10.1210/endo-128-3-1298.

Meis, J., M. Conwell, C., Durning, M., Fisher, J. & Golos, T. (1997). Pit-1/Growth Hormone Factor 1 Splice Variant Expression in the Rhesus Monkey Pituitary Gland and the Rhesus and Human Placenta 1. *The Journal of clinical endocrinology and metabolism*, (Vol. *82*). https://doi.org/10.1210/jcem.82.3.3791.

Nachtigal, M., Nickel, B. & Cattini, P. (1993). Pituitary-specific repression of placental members of the human growth hormone gene family: A possible mechanism for *locus* regulation. *The Journal of biological chemistry*, (Vol. 268).

Nickel, B. E., Kardami, E. & Cattini, P. A. (1990). Differential expression of human placental growth-hormone variant and chorionic somatomammotropin in culture. *The Biochemical Journal*, *267*(3), 653–658. Retrieved from https://www.ncbi.nlm.nih.gov/pubmed/2339980.

Norquay, D., Jin, Y., Mohan-Surabhi, R., Gietz, R., Tanese, N. & Cattini, P. (2001). A member of the nuclear factor-1 family is involved in the pituitary repression of the human placental growth hormone genes. *The Biochemical journal*, (Vol. *354*). https://doi.org/10.1042/0264-6021:3540387.

Ocampo, D. & Larhammar, D. (2018). Evolution of the growth hormone, prolactin, prolactin 2 and somatolactin family. *General and Comparative Endocrinology*, *264*, 94–112. https://doi.org/10.1016/j.ygcen.2018.01.007.

Pérez-Maya, A., Rodríguez-Sánchez, I. P., De Jong, P., Wallis, M. & Barrera-Saldaña, H. A. (2012). The chimpanzee GH *locus*: Composition, organization, and evolution. *Mammalian Genome*, *23*(5–6), 387–398. https://doi.org/10.1007/s00335-012-9392-4.

Pérez-Maya, A., Wallis, M. & Barrera-Saldaña, H. (2016). Structure and evolution of the gorilla and orangutan growth hormone *loci*. *Mammalian Genome*, *27*(9–10), 511–523. https://doi.org/10.1007/s00335-016-9654-7.

Ranke, M. B. & Wit, J. M. (2018). Growth hormone-past, present and future. *Nature Reviews Endocrinology*, *14*(5), 285–300.

Reséndez-Pérez, D., Ramírez-Solís, R., Varela-Echavarría, A., Martínez-Rodríguez, H. G. & Barrera-Saldaña, H. (1990). Coding potential of transfected human placental lactogen genes. *Nucleic Acids Research*, *18*(16), 4665–4670.

Rogers, B., Sobnosky, M. & F. Saunders, G. (1986). Transcriptional enhancer within the human placental lactogen and growth hormone multigene cluster. *Nucleic acids research*, (Vol. *14*).

Rygaard, K., Revol, A., Esquivel-Escobedo, D., Beck, B. L. & Barrera-Saldaña, H. A. (1998). Absence of human placental lactogen and placental growth hormone (HGH-V) during pregnancy: PCR analysis of the deletion. *Human Genetics*. Department of Obstetrics and Gynaecology, Frederiksberg Hospital, Denmark. FAU - Revol, A. https://doi.org/10.1007/s004390050658.

Shaw, C. J. & Lupski, J. R. (2005). Non-recurrent 17p11.2 deletions are generated by homologous and non-homologous mechanisms. *Human Genetics*, *116*(1), 1–7. https://doi.org/10.1007/s00439-004-1204-9.

Shewchuk, B., Liebhaber, S. & Cooke, N. (2002). Specification of unique Pit-1 activity in the hGH *locus* control region. *Proceedings of the National Academy of Sciences of the United States of America*, (Vol. *99*). https://doi.org/10.1073/pnas.182418199.

Shewchuk, B. M., Ho, Y., Liebhaber, S. A. & Cooke, N. E. (2006). A Single Base Difference between Pit-1 Binding Sites at the *hGH* Promoter and *Locus* Control Region Specifies Distinct Pit-1 Conformations and Functions. *Molecular and Cellular Biology*. University of Pennsylvania School of Medicine, 415 Curie Blvd., Philadelphia, PA 19103, USA. FAU - Ho, Yugong. https://doi.org/10.1128/mcb.00267-06.

Sonntag, W. E., Ramsey, M. & Carter, C. S. (2005). Growth hormone and insulin-like growth factor-1 (IGF-1) and their influence on cognitive aging. *Ageing Research Reviews*, *4*(2), 195–212. https://doi.org/https://doi.org/10.1016/j.arr.2005.02.001.

Timasheva, Y., Putku, M., Kivi, R., Kožich, V., Männik, J. & Laan, M. (2013). Developmental programming of growth: Genetic variant in *GH2* gene encoding placental growth hormone contributes to adult height determination. *Placenta*, *34*(11), 995–1001.

Trujillo, M. A., Sakagashira, M. & Eberhardt, N. L. (2006). The Human Growth Hormone Gene Contains a Silencer Embedded within an *Alu* Repeat in the 3′-Flanking Region. *Molecular Endocrinology*, *20*(10), 2559–2575. https://doi.org/10.1210/me.2006-0147.

Velegrakis, A., Sfakiotaki, M. & Sifakis, S. (2017). Human placental growth hormone in normal and abnormal fetal growth. *Biomedical Reports*, *7*(2), 115–122. https://doi.org/10.3892/br.2017.930.

Verhaeghe, J., Pintiaux, A., Van Herck, E., Hennen, G., Foidart, J. M. & Igout, A. (2002). Placental GH, IGF-I, IGF-Binding Protein-1, and Leptin during a Glucose Challenge Test in Pregnant Women: Relation with Maternal Body Weight, Glucose Tolerance, and Birth Weight. *The Journal of clinical endocrinology and metabolism*, (Vol. *87*). https://doi.org/10.1210/jcem.87.6.8569.

Walker, W., Fitzpatrick, S., Barrera-Saldaña, H., Reséndez-Perez, D. & Saunders, G. F. (1991). The Human Placental Lactogen Genes: Structure, Function, Evolution and Transcriptional Regulation. *Endocrine reviews*, (Vol. *12*). https://doi.org/10.1210/edrv-12-4-316.

Wallis, M. (2014). Molecular evolution of growth hormone. *The Biochemist*, 36 (February), 4–8. Retrieved from http://www.biochemist.org/bio/default.htm?VOL=36&ISSUE=1.

Williams, C. & Coltart, T. M. (1978). Adipose tissue metabolism in pregnancy: The lipolytic effect of human placental lactogen. *British journal of obstetrics and gynaecology*, (Vol. *85*). https://doi.org/10.1111/j.1471-0528.1978.tb15824.x.

Yoo, E. J., Cajiao, I., Kim, J.-S., Kimura, A. P., Zhang, A., Cooke, N. E. & Liebhaber, S. A. (2006). Tissue-specific chromatin modifications at a multigene *locus* generate asymmetric transcriptional interactions. *Molecular and Cellular Biology*, *26*(15), 5569–5579. https://doi.org/10.1128/MCB.00405-06.

In: A Closer Look at Hormones
Editor: Alexander V. Knyazev
ISBN: 978-1-53617-254-6

Chapter 7

PHYTOHORMONES AND ANIMAL STEROIDS IN MICROALGAE BIOMASS PRODUCTION: INDUSTRIAL APPLICATION AND WASTEWATERS EDCS REMEDIATION

Tatiana Kozlova[1,2], PhD and David Levin[2,*], PhD
[1]Instituto Timiryazev Institute of Plant Physiology RAS,
Moscow, Russia
[2]Department of Biosystems Engineering,
University of Manitoba, MB, Canada

ABSTRACT

One of the promising ways to improve the feasibility of third generation of biofuel and bioproduct economy is the integration of biomass production on the basis of maximum energy and nutrient recycling. Integration of the biomass production also offers significant

[*] Corresponding Author's Email: david.levin@umanitoba.ca.

reduction of environmental emissions due to the energy- and nutrient-recycling strategies. Interest in the use of wastewaters of different origins for the industrial production of microalgae biomass has stimulated the studies on the influence of a variety of compounds in wastewaters on microalgae physiology. Particularly, phyto- and animal hormones, which could be present in both surface and wastewaters, have been investigated on their impact on physiology of different microalgae species. The results of these studies have demonstrated clearly stimulation of certain physiological parameters of most of the microalgae species with the effect of 3 to 8-fold of increase in growth and biosynthesized molecules accumulation. Importantly, the research on the influence hormones on microalgae demonstrated simultaneously stimulation of biomass production and biomolecule yields, which is important in a view of industrial biomass production. The gained to date results highlight the possibility that hormones could be used as a tool algal biosynthesis and, particularly, for fatty acids profile manipulation in the biofuel and pharmaceutical industries. However, more intensive research has to be conducted to fill the gaps in our knowledge on the intracellular mechanisms of hormones action, the biotransformation of the hormones in algae cells and the possibility of their bioconcentration. The wide range of microalgae species has to be examined to clarify all the open questions.

Keywords: microalgae, phyto-hormones, animal steroids, integrated biomass production, wastewater treatment

1. Introduction

The production of biofuels from microalgae, with the co-production of high commercial value biochemical commodities, may be considered "third generation biofuels." One of the promising ways to improve both the feasibility of third generation of biofuels and the bioproduct economy is the integration of biomass production on the basis of maximum energy and nutrient recycling, which can reduce the costs of production of both the biofuels and the bioproducts. For example, this may be accomplished via application of industrial flue-gas carbon dioxide and nutrient-rich wastewaters to support microalgal growth. Therefore, integration of the microalgal biomass production also offers significant reduction of environmental emissions due to the energy- and nutrient-recycling strategies (Benemann, 2013).

Growing interest in the use of wastewaters of different origins for the industrial production of microalgae biomass for biofuels and bioproducts has stimulated studies on the influence of various compounds in these wastewaters on microalgae physiology. Particularly, phyto- and animal hormones, which could be present in both surface and wastewaters, have been investigated on their impact on physiology of different microalgae species (Bajguz and Czerpak, 1996a,b, 1998; Bajguz, 2002; Jobling and Owen, 2003; Park et al., 2013; Nobel et al., 2014; Kozlova et al., 2017, 2018, 2019). The results of these studies have demonstrated clearly stimulation of certain physiological parameters of most of the microalgae species investigated. The stimulatory effects for certain parameters (e.g., cell growth and synthesis/accumulation of metabolite molecules) have been observed to be as high as 3- to 5-fold (Czerpak and Bajguz, 1997; Bajguz and Piotrowska-Niczyporruk, 2014; Kozlova et al., 2017) in most cases, and as high as 7- to 8-fold for the accumulation of some fatty acids (Czerpak and Bajguz, 1997; Ozioko et al., 2015; Kozlova et al., 2019). However, the ranges of a hormone tested, growth conditions, the degree of influence of the hormones, and the most effective concentrations, vary greatly among the investigations. Moreover, some authors have highlighted species-specific sensitivities of certain microalgae to some hormones (Saygideger and Okkay, 2008; Iwamoto and Siraiwa 2012; Liu et al., 2016; Bi et al., 2018).

To date, limited research has been reported on the influence of mixed hormones on algal physiology. Nevertheless, synergistic increases in biomass production, pigments, and/or lipid accumulation have been observed in most combined-hormones experiments reported (Hunt et al., 2010; Ozioko et al., 2015; Kozlova et al., 2018). Although, under certain conditions (e.g., type of hormone, elevated hormone concentrations, duration of experiments), no additive, or adverse (antagonistic) effects of hormones were observed (Ozioko et al., 2015; Kozlova et al., 2017). This emphasizes the importance of scrupulous investigations of the chemistry of the wastewaters to be applied for microalgae biomass production, as well as the need for further research on influence of hormones, singly or in combination, found in the wastewaters on a broad-range of industrial microalgae species.

Animal steroids such as fish hormones, which are common in surface waters, as well as in some wastewaters, have been shown to have similar or even greater ability to stimulate microalgae biomass production and biosynthesis of some molecules of commercial interest. However, animal hormones fall under the list of Endocrine Disruption Compounds (EDCs), thus industrial applications of these types of hormones for biomass production are restricted. It must be demonstrated that the concentrations of these hormones cannot accumulate and/or undergo biomagnification by algae cells, or biologically transferred along the food chain. It is especially important when microalgae are grown for pharmaceutical, cosmetic, or agricultural industries applications.

On the other hand, microalgae are well known biological agents for removal EDCs from wastewaters, especially from the effluents of domestic origin (Wang et al., 2017; Liu et al., 2018), which contain considerable amounts of synthetic steroids as the products of pharmaceutical industry (Johnson et al., 2000; Shore and Shemesh, 2003). There are several studies elucidating the mechanisms and kinetics of steroid biodegradation or biotransformation by microalgal cells. The research has demonstrated rapid (within hours) uptake the dissolved hormones by algal cells, while amount of adsorbed hormone is often lesser than its intracellular concentration suggesting high rate of absorption (Koh et al., 2009; Liu et al., 2018). Thus, integration of EDCs water treatment with no-food, feed or pharmaceutical-related microalgae biomass production is a promising way to establish a low-cost feedstock biomass production on a base of environmental remediation of effluents.

Despite the interesting, and very promising, scientific data collected to date on influence of exogenous hormones on microalgal physiology, there are many gaps in our knowledge of the subject which prevents a broad application of hormones for the industrial-scale microalgae production. The most important gaps include a lack of understanding of the mechanisms and modes of action of particular exogenous hormones on an algal species, the influence of the aqueous environment on hormone lifespan, the kinetics of adsorption/absorption processes, the influence of hormone mixtures on algal species, and the interactions between exogenous and self-synthesized endogenous hormones.

Two very important assessments of the state of current knowledge on the influence of hormones on microalgae physiology need prominent clarification. It is necessary to clarify whether or not nano-concentrations of dissolved hormones cab stimulate biomass production and/or accumulation biomolecules of interest. Importantly, it is necessary to collect more data to clarify if the application of exogenous hormones stimulates microalgal biomass production and, at the same time, stimulates the yield of commercially valuable biomolecules. This is important because hormone-base simultaneously stimulation of biomass production and biomolecule yields is a major assumption that is used to justify integrated algal biomass production at the pilot and industrial scale. The conformation of these two objectives with a broad-range of industrial microalgae species can greatly contribute to the commercialization of microalgae biomass production. This chapter is focused on elucidation of several aspects of the application of hormones for microalgae biomass production for biofuels, pharmaceuticals, cosmetic and agricultural industries.

1.1. Microalgae Biomass Production for Industrial Application

The production of biofuels and high-value bio-commodities from microalgae is a growing sector of the global bioeconomy. Featuring high growth rates, an astonishing variety of synthesized biomolecules that have great economic value, and been among the most diverse organisms on our planet, microalgae are considered as the main feedstock for the "third generation biofuels." The biofuel industry, as a part of the overall bioeconomy, has been recognized since the 1970s when International Energy Agency (IEA) of the Organization for Economic Co-operation and Development (OECD) established the IEA-Bioenergy Organization.

Biofuels (biodiesel = fatty acid methyl esters) derived from microalgal triacylglycerides (TAGs) are similar, but not identical to petroleum-derived fuels molecules. The main differences are in length of carbon chains, the ratio of saturated to unsaturated carbon-carbon bonds, and the presence of oxygen atoms on the antepenultimate carbon residue. The higher proportion of saturated carbon bonds enables the fatty acid methyl esters (FAMEs) to align with one another, resulting precipitation of the compounds as the temperature drops past a critical threshold (the Cloud-Point). Thus, biodiesel derived from microalgal is mostly unsuitable for countries in the northern hemisphere that have cold temperatures, or at high altitudes.

Several technologies were developed to overcome these major challenges including subjecting the C16/C18 fatty acids to hydrothermal cracking or conversion of bioethanol to ethyl tertiary butyl ether (ETBE) (Malca and Freire, 2006; Petrus and Noordermeer, 2006). However, these techniques significantly elevate the cost of production and, consequently, the market-price of the biofuel products. Developing of new technologies, which enable more suitable biofuels from microalgal TAGs are in high demand. Among the alternative approaches is the use of genetic engineering of microalgae cells (Machida and Suzuki, 2018).

However, an entirely novel way in alteration of microalgae biosynthesis and, particularly, modification of fatty acids profiles, employs phyto- and animal hormones, which have been shown to both stimulate biosynthesis and accumulation of total neutral lipids and modify the fatty acid profiles of the accumulated TAGs in favor of FAs with short (C4 to C6) and moderate (C8 to C14) carbon chains, which are suitable for "green gasoline" and biojet fuels respectively (Salama et al., 2014; Kozlova et al., 2017, 2018, 2019).

Indeed, it is not only the biofuel industry that has great interest in industrial microalgae biomass production. The list of the commercially used biomolecules synthesized by microalgae includes antioxidants (such as β-carotene and astaxanthin), vitamins, proteins, and high-value oils (such as omega-3 and omega-6 fatty acids) that can be used for the production of cosmetics, pharmaceuticals and agricultural feeds (Olaizola,

2000; Adarme-Vega, 2012). The intense efforts of the recent decades to produce microalgal biomass at the industrial-scale has demonstrated that production biofuels can only become economically viable if high-value co-products (biomolecules) can be produced from the same biomass grown for biofuel production (Briens and Piskorz, 2008; Benemann, 2010). For example, lutein demand in United States is estimated to be nearly $150,000,000 per year (Fernández-Sevilla et al., 2010) with the value growing in recent years. The market price for, astaxanthin, synthesized by *Haematococcus pluvialis*, used as an animal fed supplement is approximately $3,000 USD per kg (Li et al., 2011), and up to $100,000 USD/kg for use in pharmaceuticals (Perez-Lopez et al., 2014). Commercially produced enzymes such as Ribulose-1,5-bisphosphate carboxylase/oxygenase (RuBisCO) and radial spoke proteins harvested from flagellated algae *Chlamidomonas* sp. and *Haematococcus* sp., play an important role in human health (Taylor et al., 2001; Colla et al., 2007) and health research (Yang et al., 2006) and also have high market price. Moreover, microalgae industrial biomass production has a great potential in CO_2 emissions and wastewaters effluents remediation since both wastes could be employed for biomass production as the sources of nutrients.

1.2. Main Challenges and Induction Strategies

Integration of biomass production based on maximum recycling of water and industrial wastes, such as flue-gas CO_2, wastewaters, and heat, greatly increases the economic feasibility by reducing the investment for both a biomass production and a plant operation. IN addition, it contributes for environmental protection and promotes sustainable industrial development. However, despite the intensive research efforts and the remarkable amount of academic, government, and industrial institutions involved in microalgal-based biofuel/bioproduct research in recent decades, industrial-scale production of microalgae still faces a number of serious challenges.

1.2.1. Challenges of Microalgae Biomass Production

The challenges that limit industrial-scale microalgal production includes: a) selection of species with elevated growth rates and which accumulate biomolecules of interest in high concentrations; b) finding effective methods of biosynthesis manipulation (induction strategies, genetic engineering); c) development of more effective methods of disease treatment; d) protection of contamination of both growing biomass and end-products; and e) improvement of engineering facilities. All the improvements required to achieve industrial-scale production must be accomplished by cost effective means.

A crucial problem, which does not rely on most of the other issues of production listed above, is the naturally incompatible enhancement of biomass productivity and yield of the biomolecules of interest (Benemann, 2013). To date, this long-standing dogma has not been achieved at industrial-scale. At the laboratory scale, relatively recent research has shown that under heterotrophic growth conditions with the excess organic carbon (acetate) and without limitation by other nutrients, microalgae cells can grow to higher cell densities than with inorganic carbon (CO_2) alone, and accumulate high concentrations of triacylglycerides (Wang et al., 2011; Fan et al., 2012). However, heterotrophic growth with organic sources of carbon, as has some disadvantages, such as the high cost of the organic carbon source and the elevated chance of the growth medium contamination by bacteria.

Another strategy to overcome this significant problem is hormone application. The most important achievement of hormone application for microalgae biomass production is that, similar to the effect of dissolved organic carbon from acetate, phyto- and animal hormones clearly promote both microalgal cell growth and accumulation of secondary metabolites (Bajgus and Czerpak, 1996 a, b; Salama et al., 2014; Kozlova et al., 2017, 2018, 2019; Babu et al., 2017; Talarek-Karwel et al., 2018). Importantly, there is a need to investigate the interactions between applied hormone and other induction methods commonly used for industrial-scale microalgal production, namely nutrient depletion and flue-gas CO2 application.

1.2.2. Induction Strategies

To date, commercial production of microalgal biomolecules is based on induction of biosynthesis/accumulation under extreme growth conditions. It is well known that lipids (TAGs) and carotenoids accumulate when microalgal cells are exposed to various adverse conditions (stressors), or combinations of these stress conditions (Jin et al., 2003; Lemoine and Schefs, 2010). The most studied and therefore extensively employed stressors can be classified into two groups: extreme growth conditions and additional physico-chemical stressors. The first group is represented by modification of light intensity and quality, temperature, pH, and nutrients depletion (Benavente-Valdes et al., 2016; Chernova and Kiseleva, 2017). The second group includes, but is not restricted to salt-stress, trace metals, herbicides, additional organic compounds (e.g., organic carbon), ultrasound, and artificial UV-irradiation (Alfalo et al., 1999; Boussiba, 2000). An outstanding tool of biosynthesis alteration is transgenics.

Among organic stressors, the influence of phyto- and animal hormones on microalgal physiology has become a topic of great interest in recent decades, for several reasons. First, it has been established that microalgae can synthesize all known phytohormones, and some animal steroid hormones, and that the role of these hormones is similar to their role in vascular plants (Stirk and Van Staden, 1996; Kiseleva et al., 2012; Romanenko et al., 2015). Second, externally applied phyto- and animal hormones have been shown to have great positive effect on growth, biomass yield, and biosynthesis/accumulation of metabolites in vascular plants. The positive effects of hormones has been established on crop species (Mandava, 1988; Ludwig-Muller, 2011; Bishop and Yokota, 2001; Xu et al., 2013). Third, there is increasing evidence that externally applied phyto- and animal hormones can have positive impacts on microalgae cell growth rate, cell density, cell size, and biosynthesis/accumulation of a variety of biomolecules in a dose-dependent manner (Bajguz and Czerpak, 1996a, b; Salama et al., 2014; Kozlova et al., 2017, 2018, 2019; Liu et al., 2016; Talarek-Karwel et al., 2018). Finally, numerous microalgal species have been shown to possess a great ability for adsorption, followed by absorption, of phyto- and animal hormones, with further biotransformation of the absorbed molecules (Rober-Kleber at al., 2003; Perrot-Rechenmann, 2010; Rock and Sun, 2005; Robert-Seilaniantz et al., 2011; Tatsuki et al., 2013; Ludwig-Muller, 2011; Baronti et al., 2000; Duong et al., 2011). Although some authors have reported positive effects of high hormone concentrations (µg/L to mg/L range) on microalgal biomass production and synthesis/accumulaton of biomolecules (Bajguz and Czerpak, 1996a,b; Julius and Theriot, 2011), other researcher have highlighted the adverse effect of high hormone concentrations on microalgal growth and biosynthesis (Bajguz and Czerpak, 1996a,b; Liu et al., 2016). These discrepancies are likely the result of different growth conditions and, most importantly, different growth medium composition. It is well known that the bioavailability of compounds, including hormones, dissolved in water depends on water chemistry, which affects competition among ions for uptake sites on algal cell membranes, and/or the binding capacity of dissolved molecules (Pollard and Rojas, 1988). Nevertheless, the use of low hormone concentration (in ng/L range) have been shown to have positive effects on microalgal growth and biosynthesis/accumulation. In addition, low hormone concentrations reduces the risk of intracellular bioaccumulation and/or biomagnification of the hormones, greatly reducing the risk of transfer of hormones through the food chain.

Thus, use of wastewaters with phyto- and/or animal hormones in nano-concentrations as growth media offers an inexpensive and safer way of using hormones to stimulate microalgal biomass production at industrial-scale. Moreover, the use of wastewaters as a source of nutrients for microalgae biomass production works toward the establishment of Industrial Integration and, concomitantly, a sustainable bioeconomy on a base of maximum energy and nutrient recycling.

1.3. Industrial Integration

A novel and powerful way to bring microalgal biofuels and bioproducts to industrial-scale production is industrial integration of the biomass production. Industrial integration has proved to be an asset in a new world development (EC, SEC 1272, 1276, 2010; Navarri, 2015). In the field of microalgal biomass production, progress in industrial integration has been made in the last two decades (McKendry, 2002; EC, 1997, 2004; DOE, 2010). Three approaches to industrial integration of microalgae have been developed: 1) Integrated production of biofuels and valuable co-products from one type of biomass; 2) Integrated production of biomass of different types for different applications; and 3) Integrated production of biomass with wastewater, or other industrial wastes treatment streams for the purpose of maximum waste recycling and end-product cross-utilisation. In this case, integration of microalgal biomass production offers significant reductions of environmental emissions due to engaging industrial flue-gas CO_2 and wastewaters.

The most developed example of an integrated system involving wastewater as a growth medium is aquaponics. In aquaponic systems, the wastewater from one type of biomass production (fish tanks) is used as a fertilizer for another type of biomass production (e.g., vegetables) while, in turn, the residues of the second biomass (e.g., microalgae) is used as a feed supplement for fish or dry fertilizer for external application (agricultural crop) (Bansal et al., 2005; Dive and Rinehart, 2010). The system was created and implemented to grow vegetables. However, it could be adapted for microalgae production. Aquaponic systems have two additional benefits for microalgae production: fish consume oxygen produced as a bioproduct from algal cell metabolism (photosynthesis), and the algal cells absorb fish hormones dissolved in the wastewater (Figure 1).

Another established system combines microalgal biomass production in photo-bioreactors (PhBRs) with industrial-scale hydroponic facilities, when reactors are accommodated within greenhouses, which allows recycling of hydroponic effluents and of sharing artificial lighting and heat (Roquette GmbH, Frankfurt, Germany). In this integrative system, microalgae utilize nutrients and phytohormones from hydroponic effluents.

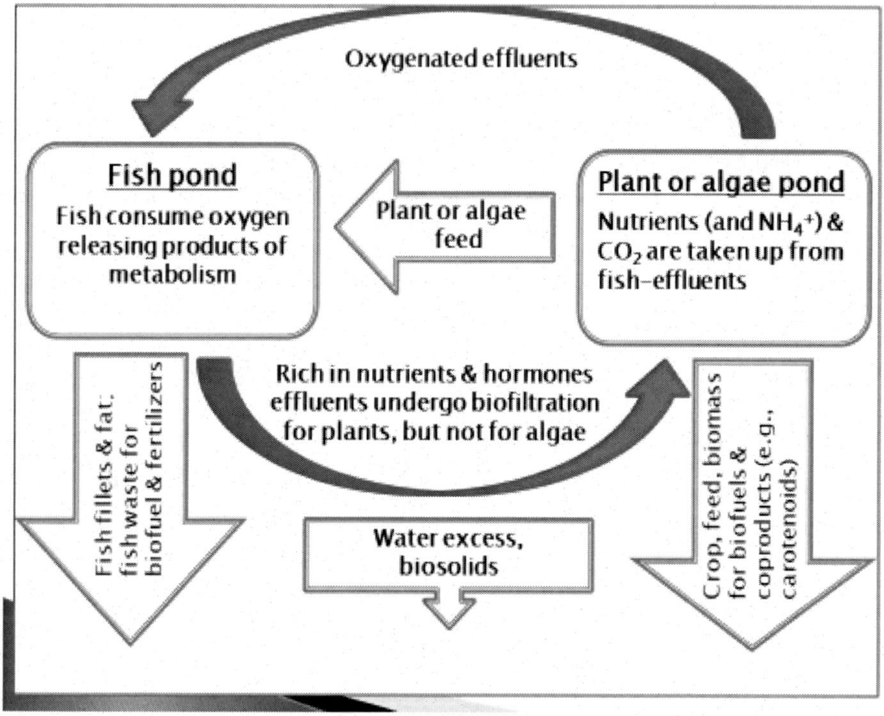

Figure 1. Profitability of an integrated system for plant/microalgae and herbivore/mesocarnivore fish biomass production.

1.3.1. Wastewaters as Growth Media

The application of wastewaters for microalgae biomass production has been proved greatly advantageous for microalgal biomass production (Kim et al., 2007; Kong et al., 2010; Dalrymple et al., 2013), as well as for the biosynthesis of neutral lipids (Kim et al., 2007; McGinn et al., 2012; Han et al., 2016) and carotenoids (Kang et al., 2006; Kim et al., 2007; Safafar et al., 2015). For example, swine hormones in agricultural wastewater have been shown to have a positive effect on algal growth and lipid yields (Zhang et al., 2013), resulting in one of the highest known algal biomass productivities (8 g/L dcw) (Kim et al., 2007). The effluent hormones in the swine wastewater are free-of-cost inducers, which reduce significantly the costs of production.

The challenges of wastewater applications include the elevated possibility of contamination either by bacterial and planktonic wastewater communities, or by toxic compounds (metals, pharmaceuticals, herbicides, detergents, etc.). Consequently, complexity of wastewater chemistry raises the need of qualitive and quantitively analysis for dissolved compounds with further investigation of the influence of the compounds, and their mixtures, on algal physiology in a species-specific manner. Moreover, to resolve the problems of contamination, additional wastewater treatments (filtration, sanitation) are necessary prior using the wastewater as an algae culture media. Additional treatments, however, increase the cost of wastewater application and may eliminate the positive effects of some organic compounds of the wastewaters, such as enzymes and hormones. Thus, the development of novel and effective methods for the improvement of wastewater application to grow microalgae biomass for biofuels and bioproducts will enable its use in industrial-scale production (McGinn et al., 2011).

Some types of wastewaters, such as from fish-farms and/or hydroponic facilities have some advantages compared to agricultural and municipal effluences. Fishery and hydroponic wastewaters normally have relatively low chemical oxygen demand (COD), biological oxygen demand (BOD), and turbidity/total suspended solids (TSS) (Rakocy et al., 2005). Effluents from cold-water fish-farming (e.g., salmonid) also have low HPC (heterotrophic plate counts) due to relatively low organics and low water temperature (4 to 15°C). This reduces the need for pretreatment of the wastewaters prior to pumping it to the algae production units. Importantly, fishery and hydroponic effluents normally contain hormones in nano-concentration compared to high concentrations (μM to hundreds of mM) of domestic effluents. Nano-concentrations, especially of sex hormones from fishery wastewater have been considered as a safer loading, which will be described further.

Indeed, there still is a rather scattered knowledge of the influence of organic compounds of wastewaters on microalgae (or plant) physiology. Aqueous environments facilitate the interactions of metabolites released by the growing organisms of an integrated recycling system. To date, the combined effect of dissolved organic compounds and their products of degradation on microalgae physiology needs extensive investigation.

1.3.2. Routes of Hormones Exposure

Since microalgae can synthesize phytohormones and some animal steroids (Tarakhovskaya et al., 2007; Kiseleva et al., 2012; Stirk et al., 2013), their metabolic pathways are accomplished with the same enzymes, which are found in similar pathways in plants and animals, including humans (Brzostek et al., 1998, Li et al., 1997, Rosati et al., 2003; Tsavkelova et al., 2006; Kiseleva et al., 2012). In natural ponds, phytohormones synthesized by algae and plants are secreted in trace concentrations into the water (Nobel et al., 2014; Hirsch et al., 1989). Aquatic faunae contribute to the loading of sex and stress steroids in aqueous environments. Agricultural wastewaters, such as those derived from hydroponic cultivation in greenhouses, and run-off water from crop fields treated with plant growth regulating substances or hormone-herbicides (Khripach et al., 2000; Cobb, 1992; Grossmann, 2010), contribute significantly high concentrations of phytohormones, reaching hundreds of μM, to these natural sources (Dalrymple et al., 2013; Lopez et al., 2005; Tate et al., 2013). It is worth mentioning that agricultural wastewaters contain both, naturally synthesized and synthetic hormones, which could have different impact on microalgal physiology (Yu et al., 2015; Hunt et al., 2010; Du et al., 2017). The main sources of synthetic hormones are growth-regulators, regulators of sexual activity, feed efficiency (animals) or nutrients intake (plants) promotors, and drugs applied for disease treatment.

Relatively low concentrations of phytohormones, in the nM to μM range, are found in hydroponic system effluents, despite the intensive culturing of plants coupled with the fact that recycled water in such a system tends to accumulate metabolites released by growing plants (Moya et al., 1995; Hartung, 2010; Dalrymple et al., 2013; Mota et al., 2014; Nobel et al., 2014). In fish farm effluents, the concentrations of hormones are more variable and greatly depend on type of farming: raceway tanks vs. hatchery tanks. Vermeirssen et al. (1996) suggested that free and conjugated forms of fish steroids enter aquatic environments mostly via a respiratory route (gills), from urine and bile. For instance, one of the most common steroid 17β-estradiol (E2) efflux into water is mostly (up to 90%) through gills (Hala et al., 2012), and E2 (as well as other sex hormone conjugates) are easily bio-transformed into their free forms (Baronti et al., 2000; Duong et al., 2011). About 40% of synthesized 17,20β-dihydroxy-4-pregnen-3-one (17,20β-P) is released directly into water through gills and with urine (Vermeirssen et al., 1996), which supports the main function of 17,20β-P as a pheromone of sexual engagement (Joannou and Reeder, 1995; Nagahama et al., 2008).

1.3.3. Hormone Concentrations in Effluents and Surface Waters

Concentrations of phytohormones in surface water normally are just above the detectable level not exceeding 5-10 ng/L. The actual loading of phytohormones in hydroponics wastewaters, such as brassinolides and natural auxins, has been demonstrated to be in the range of 200-300 nM or lower (Moya et al., 1995; Khripach et al., 2000; Savidov, 2004; Dalrymple et al., 2013; Nobel et al., 2014), although ABA was found in hydroponic wastewater at concentrations as high as 500 μM (Lopez et al., 2005). Agricultural run-off water from the fields treated with phytohormones and effluents from hormone production factories can contain phytohormones in mg/L. Concentrations of animal steroids in surface waters are also well studied and reported to be in the range of 1 to 200 ng/L E2 in North America and Europe (Pojana et al., 2007; Zhou et al., 2011; Zhang et al., 2011; Esteban et al., 2014; Liu et al., 2018). Concentrations of 17,20β-P in surface waters are approximately 5-10-fold higher (Hala et al., 2012; Mota et al., 2014) (Table 1).

The reproductive tanks of fish hatcheries farm contain high concentrations of steroid hormones, reaching hundreds of μg/L (Silva et al., 2012; Jobling and Owen, 2003). Remarkably, this type of wastewater normally considered as "clean" and could be discharged into natural ponds with no water-treatment applied. Different fish species can release up to 40-50 ng/g/h of 17,20β-P, as was shown in studies conducted on *Carassius auratus* (Stacey and Sorensen et al., 2005) and rainbow trout (Barry et al.,

1997), where hundreds of μg/L of this hormone were detected at maximum fish density in a spawning tank. The ratio of 17,20β-P and E2 in a reproduction tank varied from about 4:1 to 20:1, depending on fish species and their age (Hala et al., 2012).

Table 1. Concentration of hormones in natural and wastewaters

Site of sampling	Surface water	Underground water	Domestic/ livestock runoff effluents	Fishery wastewater	Hydroponics	Sediments
Phytohormones						
ABA	(21); (22); (49); (52)	n-f	(46); (49)	n-f	(49); (57)	n-f
IAA	(7); (11); (15);(19); (50)	(2); (24); (51); (65)	(11); (12); (19); (46); (61); (65)	(56)	(12); (19); (46); (49); (57)	(7); (11); (15); (19); (50)
BRs	(29)	n-f	(49)	n-f	(29); (49)	n-f
Animal steroids						
E1	(14); (28); (37); (55); (58); (62); (70)	(8); (36); (37); (38); (62);	(16); (38); (58)	(47); (48); (68)	(28); (47); (48);	(38); (58); (62)
E2	(10); (26); (27); (28); (36); (37); (38); (54); (62); (63); (69)	(1); (13); (27); (36); (37); (62)	(1); (13); (16); (38); (58)	(23); (27); (41); (47); (48); (59); (68)	(28); (47); (48)	(9); (13); (27); (38); (54); (58); (62)
EE2	(5); (14); (26); (30); (40); (54); (67)	(18); (26); (40); (37); (66)	(6); (40); (67)	(40)	n-f	(26); (39);(54); (62)
Progesterone	(20); (25); (33); (34); (38); (43); (48); (64); (69)	(4); (20); (38); (45); (64);	(3); (35); (38); (42); (53)	(3); (23); (26); (41); (47); (48); (59); (60)	(31); (47); (48);	(17); (38); (43); (44)

Note: n-f – data not found; see Table 1 References.

The amount of a hormone in fishery wastewater can be estimated knowing fish density per volume of water, the rate of hormone production, the rates of hormone release into the water, and the rate of reverse uptake by fish, although the last point is not applicable for aquaponic systems, where water from fish tanks flows out to microalgae or plant production facilities. The concentrations of animal hormones, such as the steroid E2 or its synthetic analog17α-ethynyl-estradiol (EE2) in domestic or swine-farms wastewaters can be as high as 350 μg/L (Johnson and Williams, 2004; Dalrymple et al., 2013; Owen and Jobling, 2013) and can bear significant amount of bacteria (Table 1).

Synthetic hormones are normally more toxic to aquatic inhabitants. Although in natural habitat EE2, being a common human medication, occurs just in small quantities of 0 to 35 ng/L (Stachel et al., 2003; Vethaak et al., 2005; Chen et al., 2007; Pojana et al., 2007), this synthetic estradiol is one of the most dangerous water pollutants and it can affect not only animals, but also decrease photosynthetic processes in plants (Perron & Juneau, 2011) (Table 1).

Notably, detectible concentrations of phyto- and animal hormones have been found in well-water, which often is used for microalgae medium preparation, for hydroponics water-supply, and fishery raceways (Aris et al., 2014; Adeel et al., 2017). In underground water, steroid concentrations can reach the 10-100 ng/L range for estrone (E1) and 1-10 ng/L for estrogen (E2) (Kjaer et al, 2007). The prevalence of estrone concentrations may also be seen in surface and wastewaters and primarily is due to prompt transformation of E2 and 17α-ethynyl-estradiol, EE2 to E1 (Haiyan et al., 2007).

2. HORMONES APPLICATION FOR BIOMASS PRODUCTION

The application of hormones for microalgae biomass production is a novel approach of induction of microalgae biomass production, and stimulation of biomolecule synthesis and accumulation. Numerous reports demonstrated positive effect of wide-range of phyto- and animal hormones on microalgae growth, although the mechanisms of action of exogenous hormones on microalgae physiology, of interactions between endogenous and exogenously applied hormones, and of induction of biomolecule synthesis and accumulation are not well studied.

It is logical that hormones have great impact on algal physiology, since phytohormones and some animal steroids are known to enhance growth, organ differentiation, and stress tolerance in vascular plants (Osborne and McManus, 2005; Peleg and Blumwald, 2011; Davies, 2013; Vidhyasekaran, 2015). Most of the phytohormones essential for plant metabolism are synthesized by algal cells (Stirk and Van Staden, 1996; Kiseleva et al., 2012; Romanenko et al., 2015), but not many studies about their role in algal physiology and metabolism, or their effects on algal biosynthesis and stress tolerance, have been reported.

None-the-less, hormones have been shown to influence microalgae physiology in a number of important ways. Hormones have been shown to: a) simultaneously induce biomass production and valuable biomolecules accumulation in grown biomass; b) alter fatty acid profiles in favour of short and/or moderate-length carbon chain fatty acids (demonstrated on laboratory scale) (Salama et al., 2014; Kozlova et al., 2017, 2018, 2019; Talarek-Karwel et al., 2018); c) reduce the lag phase of algal growth. This has been reported for auxin (Hunt et al., 2010), EBL (Talarek-Karwel et al., 2018), mixtures of IAA and EBL (Kozlova et al., 2018), and E2 (Kozlova et al., 2019).

The first point above gives a unique opportunity to amplify the yield of the molecules of interest bringing microalgae industry to competitive level compared to other sources of energy and nutrients. The second point is important for algal biofuel industry making the biosynthesized oils more suitable for fuels production reducing the cost of the crude oil downstream processing. The third point promotes both greater biomass yield and its rate of production, so the maximum biomass yield can be harvested earlier.

Addressing the question of the safe and inexpensive use of hormones for algal biomass production, nano-concentrations of applied hormones must be used, since low concentrations are expected in hydroponic effluences and other suitable for microalgae wastewaters (Moya et al., 1995; Hartung, 2010; Dalrymple et al., 2013; Nobel et al., 2014). Successful application of phytohormones in nano-concentrations for microalgae biomass and valuable biomolecule production was confirmed in several studies (Nobel et al., 2014; Salama et al., 2014; Kozlova et al., 2017, 2018; Talarek-Karwel et al., 2018). However, application of hormones on industrial-scale remains at an early stage of development.

Another problem is the distinction between laboratory batch-scale, pilot-scale, and industrial-scale production of microalgal biomass induced by hormones. Until now, the vast majority of experimental work has been conducted in batch experiments at the laboratory bench-scale. The major advantages of this is that the experimental set-up allows for complete control of growth conditions and aseptic maintenance of the experiment. The disadvantages of bench-scale experiments is that they are disconnection from the reality of industrial production, where wastewater input is constant or in feed-batch mode. The one-time application at the beginning of bench-scale batch experiments has a high possibility of mitigation of the hormone impact on the algae. In fact, phytohormone suppression of chlorophyll-a and total carotenoind and lipid accumulation over time has been noted by several investigators in bench-scale batch experiments (Tatkowska and Buczek, 1980; Kozlova et al., 2017; Talarek-Karwel et al., 2018). This attenuation of the positive effect of phytohormones on intracellular pigments accumulation was reported to be even stronger when a mixture of two hormones (EBL and IAA) was examined with *Scenedesmus quadricauda* cells (Kozlova et al., 2018).

Thus, experimental work at the laboratory bench-scale with fed-batch and/or constant-flow experimental designs are very important for the assessment of hormones on microalgal physiology and metabolism.

2.1. Phytohormones to Microalgae

Many studies have reported positive stimulatory effects, in a concentration-dependent manner, on algal growth parameters (Bajguz and Czerpak, 1998; Park et al., 2013; Nobel et al., 2014; Kozlova et la., 2017, 2018), bioaccumulation of chlorophyll-a and chlorophyll-b (Bajguz and Czerpak, 1998), carotenoids (Kozlova et la., 2017, 2018), protein (Taylor et al., 2001; Colla et al., 2007;), and lipids (Salama et al., 2014; Du et al., 2017; Kozlova et la., 2017, 2018) (Figure 2). It is difficult to select the most powerful inducers among the phytohormones, since the investigations have been conducted on different algal species and because of differences in growth conditions.

The data gained to date suggest that any phytohormone can have a different degree of influence on the physiological aspects of a particular algal species grown under constant ambient conditions. For instance, the auxin, 3-indoleacetic acid (IAA), was found to be one of the most powerful phytohormones for induction of algal growth and biomass yield, as well as pigment (chlorophyll, carotenoids) accumulation (Park et al., 2013; Nobel et al., 2014; Ozioko et al., 2015; Liu et al., 2016; Babu et al., 2017; Kozlova et al., 2017). However, the positive effect of IAA on lipid accumulation was weaker (Salama et al., 2014; Kozlova et al., 2017) or not evident (Park et al., 2013; Liu et al., 2016). Thus, the collection of sufficient data, with modeling work, is necessary for accurate prediction of the effects of a single hormone, or a mixture of hormones, on one or another metabolic pathway of a particular algae species.

The ability of phytohormones to promote microalgal growth and biosynthesis is dose dependent. The most effective concentrations of phytohormones vary somewhat in the literature. For example, IAA was reported to be most effective at concentrations between 10-3 and 100 μM by several authors (Park et al., 2013; Czerpak et al., 1994, 1999; Piotrowska et al., 2008), including one of the pioneering papers of Pratt (1938).

Even lower concentrations, between 10 and 100 nM, have been demonstrated to be most effective to stimulate an increase in cell density, biomass production, and pigment and lipid accumulation when IAA was applied to *S. quadricauda* cultures (Kozlova et al., 2017) (Figure 2A). Nobel et al. (2014) observed an inhibitory effect of IAA at concentrations of 0.1 μM and higher. Certainly, morphological and physiological differences between microalgae species and variations in algae growth conditions are responsible for the differences in outcomes of experimental work.

Several studies have investigated the influence of extremely high concentrations of plant growth regulators on microalgae physiology. Unexpectedly, the authors did not observed an adverse effect at very high concentrations of Auxins, Gibberellin, Abscisic acid, and Kinetin applied in mg/L to *Chlamydomonas reinhardtii*, *Chlorella pyrenoidosa*, and *S. quadricauda* (Hunt et al., 2010; Park et al., 2013; Liu et al., 2016). Nevertheless, such high concentrations can only be achieved in industrial production units by the addition of manufactured hormones into growth medium, which is very costly and most likely not economically viable. Thus, application of naturally occurring hormones from fishery, hydroponic, and other suitable wastewaters could be one of the solutions to reduce the cost of industrial scale microalgal production. The influence of phytohormones on microalgae physiology must consider the following key hormones and their combinations.

2.1.1. Auxins

In vascular plants, auxins regulate cell growth and elongation, as well as tissue and organ differentiation (Teale et al., 2006; Ludwig-Muller, 2011). Auxins of natural and synthetic origins were intensively tested on different algae species including *Euglena gracilis Klebs* (Nobel et al., 2014), *C. reinhardtii* (Park et al., 2013), *Ch. pyrenoidosa* (Czerpak et al., 1994, 1999; Liu et al., 2016), *Chlorella sorokiniana* (Hunt et al., 2010), and *S. quadricauda* (Liu et al., 2016, Kozlova et al., 2017). The results of these studies agree on the positive effects of different auxins on microalgae cell division (Park et al., 2013; Piotrowska and Bajguz, 2014; Nobel et al., 2014; Kozlova et al., 2017), cell size (Park et al., 2013; Nobel et al., 2014; Kozlova et al., 2017), biomass production (Czerpak and Bajguz, 1997; Liu et al., 2016; Babu et al., 2017; Kozlova et al., 2017), protein (Park et al., 2013; Piotrowska and Bajguz, 2014), lipids (Liu et al., 2016; Babu et al., 2017; Kozlova et al., 2017), and pigment biosynthesis (Czerpak and Bajguz, 1997; Park et al., 2013; Piotrowska and Bajguz, 2014; Ozioko et al., 2015; Kozlova et al., 2017) (Figure 2).

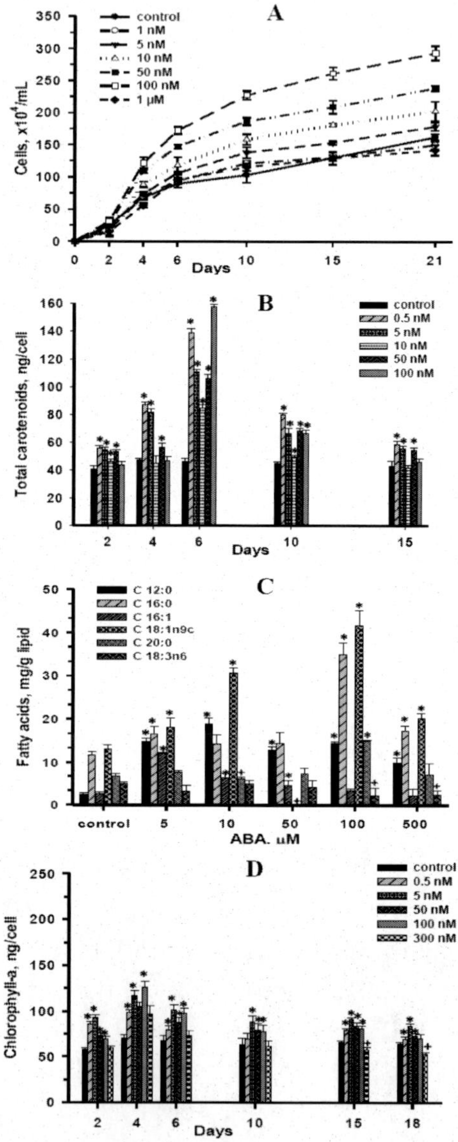

Figure 2. Examples of phytohormones influence on microalgae (*Scenedesmus quadricauda*) physiology: A) auxin IAA induces cell growth; B) EBL induces carotenoids accumulation; C) ABA induces FAs accumulation; D) BL induces chlorophyll-a accumulation. Statistically significant differences from the controls ($P < 0.05$) are indicated, where * - significantly greater than the control and $^{+}$ - significantly lower than the control. Data shown as the mean ± SD, n is 4 to 15 (Kozlova et al., 2017).

Indole-3-propionic acid (IPA), indole-3-acetic acid (IAA), indole-3-butyric acid (IBA) were found to be the most effective auxins for stimulation of microalgal growth and biosynthesis (Liu et al., 2016; Piotrowska and Bajguz, 2014). It is of interest to note that the most effective concentrations of synthetic auxins is usually in higher range than the effective concentration range of natural auxins. Synthetic hormones have been reported to pose a high risk of intracellular accumulation in microalgal cells since the concentration factor is normally higher for synthetic analogues of hormones (Larsson et al., 2000), which may be an important reason to avoid using synthetic auxins (and other synthetic hormones) as inducers of microalgae biomass production.

Discrepancies in the reported impacts of auxins on microalgal physiology are likely a consequence of species-specific responses and varied growth conditions. As a result, the most effective hormone concentrations are inconstant across many studies. For example, Nobel et al. (2014) observed stimulation of *E. gracilis* growth, biomass yield, and cell size at 1 nM and 10 nM IAA, and physiological inhibition at higher IAA concentrations, while studies with *S. quadricauda* (Kozlova et al., 2017) demonstrated a clear stimulation of cell growth, biomass yield, pigments and lipid accumulation at IAA concentrations from 5 to 100 nM. Liu et al. (2016) demonstrated induction of *S. quadricauda* growth and biomass production at much higher IAA concentrations (228 and 342 μM), despite the fact that such concentrations were corresponded toxic by some authors (Piotrowska and Bajguz, 2014; Kozlova et al., 2017).

The reported positive effect of auxins in general, and particularly for IAA, on lipid accumulation by microalgae are also inconsistent. Recent studies conducted on *S. quadricauda* (Liu et al., 2016; Kozlova et al., 2017), *S. obliquus* (Salama et al., 2014), and *Ch. sorokiniana* (Babu et al., 2017) demonstrated a clear induction of lipid production by IAA. However, Park et al. (2013) found no effects of IAA on fatty acid accumulation by *C. reinhardtii* cells. Liu et al. (2016) also reported that IAA had no effect on fatty acid accumulation by *Ch. pyrenoidosa*. Yet, it is important to note, that both experiments (Park et al., 2013; Liu et al., 2016) used very high IAA concentration, up to 5 mg/L. At these concentrations,

IAA likely would have no-effect, or would have adverse effects, on neutral lipid accumulation. A possible explanation of the mechanism of fatty acid profile alteration by auxins may be that they can affect activity of acyl-ACP thioesterase (FAT), the enzyme that determines the final chain length of the fatty acids in algae (Borowitzka, 1988).

As indicated above (Section 1.2.2), hormone bioavailability and uptake but algal cells is dependent on water chemistry and competition among ions for uptake on membrane sites. Auxins may directly alter algae membrane permeability, since auxin has been shown to be capable of activating a proton pump ATPase in the plasma membrane (Rober-Kleber at al., 2003; Perrot-Rechenmann, 2010) in vascular plants. It could be also true for microalgae, since there are some similarities between vascular plants and algae in their response to exogenous hormones. For example, the ability of IAA to enlarge microalgae cells is consistent with the main function of auxin in vascular plants to regulate cell growth and promote cell elongation. The similarities and differences in the responses to a hormone between vascular plants and microalgae need further clarification.

2.1.2. Brassinosteroids

Brassinosteroids (BRs) are less studied for their impact on microalgal growth and biosynthesis. However, a great effort to elucidate the influence of BRs on the physiology of microalgae and some aquatic plants has been made by researchers in the Department of Plant Biochemistry and Toxicology, Institute of Biology, University of Bialystok, Poland. The most studied BRs, 24-epibrassinolide (EBL) and brassinolide (BL), exhibit similar stimulatory effects on different algae species. However, the impacts of EBL was reported to be stronger by some authors (Kozlova et al., 2017), while other researchers have reported that BL has a stronger effect on algal growth and biomolecule accumulation (Bajguz, 2002).

A vast spectrum of functions have been found for BRs in vascular plants, including cell division and growth, cell wall formation, regulation of photosynthesis and metabolite accumulation (Vidhyasekaran, 2015). Both EBL and BL have been shown to work synergistically with IAA to elongate cells and promote tissue differentiation in vascular plants.

Moreover, like ABA, BRs work to protect plant tissues from environmental stresses (Clouse and Sasse, 1998; Cano-Delgado et al., 2004) and pathogens (Vidhyasekaran, 2015). In microalgae, both BRs have clear inductive effect on growth and biomass production, enhancing both parameters by 2- to 3-fold (Bajguz and Czerpak, 1998; Bajgus, 2000; Bajguz, 2002; Kozlova et al., 2017; Talarek-Karwel et al., 2018). Yet, despite of the fact that BL plays a role in cell elongation in plants, it was shown to have an adverse effect on cell size (unlike EBL) (Kozlova et al., 2017).

Reports on the impacts of EBL and BL on biomolecule accumulation are very diverse. Yet, both EBL and BL have been shown to stimulate pigments production (Bajguz and Piotrowska-Niczyporuk, 2013; Kozlova et al., 2017) (Fig 2B, D), monosaccharide production (Bajguz and Czerpak, 1998; Bajguz and Piotrowska-Niczyporuk, 2013), as well as fatty acid accumulation and protein content (Bajguz, 2000; Bajguz, 2002; Bajguz and Piotrowska-Niczyporuk, 2013; Kozlova et al., 2017; Talarek-Karwel et al., 2018). Importantly for microalgae biofuel production, BRs were found to inhibit the formation of reactive oxygen species by suppressing oxidative damage of synthesized metabolites, particularly lipids (Talarek-Karwel et al., 2018).

2.1.3. Abscisic Acid

Abscisic acid (ABA) plays special role in plants and algae physiology. All the functions of this hormone are tightly linked to ambient conditions of an organism. For instance, ABA is an important two-way growth regulator, adjusting plant growth to environmental conditions such as cold or hot weather, water deficit or its excess, the impact of pathogens, etc. (Schroeder et al., 2001; Pastori and Foyer, 2002; Tuteja, 2007). In microalgae, ABA has been found to increase growth parameters and biomolecule accumulation (Park et al., 2013; Nobel et al., 2014; Wu et al., 2017; Kozlova et al., 2017), which was demonstrated with number of species, such as *C. reinhardtii, Chlorella sp., E. gracilis, S. quadricauda, and Dunaliella salina*. However, it has been reported that ABA has a weaker effect on microalgae physiology and an adverse effect on cell size compared to other phytohormones, such as auxins and BRs (Hartung, 2010; Park et al., 2013; Kozlova et al., 2017)..

A number of studies have reported a significant increase in fatty acid yields in the presence of ABA, despite the fact that cell growth was not affected by this hormone (Park et al., 2013; Wu et al., 2017; Kozlova et al., 2017; Babu et al., 2017) (Figure 2C). As with auxins and other phytohormones, the majority of research on ABA was conducted using very high hormone concentrations from hundreds of μM to tens of mM. For example, Park et al. (2013) reported that ABA demonstrated great ability to stimulate an increase in starch and protein content of *C. reinhardtii* at very high concentrations: 100 to 10,000 μg/L. Hartung (2010) very appropriately remarked on the marginal meaning of such experiments, which do not represent physiological relevance. Moreove, as mentioned previously, such concentrations are too expensive for application on the industrial scale of microalgae biomass production.

The weak ability of ABA to stimulate *S. quadricauda* growth could be explained by taking into account that the prime function of ABA in vascular plants is as a protective mediator against environmental stresses and pathogens. Remarkably, this function of ABA was confirmed in microalgae (Hirsch et al., 1989; Cowan and Rose, 1991; Khasin et al., 2017), although to a much weaker extent compared to the effects in vascular plants (Hartung, 2010). In the laboratory conditions of batch experiments, when only ABA concentrations are variables, no environmental stresses present. This, at least partially, can explain moderate induction of algae growth and some biosynthetic activity by ABA.

Cowan and Rose (1991) suggested that ABA is likely a regulator of carotenogenesis in microalgae, particularly in response to an environmental stress. Indeed, ABA has been shown to be connected with carotenogenesis in vascular plants, where ABA is considered as a catabolite of neoxanthin and xanthoxin (Zeevaart and Creelman, 1989; Parry et al., 1990). Yet, the metabolic pathway of ABA is different in microalgae and cyanobacteria. It is debatable whether or not secondary carotenoids (xanthophylls) are involved in the biosynthesis of ABA, or the hormone could be directly synthesized from farnesyl-diphosphate (Hartung, 2010). It also was highlighted in some reports that 9'-cis-neoxanthin, which is the direct precursor of abscisylaldehyde, could not be detected in some algae species (Bopp-Buhler et al., 1991). This is a good illustration of the mechanism of species-specific hormone metabolism in microalgae. The connection of intracellular ABA metabolism and exogenous hormone influence on algae physiology needs an extensive study for accurate design of ABA application on industrial scale.

2.1.4. Gibberellin

Gibberellic acid, also known as Gibberellin A3, GA, and GA_3, is a hormone found in plants, fungi and algae. GA is a pentacyclicditerpene acid that regulates growth and elongation of plant cells and development of plant tissues. GA stimulates the cells of germinating seeds to produce mRNA molecules that code for hydrolytic enzymes (Camara et al., 2015).

Importantly for microalgae biomass production, GA regulate cells dormancy under adverse conditions and cells senescence (Hedden and Sponsel, 2015). Low concentrations of GA (100 nM) have been shown to have stimulatory effects on cell division and cell mass accumulation (as measured by dry cell mass) in *S. quadricauda*, but did not stimulate an increase in cell specific protein concentration (Buczek et al., 1975). Higher concentrations GA had significant stimulatory effects on cell mass production of *Chlamydomonas reinhardtii* cultivated for biodiesel production. One ppm (1 mg/L) GA increased maximum cell density approxmately 1.6-fold, but had no effect on FAME yield. Slightly lower concentrations of GA (0.8 mg/L) GA increased protein and chlorophyll content of *Ch. reinhardtii* cells, but had no effect on starch concentrations (Park et al., 2013). High concentrations GA in combination with other biochemical stimulants were shown to stimulate cell mass production. For example, 10 mg/L of GA plus 5 mg/L 1-naphthaleneacetic acid (NAA) stimulated call mass production of *Chlorella sorokiniana* by 138%, while 10 mg/L GA plus 5 mg/L NAA, plus 1 mg/L zeatin stimulated call mass production of *Ch. sorokiniana* by 170% (Hunt et al., 2010). As indicated in

Section 2.1.1, discrepancies in the effects of GA are likely due to species-specific responses to this hormone.

The mechanism(s) of how GA stimulates cell mass production are not well understood. However, an interesting study of the effect of GA on the brown seaweed, *Fucus vesiculosus*, demonstrated a clear effect of GA on the photosynthetic activity of *F. vesiculosus* embryonic cells. One (1) mg/L GA had a significant stimulation effect on the maximum quantum yield (F_v/F_m) relative to untreated control cells (Tarakhovskaya et al., 2013), suggesting that GA stimulates at least some elements of the photosynthetic machinery, such as maximum quantum yield of PSII in plants, including seaweed, and by inference, microalgae.

2.1.5. Phytohormones Crosstalk: Benefits for Production

When wastewater applied to microalgae production units, the influence of mixed dissolved phytohormones on microalgae physiology is expected.

In nature, phytohormones rarely, if ever, work as single agents. Inside living cells of vascular plant, phytohormones (synthesized by cells or absorbed from the environment) cross-talk with each other via signal transduction cascades of metabolic pathways forming a signaling networks (Sasaki et al., 2001; Verhage et al., 2010; Harrison, 2012; Yang et al., 2013; Vidhyasekaran, 2015). This network is not constant in time, but depends on environmental conditions, such as temperature, soil and/or water chemistry, and the presence of pathogens (Vidhyasekaran, 2015; Robert-Seilaniantz et al., 2011; Thaler et al., 2012; Alazem and Lin, 2015). Cross-talk between phytohormones can have different courses of action, resulting in synergistic, additive, or antagonistic effect on plant or microalgae physiology.

Combinations of phytohormones are known to enhance growth and stress tolerance of vascular plants, but the effects of phytohormone combinations on microalgae physiology are not well studied. In vascular plants, some phytohormones support each other by working singularly, and/or together, on the same physiological functions. For instance, abscisic acid (ABA) and auxin both regulate cell growth and elongation. However, ABA is also an important two-way growth regulator, that functions to adjust plant growth to environmental conditions and the impact of pathogens (Schroeder et al., 2001; Pastori and Foyer, 2002; Tuteja, 2007), while auxins also regulate tissue and organ differentiation (Muller, 2011). Brassinosteroids (BRs) can work synergistically with auxins to elongate cells and promote tissue differentiation, and with ABA as an environmental stress remediator (Clouse and Sasse, 1998; Karnachuk et al., 2002; Cano-Delgado et al., 2004; Vidhyasekaran, 2015).

The mechanisms of phytohormone signaling network formation between exogenously applied phytohormones, the relevant metabolic pathways, and the influence of these networks on the physiology of a particular microalgal species are not yet understood. A few phytohormone cross-talk experiments conducted on different algae species have provided some insights into the influence of different plant hormone combinations on microalgae physiology. Hunt et al. (2010) reported significant increases in biomass and chlorophyll accumulation, but not in lipid or protein production, when *Ch. sorokiniana* was exposed to combined phytohormones from different taxa (auxin, gibberellin and cytokinin), while no effect was confirmed with a mix of different auxins (Vance, 1987; Hunt et al., 2010). Nobel et al. (2014), examined different combinations of two-, three-, four-, and five-hormones (ABA, IAA, and three cytokinins), concluded that any combination of phytohormones produced better induction of *E. gracilis* growth compared to each hormones.alone. Importantly, the hormones were tested at nano-concentrations in this work.

Clear synergistic relationship between auxin and brassinosteroids was observed in two recent studies (Bajguz and Piotrowska-Niczyporuk, 2013; Kozlova et al., 2018). Bajguz and Piotrowska-Niczyporuk (2013) demonstrated synergistic increases in cell growth, protein, monosacchride, and total chlorophyll content in *Chlorella vulgaris* cells for all examined combinations of IAA and BL, although they examined relatively high concentrations of the hormones (1-100 μM).

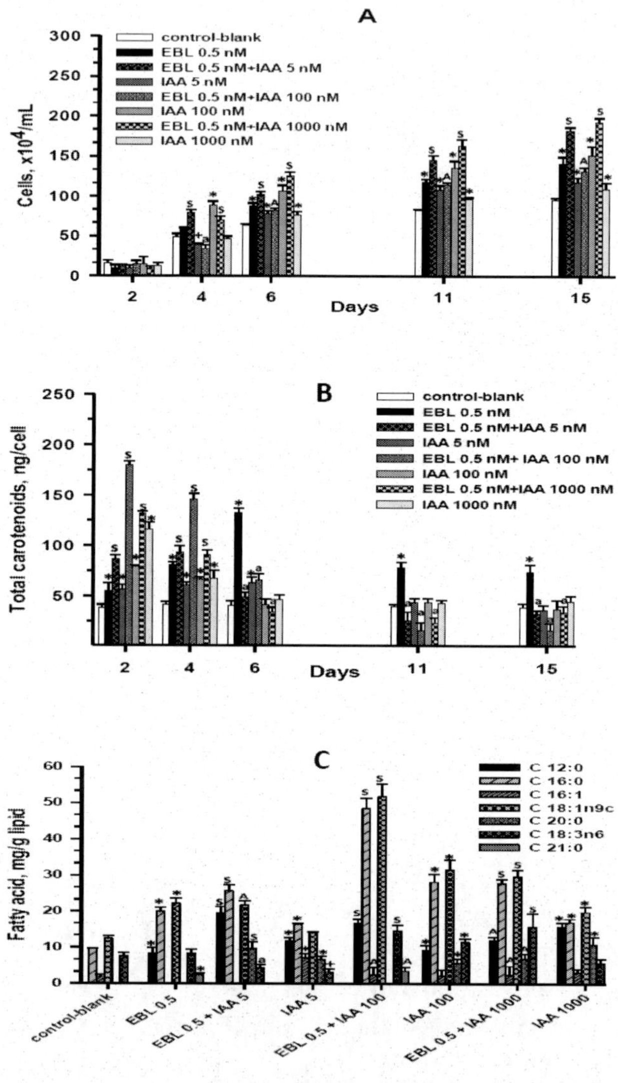

Figure 3. Examples of mixed hormones influence on microalgae (*Scenedesmus quadricauda*) physiology. 0.5 Nm EBL was mixed with 5, 100 and 1000 nM IAA, where the effect of the mixtures (columns with patterns) were compared to the single hormones (columns with no patterns) used for the mixtures preparation and to the controle-blank (no hormones added). A) influence on algae growth; B) influence on carotenoids accumulation; C) influence on FAs accumulation and their profile. Statistically significant differences are indicated: *, for single hormones, significantly greater values compared with the control-blank ($P < 0.05$); for mixed hormones, S = synergistic effect; A = additive effect; a = antagonistic effect. Data shown as the mean ± SD, n = 12 (Kozlova et al., 2018).

Kozlova et al. (2018) confirmed synergistic work of IAA with EBL on *S. quadricauda* growth, chlorophyll-a, total carotenoids, and FA accumulation applying both hormones in nano-concentrations. Importantly, this work demonstrated that some studied IAA and EBL combinations synergistically promoted both greater biomass yield and its rate of production, decreasing the lag phase of algae growth and this was stronger than in the single-hormone trials (Kozlova et al., 2017).

Ozioko et al. (2015) also observed clear synergistic effects of auxin mixed with GA on biomass and chlorophyll-a production in *Ch. sorokiniana* cells. The excessive activity of auxins in the signaling network of high plants was notes by several authors (Stepanova et al., 2008; Swarup et al., 2002; Rock and Sun, 2005; Vidhyasekaran, 2015) and likely is due to the fact that auxin homeostasis has multiple pathways in vascular plants (Hagen et al., 2004). The activation of the auxin pathway depends on particular environmental factors, which trigger certain combinations of genes and enzymes (Hagen et al., 2004).

Moreover, auxins have high binding capacity and a tendency to form conjugated conformations, which facilitate their involvement in phytohormone cross-talk networks (Rock and Sun, 2005; Robert-Seilaniantz et al., 2011; Tatsuki et al., 2013; Vidhyasekaran, 2015), although formation of conjugates is reported to be different in microalgae (Stirk et al., 2013). With brasinosteroids, auxins exhibit a strong one-way metabolic relationship, as was proposed by Davies (2004) and others (Nakamura et al., 2003; Reid et al., 2004; Vert et al., 2008).

The theory states that only BR can affect IAA biosynthesis and activity in plants, but not *vice versa*. Indeed, the ability of exogenous BRs to synergistically enhance endogenous IAA accumulation and/or to elevate plant sensitivity to auxin was demonstrated in vascular plants (Katsumi, 1985; Sala and Sala, 1985; Sasse, 1999; Nakamura et al., 2003). Remarkably, similar patterns could be seen in microalgae *Ch. vulgaris* and *S. quadricauda* (Bajguz and Piotrowska-Niczyporuk, 2013; Kozlova et al., 2018). However, the complex outcome of the combinations of BL with IAA on growth and metabolites accumulation in *S. quadricauda* (Kozlova et al., 2018) (Figure 3) suggests that the mechanism of EBL and IAA crosstalk is likely specific to targeted biosynthetic pathways. This assumption is supported by the work of Bajguz and Piotrowska-Niczyporuk (2013), who demonstrated that, in some circumstance, auxin could also increase the activity and effectiveness of BRs in a synergistic manner, which is in agreement with opinions of Bao et al. (2004) and Nemhauser et al., (2004).

The molecular mechanisms of cross-talk has been elucidated for several phytohormones. For instance, Nakamura et al. (2003) and Vert et al. (2008) clearly demonstrated the molecular mechanism of BRs and IAA cross-talk, suggesting that the signaling pathways of both hormones are capable of triggering the transcriptional system of IAA, and BRs work to induce the early auxin-inducible genes. De Vleesschauwer et al. (2012) demonstrated negative cross-talk of BRs with salicylic acid (SA) and GA pathways, and found a dramatic change in BRs function from protection of plants from biotic stress to immune suppression in rice (*Oryza sativa*). Certainly, the molecular mechanisms of signaling networks between different hormones mixtures needs further intensive investigations. Moreover, the most effective phytohormone combination is plant/algae species-specific. Thus, the optimal combination of plant growth regulators treated to paddy water was determined for rice as GA and ABA, although their ratios depend on water chemistry and density of rice seedlings (Moya et al., 1995). For commercially used algae species, the optimal combination of phytohormones is yet to be determined.

By any mean, the consequences of mixed hormone experiments depend on the concentrations of the specific hormones in mixture, as well as on ambient conditions or on the specific plant or algae species under investigation (Hunt et al., 2010; Peleg and Blumwald, 2011; Thaler et al., 2012; Van der Does et al., 2013; Kozlova et al., 2018). Moreover, in term of wastewater application for microalgae biomass production and dependently on the wastewater origin, there is a high probability that animal hormones can interact with phytohormones. There is lack of data on the subject of interactions of animal and plant hormones in aqueous solutions.

2.2. Animal Hormones and Microalgae

The great importance of investigating the effect of animal hormones on microalgal physiology is derived from the fact that not only wastewaters, but also surface and ground waters, of many regions contain detectable level of steroid hormones (Julius et al., 2007; Silva et al., 2012; Aris et al., 2014). The use of wastewaters containing animal hormones as a growth medium for microalgae production, raises question about the influence of dissolved animal steroids on microalgae physiology and metabolism. Moreover, the possibility of steroid hormone bioconcentration and/or biomagnification in microalgal cells, followed by the possibility of their transfer via food chain, is an alarming issue which needs intensive investigation, because natural and especially synthetic hormones are part of the extremely hazardous group of chemicals named Endocrine Disruption Compounds (EDCs).

Reproductive and stress hormones are among the organic compounds often found in agricultural and domestic wastewaters. Wastewaters have gained a reputation as potent inducers of crop plant growth and biomolecule accumulation (McMurtry et al., 1997; Takeda et al., 1997; Rakocy et al., 2004, 2007; Bansal et al., 2005; Bailey and Ferrarezi, 2017), because they contain all macro- and micronutrients required for successful growth of higher plants (Xu and Shen, 2011). For instance, fishery wastewaters have been utilized successfully as a source of nutrients in integrated aquaponics systems and have had positive effects on hydroponic crop quality and biomass production (Zweig, 1999; Rakocy et al., 2004).

In addition to nutrients dissolved in fishery wastewater, fish hormones play an important role in the promotion of vascular plant physiology. Sex steroids have demonstrated great ability in promoting shoot and root growth, enhancing the photosynthetic rate, the accumulation of carbohydrates and proteins, as well as flower and ovary formation in vascular plants (Ylstra et al., 1995; Bhardwaj and Thukral, 2000; Janeczko et al., 2003; Upadhyay and Maier, 2016).

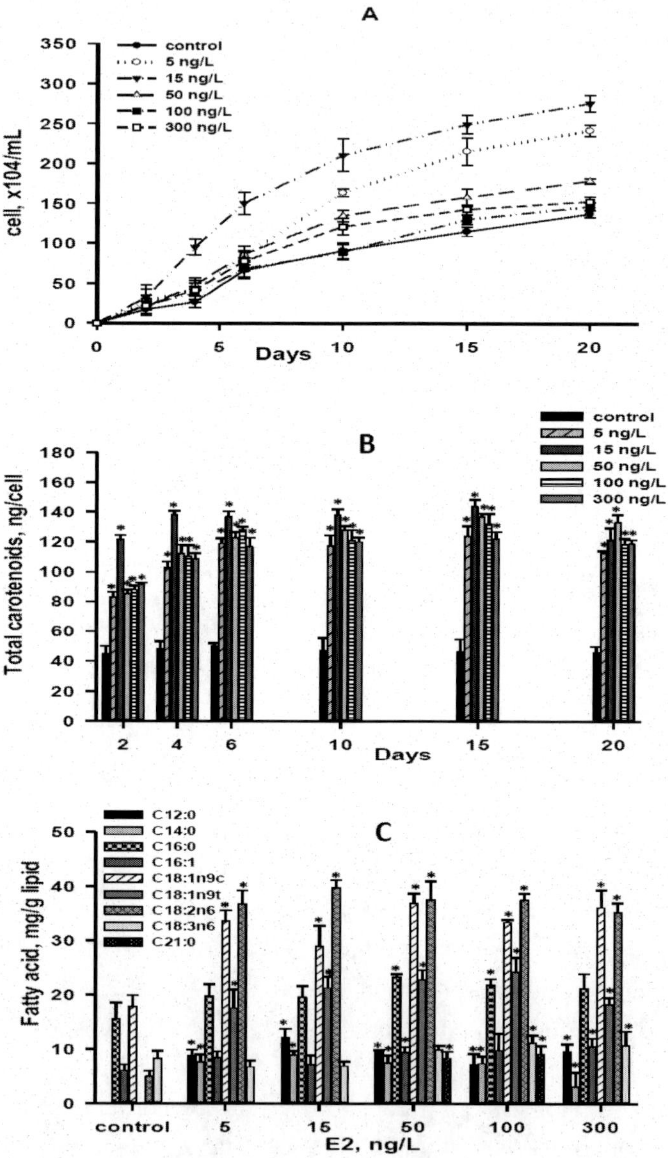

Figure 4. Examples of the influence of 17β-estradiol (E2) on microalgae (*Scenedesmus quadricauda*) physiology. A) influence on algae growth; B) influence on carotenoids accumulation; C) influence on FAs accumulation and their profile. Statistically significant differences of treated versus controls ($P < 0.05$) are indicated. (Kozlova et al., 2019).

Yet, for most vascular plants, animal effluents that are rich in ammonium ions have to be biologically treated to convert the ammonium to nitrate and nitrite. Such a treatment eliminates almost all the hormones dissolved in water. Opportunely, for microalgal biomass production there is no need in such a treatment, since microalgae can preferably consume ammonium ions as a source of nitrogen (Boussiba and Gibson, 1991; Graham and Wilcox, 2000; Perez Garcia et al., 2011).

Animal reproductive and stress hormones have been reported to induce microalgae growth (Bajgus and Czerpak, 1996a,b; Czerpak and Szamrej, 2000; Kozlova et al., 2019) and production of valuable molecules, such as carotenoids (Bajgus and Czerpak, 1996a; Czerpak et al., 2001b; Kozlova et al., 2019), proteins (Bajgus and Czerpak, 1996b; Czerpak and Szamrej, 2000), soluble sugars (Bajgus and Czerpak, 1996b; Czerpak and Szamrej, 2000), and lipids (Kozlova et al., 2019). Similar to phytohormones, high concentrations of steroids are toxic to algae and bring a risk of intracellular hormone accumulation (Liu et al., 2018). Certainly, the current knowledge on the influents of animal hormones on microalgae growth and metabolism suggests that hormones could be used as a tool for algal biosynthesis and, particularly, for fatty acid profile manipulation in the biofuel, agricultural, pharmaceutical and cosmetic industries. However, knowledge of the mechanisms of steroid action, their most effective concentrations, the biotransformation of steroids in microalgal cells, and the possibility of hormones bioconcentration, require further intensive investigation on wide-range of microalgal species to clarify the feasibility of animal steroids application for microalgae biomass production.

2.2.1. Sex Hormones

The use of fishery wastewater for microalgae production provides an opportunity to employ the nano-concentrations of dissolved fish steroid hormones as free-of-cost inducers of algal biomass production and valuable biomolecules. Even when surface water or ground water is considered t for growth of microalgae, substantial concentrations of animal steroids may be present (Table 1).

Regardless to the source of water taken for microalgae production (surface, well, or wastewater from farms), two major groups of animal sex steroids are expected to be dominant: estrogens and progesterones. Despite the fact that ratio of estrogens and progesterones in water is estimated as 1:4 to 1:20 (Hala et al., 2012), the hormonal activity and influence of estrogens on aquatic biota, even in very low concentrations, is extremely high (Lai et al., 2002a; Joblin and Owen, 2003; Silva et al., 2012; Adeel et al., 2017). The steroid hormone activity is even more important to consider when synthetic hormones are present in the water, since it is known that estrogenic activity of synthetic steroids, such as EE2, is about 100 to 7,000 times (depending on water physico-chemical parameters and a species-receiver) greater than the natural analogue (E2) (Joblin and Owen, 2003; Nendza and Wenzel, 2006).

Numerous studies reported positive influence of animal steroids on vascular plant growth, organs development, biomass production and metabolites accumulation. Upadhyay and Maier (2016) reported that E2 induced shoot and root growth and enhanced the photosynthetic rate, and carbohydrate and protein accumulation, in *Arabidopsis sp.* treated with low hormone concentrations (10 and 100 nM). Research conducted on number higher plants, including aquatic, demonstrated great ability of E2 to promote flowering and ovary formation (Czygan, 1962; Ylstra et al., 1995; Bhardwaj and Thukral, 2000; Janeczko et al., 2003).

The mechanism(s) of exogenous animal steroid influence in aqueous environments have some specifics. When applied to growth media, steroids are taken-up rapidly by algal cells (absorption > adsorption), with about 90% uptake in 48 hours (Liu et al., 2018). Biodegradation occurs over 4-6 days, resulting in intracellular accumulation of the steroidal catabolites, of both known (e.g., E2 transfers to E1) and unknown nature (Shi et al., 2010; Liu et al., 2018). The influence of these catabolites on microalgae growth and biosynthesis needs intensive investigation. It was suggested that the catabolites of E2 degradation work similarly to the originally applied estradiol in the promotion of carotenoids and chlorophyll accumulation by *S. quadricauda* cells (Kozlova et al., 2019).

A few studies conducted on the influence of animal sex hormones on microalgal physiology have demonstrated strong abilities of estrogens and progesterones to increase growth and biomass production (Bajguz and Czerpak, 1996a, b; Julius et al., 2011; Beecher, 2013; Kozlova et al., 2019), although in the species-specific manner (Perron and Juneau, 2011). Moreover, it was reported that, similar to the influence of certain phytohormones, E2 greatly reduces the lag phase of algal growth (Bajguz and Czerpak, 1996a; Kozlova et al., 2019), which is clearly seen at the most effective concentration (Figure 4A). Indeed, E2 was found to be responsible in promotion of the early growth stages of aquatic organisms such as fish juveniles (Liu et al., 2012), plant shoots, buds and young roots (Janeczko and Skoczowski, 2005; Lino et al., 2007), and it seems also be true for microalgae.

Estrogens have shown very positive effects on pigment accumulation by microalgae, although total carotenoid accumulation was affected greater than chlorophyll accumulation (Bajguz and Czerpak, 1996a; Kozlova et al., 2019) (Figure 4B). Similar responses were observed in vascular plants, where exogenously applied E2 was shown to greatly induce pigment accumulation, simultaneously increasing intracellular content of GA and auxin (Kopcewicz and Porazinski, 1974). Remarkably, animal sex hormones can be synthesized by many species of high plants and microalgae. Steroid synthesis by primary producers depends on the physiological stage of the organism. For example, Janeczko (2012) detected estrogen and progesterone at concentrations 2 to 3-times greater in flower buds compared to developed flowers. Lino et al. (2007) found progesterone in concentrations that are 40-times greater in immature green tomatoes compared to ripe, red tomatoes, suggesting that phytosteroids play important role in carotenogenesis in plants. Furthermore, it was highlighted that both estrogens and progesterones directly and indirectly, via activation of phytohormone synthesis, affect photosynthesis and carotenogenesis in algal cells (Grossmann, 2010; Kozlova et al., 2019). Indeed, it was demonstrated early that estrogen has drastic effect on photosystem II (PSII) energy fluxes (Perron and Juneau, 2011).

Importantly for the biofuel and bioproducts industry, two sex steroids, estradiol (E2) and progesterone (17,20β-P), were shown to significantly induce lipid production and alter FA profiles in a recent study conducted on *S. quadricauda* (Kozlova et al., 2019) (Figure 4C). In this study, it was suggested that, similarly to phytohormones, animal sex steroids could be used as a tool for fatty acid profile manipulation in the biofuel and pharmaceutical industries. Mechanism of FAs profile alteration by animal sex hormones may be linked to chemical transformation of steroids during the exposure resulting in their hydroxylation, oxidation, and hydroxidation, which further results in the formation of conjugates (e.g., glycosides) (Pollio et al., 1994; Luo et al., 2014; Sui et al., 2015). The mechanism of transformation is subject of physico-chemical characteristics of water and cell plasma, hormone concentrations, and algae density. Therefore, this mechanism likely interferes with FAs formation, since both steroids and FAs have ketone and hydroxyl groups in their structure (Zhong-han et al., 1994; Finsterbusch et al., 1999, Janeczko and Skoczowski, 2005).

Estrogens and progestogens *were* detected in seeds (e.g., wheat, rice, corn), vegetable oils, and plant rhizomes (potatoes) in high concentrations of 5.1 ng/g wwt (Hartmann et al., 1998; Lino et al., 2007). Hence, it is possible that these steroid hormones are involved in regulation of lipogenesis and/or in lipid deposition in higher plants. Indeed, estrogens and progestogens can be synthesized by plant cells where their metabolic pathways are accomplished with the same enzymes, which are found in similar pathways in animals, including humans (Brzostek et al., 1998, Li et al., 1997, Rosati et al., 2003). Moreover, estrogen receptors were detected in many species of higher plants (Janik and Adler, 1984). The presence of the receptors and appropriate enzymes for animal steroid metabolism was proposed to play a role in steroid actions in plants (Bhardwaj and Thukral, 2000; Milanesi et al., 2001). This may be similar in algal cells.

2.2.2. Stress Hormones

It has been well established that some microalgae and fungi contain biosynthesized steroid hormones from groups of andro- and estrogens, corticosteroids (mainly glucocorticoids), brassinosteroids, and ekdysteroids. Animal steroids exogenously applied to plant organisms mostly undergo intracellular chemical transformation - biotransformation, and become typical phytosteroids, which show large and varied biological activity. Stress steroids mineralocorticoids, such as aldosterone (11-deoxycortico sterone), and some hydroxyl and methyleno- derivatives can inhibit germination of seeds, development of generative organs, vegetative growth and biosynthesis of nucleic acids (Czerpak and Szamrej, 2003). The characteristic feature of glucocorticoids is the oxygen atom at carbon C-11 of the cyklopentanoperhydrophenanthrenic skeleton. Differences in chemical structures of sex hormones and corticoid hormones can greatly influence their biological and biochemical activities. In the microalgal species, *Chlorella vulgaris*, both mineralocorticoid aldosterone and glucocorticoid prednisolone (at concentrations of 5 □M each) stimulated significant increases in fresh weight cell mass (239% and 206%), dry cell weight (134% and 110%, water-soluble protein content (267% and 296%), and reducing sugars (375% and 333%), compared with untreated controls (Czerpak and Szamrej, 2000). As it was reported for animals and plants (Li et al., 1997; Aoyama and Chua, 1997; Pazirandeh et al., 2002; Revollo and Cidlowski, 2009), mineralo- and glucocorticoids might play an important role in microalgae' immune system.

2.3. Challenges of Hormone Application to Microalgae

The major challenges of hormone applications for induction of microalgae biomass production may be summarized as following: 1) the high cost of manufactured hormones; 2) toxicity of hormones and their combinations, and possibility of hormone bioconcentration and/or biomagnification; 3) insufficient knowledge on the influence of hormones and the products of their degradation on the industrially employed microalgae species; and 4) a lack of experimental data on feed-batch and continuous input of dissolved hormones.

As discussed for phyto- and animal hormones, the majority of the research reported in the literature has been conducted at concentrations higher then that expected in hydroponics, fishery wastewaters, or sometimes even higher than in agricultural run-off effluents (Julius et al., 2007; Perron and Juneau, 2011). First disadvantage of high concentrations is elevated cost of such a biomass production, since the concentrations in hundreds of μM or mM can only be achieved via the used of manufactured synthetic hormones.

The second disadvantage is very important, but not often discussed by scientific society. Hartung et al. (2010) declared this problem as: "hormone concentrations for an experiment have to be selected considering the physiology of the hormone in a particular algae species." In other words, the relationship between internally synthesized hormone and its exogenous concentration must be considered (e.g., permeability of algal cell membrane for the hormone, its intracellular transformation) when choosing the range of hormone concentrations in an experiment. High hormone concentrations often greatly, and critically, surpass the physiological relevance of a hormone uptake making questionable the outcomes of the experimental work.

Furthermore, high concentrations of hormones (in μM and mM) brings us to the second issue of hormone application. It can lead to the adverse effect on microalgae physiology and can cause bioconcentration of applied hormone in algae cells (Lai et al., 2002b) contaminating the grown biomass and the end-products. This problem is most important when animal steroids are designed as the inductors of production, which raises the concern of the suitability of the biomass for agriculture (animal feed) and pharmaceutical industry.

The toxicity of estrogens is well studied and the 50% Effective Concentration (EC_{50}) 48 hour value lies between 1 and 5 mg/L E2 for aquatic (Lemnaceae) and terrestrial plants (Heftmann, 1975; Mohsen et al., 1987; Bhardwaj and Thukral, 2000; Oost et al., 2016; Liu et al., 2018) depending on the species and water chemistry, which gives an LOEC value lower than 1 mg/L. However, in a study conducted with the fungus *Pythium periplocum*, a significant decrease in sexual reproduction was observed at much lower E2 concentrations of 272 μg/L, while 900 μg/L E2 was lethal for this species (Hendrix & Guttman, 1968).

Quite a wide-range of toxicities of estrogens may be found in the literature for microalgae. E2 was reported to have the lowest observed adverse effect on growth of *Chlorella vulgaris* cells, as well as on carotenoid and chlorophyll accumulation, at a concentration of 272 μg/L (Bagjuz and Cserpak, 1996a,b). Synthetic estrogen is more toxic. Balina et al., reported LOEC at 4-8 μg/L, while EC_{50} in her study was 80 μg/L when EE2 was applied to *Desmodesmus communis* culture. However, Beecher (2013) reported a high toxic effect of natural E2 on *Pseudokirchneriella subcapitata* growth at low concentrations of 500 ng/L. It is one of the lowest LOEC for E2 toxicity reported for an algal species.

Data on progesterone toxicity to algae and plants is very limited, although Sorensen et al. (1994) suggested that olfactory system of male goldfish can react to waterborne 17,20β-P at concentrations as low as 3 ng/L, but no diverse effect of such small concentrations of the steroid was reported. Kozlova et al. (2019) did not observed any diverse effect of 17,20β-P on *S. quadricauda* in the range of 1 to 1000 ng/L.

Reports on natural phytohormones toxicity are also limited. However, Grossmann (2010) emphasized that high IAA concentrations acts as synthetic auxin-herbicides initially stimulating cell divisions and biosynthetic activity against a background of rapid ABA accumulation, activation of membrane ion channels, and proton pump H+-ATPase in the plasma membrane, in the first couple of hours of exposure. However, in the following 5-10 hours, abnormal processes began: total chlorophyll levels are greatly increased in conjunction with a reduction of transpiration and carbon fixation, which further leads to dysfunction of metabolism.

Indeed, several authors observed adverse effect of phytohormones on microalgae, although the toxic concentrations are varied through the reports. For example, Nobel et al. (2014) observed an inhibitory effect of IAA at concentrations of 10^{-7} M and higher, while Park et al. (2013) did not observe any inhibition at a higher concentration of 10^{-4} M IAA. A possible explanation of these differences lies in morphological and physiological differences between microalgae species and in dissimilarities in algal cell growth conditions. BRs, ABA, GA, and other phytohormones also demonstrated toxic effects on microalgae physiology at high concentrations. BRs inhibited cell size and chlorophyll-a (Kozlova et al., 2017). BL at concentrations of 10^{-7} to 10^{-6} M inhibited growth of *Ch. vulgaris* (Bajguz and Czerpak, 1998). Remarkably, BRs demonstrated adverse effects on growth (Bajguz and Czerpak, 1998), cell size, and accumulation of chlorophyll-a and FAs (Kozlova et al., 2017) at low concentrations of 0.1-1 μM and 50-300 μg/L, respectively. ABA was toxic to biomass production, cell size, and pigment accumulation at 100 to 500 μg/L (Kozlova et al., 2017). Interestedly, Du et al. (2017) observed toxic effects of ABA on growth of *Ch. pyrenoidosa* cells at 100 to 500 μg/L, then induction of growth at 1000 to 10 000 μg/L, followed again by adverse effects at >10 000 μg/L, while only very high concentrations of 50,000 μg/L were adverse for carotenoid accumulation. High ABA concentrations of ≥ 10,000 μg/L were confirmed to be toxic for lipid accumulation by two algae, *Ch. vulgaris* ZF and *Chlorella*. sp FACHB-31 (Wu et al., 2017).

In any case, the concentrations of a hormone in water designed for microalgae production must be lower than the Lowest Observed Effective Concentration (LOEC) for different plants and algal species to avoid not only toxic effects on algal physiology, but also the possibility of accumulation of the hormones in algal cells. Although, for many microalgae species this value must be determined in the designed for experiments and/or biomass production media, taking in account its physico-chemical parameters.

There are some reports that microalgae can bioconcentrate in their cells and subsequently biomagnify hormones into higher trophic levels at high (> 1000 μg/L) waterborne exposures (Pflugmacher et al., 1999; Sijm et al., 1998). This assumption is supported by the fact that microalgae have high membrane permeability for hormones, and often it occurs via a simple diffusion pathway (Williams and Stancel, 1996). However, the risk of bioconcentration is considered to be low for naturally occurred hormones in nano-concentrations, since the concentration factors of natural hormones in algae are significantly lower than those reported for synthetic hormones (Larsson et al., 2000). Lai et al. (2002b) demonstrated that the bioconcentration of estrogens in algal cells might occur via an active binding to receptors or enzyme mechanisms, followed by intracellular biotransformation. Notably, the authors confirmed that biotransformation and biodegradation of intracellular estrogens would work well up to the saturation point, which is 400 g/L/g wwt or greater. Such a high concentration is very unlikely to observe in microalgae production facilities.

Thus, despite the gained data on the positive influence of phytohormones, their combinations, and animal steroids on microalgae biomass production and synthesis of valuable biomolecules, there are several gaps in our understanding of the impact of both phyto- and animal hormones on various features of microalgae physiology. More research on dose-depended risk of algal end-products contamination by applied hormones has to be conducted to clarify the suitability of the biomass grown on effluents from fishery or livestock farms for agriculture (animal feed), cosmetic, and pharmaceutical industry use.

3. Integrative Approach of Water Treatment: EDCs Water Treatment and Algae Biomass for Biofuels and Fertilizers

Microalgae is one of the most demanded and well established agents for secondary and/or tertiary water treatment. Different microalgae species had been successfully employed for nutrients (Shi et al., 2007; Park et al., 2011; Rawat et al., 2013), metals and some toxic organics such detergents and herbicides (Cragg et al., 2011; Zhou et al., 2013; Sun et al., 2014) removal. It was proved that microalgae were capable of easily cleaning wastewater with organic loadings of 100-150 kg BOD_5/ha/day while nutrient assimilation rates reached 600 kg CO_2/ha/day, 24 kg N/ha/day, and 3 kg P/ha/day giving biomass productivity up to 30 g/m^2/day (Cragg et al., 2011). Thus, the use of microalgae for secondary and/or tertiary water treatment not only "polishes" the effluents out of residual organics, but also decreases the release of the gases produced by bacterial communities to the atmosphere (Benemann, 2013), and offers algal biomass, which could be used for biofuels and agricultural (fertilisers) industries.

Common technologies of elimination of EDCs from wastewaters include physical sorption onto biosolids of water treatment tanks or inorganic absorbents, membrane filtration, biological treatment with bacteria of activated sludge and others. However, it was recorded that all the common methods provide only partial removal of EDCs (Kuch and Ballschmiter, 2001; Jeannot et al., 2002; Liu et al., 2009). Therefore, the combination of primary treatment by common technologies with the additional treatment employing microalgae is a promising way to significantly reduce the EDCs and phytohormones-pesticides from domestic and agricultural effluences.

In EDCs water treatment the gained scientific data and industrial experience on the role and the strength of microalgae to remove and degrade steroids of wastewater are somewhat debatable. Many authors stated a great potential of microalgae in removing EDCs from wastewaters, first of all of domestic origin (Lai et al., 2002b; Shi et al., 2010; Wang et al., 2017; Liu et al., 2018). Another authors declared that not any algae can successfully eliminate and biodegrade EDCs from water, as it was shown for the synthetic steroid EE2, one of the major and most hazardous hormone abundantly found in domestic and some agricultural effluents (Della Greca et al., 2008).

The inequality of the opinions on the subject could relate to the differences in ambient conditions of the treatments: physico-chemical parameters of water, algae density and their membrane permeability, EDCs concentrations and their physical features. Indeed, EDCs could be neutral and ionized molecules. In wastewater, both forms of the molecules are likely present. However, the mechanism of adsorption and crossing the cell membranes for these forms of EDCs is different and differently affected by surrounded ware chemistry, which was described for aquatic plants (Collin et al., 2006; Shi et al., 2010; Malchi et al., 2014) and might be similar for microalgae.

Numerous studies conducted on algae species attempted to elucidate the mechanism and the kinetics of steroid removal (including passive and active actions) and their biodegradation by microalgae cells. The research demonstrated rapid (within hours) uptake the dissolved hormones by algae cells. Notably, the amount of adsorbed hormone is often less than its intracellular concentration suggesting high rate of absorption (Koh et al., 2009; Liu et al., 2018). The corresponded percent of intracellular biodegradation (BDP) and the speed of action vary among the studies. For natural estradiols it takes from 3 hours to 4-5 days to be removed and biotransformed, which depends on algal species, water chemistry, lighting, the hormone concentration to algae density ratio, and the presence of bacteria (Shi et al., 2010; Li et al., 2010; Norvill et al., 2017; Liu et al., 2018). Importantly, it was reported that E2 likely cannot be transferred via the food chain (Beecher, 2013).

Thus, the integration of EDCs water treatment with no-food, feed or pharmaceutical related microalgae biomass production is a promising way to establish a low-cost feedstock biomass production on a base of environmental remediation of effluents.

CONCLUSION

In recent decades, the attention of numerous investigators has been focused on the viability of the industrial-scale of microalgae production. One of the promising ways to improve the economic feasibility of third-generation biofuels and bioproducts is integration of the biomass production with maximum energy- and nutrient recycling applying an advanced technology of the use of naturally occurred in wastewater hormones. Integration of the biomass production also offers significant reduction of environmental emissions.

Importantly, nano-concentrations of phyto- and animal hormones commonly occurring in agricultural effluents have been proved to significantly induce microalgae productivity. What is the great importance, phyto- and animal hormones demonstrated clear ability to simultaneously induce both microalgae biomass production and valuable molecules accumulation. It opens the door for the opportunity to bring microalgae biomass production to the industrial scale.

The gained to date results highlight the possibility that hormones could be used as a tool for fatty acids profile manipulation in the biofuel and pharmaceutical industries.

However, as it was highlighted in this chapter, more intensive research has to be conducted to fill the gaps in our knowledge on the intracellular mechanisms of hormones action and their most effective concentration, the biotransformation of the hormones in algae cells and the possibility of their bioconcentration, and, hence, on dose-depended risk of algal end-products contamination. The last but not the least, wide range of microalgae species has to be examined to clarify all the above questions.

REFERENCES

Adarme-Vega T. C., Lim D. K. Y., Timmins M., Vernen F., Li Y., and P. M. Schenk, 2012. Microalgal biofactories: a promising approach towards sustainable ω-3 fatty acid production, *Microb. Cell Factories* 11:96-109.

Adeel M., Song X., Wang Y., Francis D., and Y. Yang, 2017. Environmental impact of estrogens on human, animal and plant life: A critical review, *Environ. Internat.* 99:107–119.

Aflalo C., Bing W., Zarka A., and S. Boussiba, 1999. The effect of the herbicide glufosinate (Basta) on astaxanthin accumulation in the green alga *Haematococcus pluvialis*, Z. Naturforsch., C: *Bio Sci.,* 54: 49–54.

Alazem M., and N. S. Lin, 2015. Roles of plant hormones in the regulation of host-virus interactions, *Mol Plant Pathol*, 16:529–540.

Aoyama T., and N. H. Chua, 1997. A glucocorticoid-mediated transcriptional induction system in transgenic plants, *Plant J.* 11(3):605-12.

Aravantinou A. F., Theodorakopoulos M. A. and I. D. Manariotis, 2013. Selection of microalgae for wastewater treatment and potential lipids production, *Bioresour. Technol.* 147:130–4.

Aris A. Z., Shamsuddin A. S., and S. M. Praveena, 2014. Occurrence of 17α-ethynylestradiol (EE2) in the environment and effect on exposed biota: a review, *Environ. Internat,* 69:104–119.

Babu A. G., Wu X., Kabra A. N., and D. P. Kim, 2017. Cultivation of an indigenous *Chlorella sorokiniana* with phytohormones for biomass and lipid production under N-limitation, *Algal Res.* 23:178–185.

Bailey D. S. and R. S. Ferrarezi, 2017. Valuation of vegetable crops produced in the UVI Commercial Aquaponic System, *Aquacul. Reports* 7:77–82.

Bajguz A. and R. Czerpak, 1996a. Metabolic activity of estradiol in *Chlorella vulgaris* Beijerinck (*Chlorophytaceae*) Part 1. Content of photosynthetic pigments, *Pol Arch Hydrobiol* 43:421-426.

Bajguz A. and R. Czerpak, 1996b. Metabolic activity of estradiol in *Chlorella vulgaris* Beijerinck (Chlorophyceae) Part II. Content of the cellular sugar and protein accumulation, *Pol. Arch. Hydrobiol.* 43:427-430.

Bajguz A. and R. Czerpak, 1998. Physiological and biochemical role of brassinosteroids and their structure-activity relationship in the green alga *Chlorella vulgaris* Beijerinck (Chlorophyceae), *J. Plant Growth Regul.* 17:131–139.

Bajguz A., 2000. Effect of brassinosteroids on nucleic acids and protein content in cultured cells of *Chlorella vulgaris*, *Plant Physiol. Biochem.* 38: 209–215.

Bajguz A., 2002. Brassinosteroids and lead as stimulators of phytochelatins synthesis in *Chlorella vulgaris*, *J. Plant Physiol.* 159:321–324.

Bajguz A., 2011. Suppression of *Chlorella vulgaris* growth by cadmium, lead, and copper stress and its restoration by endogenous brassinolide, *Arch. Environ. Contam. Toxicol.* 60:406. doi:10.1007/s00244-010-9551-0.

Bajguz A. and A. Piotrowska-Niczyporuk, 2013. Synergistic effect of auxins and brassinosteroids on the growth and regulation of metabolite content in the green alga *Chlorella vulgaris* (Trebouxiophyceae), *Plant Physiol. Biochem.* 71:290-297. http://dx.doi.org/ 10.1016/j.plaphy. 2013.08.003.

Bajguz A., and A. Piotrowska-Niczyporuk, 2014. Interactive effect of brassinosteroids and cytokinins on growth, chlorophyll, monosaccharide and protein content in the green alga *Chlorella vulgaris* (Trebouxiophyceae), *Plant Physiol. Biochem.* 80:176–83.

Bansal V., Savidov N. and J. Rakocy, 2005. The effects of aquaponics and hydroponics nutrient solutions on shoot and root growth of six greenhouse crops, *Austral. J. of Crop Sci.* AJCS 4(2): 85-90. ISSN: 1835-2707.

Bao F., Shen J., Brady S. R., Muday G. K., Asami T., and Z. Yang, 2004. Brassinosteroids interact with auxin to promote lateral root development in *Arabidopsis*, *Plant Physiol.* 134: 1624-1631.

Barry T. P., Riebe J. D., Parrish J. J., and J. A. Malison, 1997. Effects of 17α, 20β-Dihydroxy-4-pregnen-3-one on cortisol production by Rainbow Trout interrenal tissue *in vitro*, *Gener. and Comparat. Endocrin.* 107(2):172-81.

Baronti C., Curini R., D'Ascenzo G., Di Corcia A., Gentili A., and R. Samperi, 2000. Monitoring natural and synthetic estrogens at activated sludge sewage treatment plants and in receiving river water, *Environ. Sci. Technol.* 34:5059–5066.

Beecher L. 2013. *Assessment of 17β-estradiol removal from wastewater via abiotic and biotic routes and potential effects on food chain pathways*. A Dissertation Presented to the Graduate School of Clemson University: 152pp.

Benemann J., 2010. Photobioreactors vs. ponds for microalgae biomass production, *Aquacult. Soc. Annual Meeting, World Aquaculture*, San Diego, CA, March 3. https://www.was.org/documents/MeetingPresentations/AQ2010/AQ2010_0027

Benemann J., 2013. Microalgae for biofuels and animal feeds, *Energies.* 6:5869-5886.

Bhardwaj R., and A. K. Thukral, 2000. Effects of steroids on growth and biochemical parameters of maize plants. *Conference materials of the 12th Congress of the Federation of European Societies of Plant Physiology*. 21-25 August 2000, Budapest, Hungary, p 82.

Bi R., Zeng X., Mu L., Hou L., Liu W., Li P., Chen H., Li D., Bouchez A., Tang J., et al., 2018. Sensitivities of seven algal species to triclosan, fluoxetine and their mixtures. *Sci. Rep.* 8:15361. doi: 10.1038/s41598-018-33785-1.

Bishop G. J., and T. Yokota, 2001. Plants steroid hormones, brassinosteroids: current highlights of molecular aspects on their synthesis/metabolism, transport, perception and response, *Plant Cell Physiol.* 42:114–120.

Bopp-Buhler M. L., Wabra P., Hartung W., and H. Gimmler, 1991. Evidence for direct ABA synthesis in *Dunaliella* (Volvocales). *Cryptogamic Botany* 2/3:192–200.

Borowitzka M. A., 1988. Fats, oils and hydrocarbons. In, *Micro-algal biotechnology*, edited by M. A. Borowitzka and L. J. Borowitzka, Cambridge University Press, Cambridge: 257-287.

Boussiba, S., Gibson, J., 1991. Ammonia translocation in cyanobacteria. FEMS *Microbiol. Lett.* 88 (1), 1e14.

Boussiba S., 2000. Carotenogenesis in the green alga *Haematococcus pluvialis*: Cellular Physiol. and Stress Responce, *Physiol. Plantar.* 108(2):111–117.

Briens C., Piskorz J., et al., 2008. Biomass valorization for fuel and chemicals production. A Review. *Internat. J. of Chemical Reactor Engineer*. 6(6):2-9.

Brzostek A., Popławski T., Murooka Y., and J. Dziadek, 1998. Genes and enzymes involved in conversions of steroids (in Polish). *Biotechnologia* 2: 65-85.

Camara M. C., Lopes da Silva A. L. Rodrigues C., Vandenberghe L. 2015. General Aspects and Applications of Gibberelins and Gibberellic Acid in Plants. p. 1-21 In, Hardy, J. (Org.). Gibberellins and Gibberellic *Acid: Biosynthesis, Regulation and Physiological Effects*. 1st ed. Hauppauge: Nova Science Publishers.

Cano-Delgado A., Yin Y., Yu C., Vafeados D., Mora-Garcia S., Cheng J. C., et al., 2004. BRL1 and BRL3 are novel brassinosteroid receptors that function in vascular differentiation in Arabidopsis, *Development* 131:5341–5351.

Chen C. Y., Wen T. Y., Wang G. S., Cheng H. W., Lin Y. H., and G. W. Lien, 2007. Determining estrogenic steroids in Taipei waters and removal in drinking water treatment using high-flow solidphase extraction and liquid chromatography/tandem mass spectrometry, *Sci Total Environ.* 378:352–65.

Cobb A., 1992. *Herbicides and plant physiology, Auxin-type herbicides*, Chapman Hall: 82–106.

Colla L. M., Reinehr C. O., Reichert C. and J. A. Vieira Costa, 2007. Production of biomass and nutraceutical compounds by *Spirulina platensis* under different temperature and nitrogen regimes, *Biores. Technol*. 98(7):1489-93.

Collins C., Fryer M., and A. Grosso, 2006. Plant uptake of non-ionic organic chemicals, *Environ. Sci. Technol*. 40:45–52.

Collins C. D., Martin I., and W. Doucette, 2011. Plant uptake of xenobiotics. In: Schroder, P., Collins, C. D. (Eds.), *Organic Xenobiotics and Plants: From mode of action to Ecophysiology*. Springer, Heidelberg, pp. 3–16.

Clouse S. D., and J. M. Sasse, 1998. Brassinosteroids: essential regulators of plant growth and development. *Annu Rev Plant Physiol Plant Mol Biol* 49:427–451.

Cowan A. K., and P. D. Rose, 1991. Abscisic acid metabolism in salt-stressed cells of *Dunaliella salina, Plant Physiol.* 97:798-803.

Craggs R. J., Heubeck S., Lundquist T. J., and J. R. Benemann, 2011. Algal biofuels from wastewater treatment high rate algal ponds, *Water Sci. Technol.* 63:660–665.

Czerpak R., Bajguz A., Bialecka B., Wierzcholowska L. E., and M. M. Wolanska, 1994. Effect of auxin precursors and chemical analogues on the growth and chemical composition in *Chlorella pyrenoidosa*, Chick. *Acta Soc. Bot. Pol.* 63:279-285.

Czerpak R., and A. Bajguz, 1997. Stimulatory effect of auxins and cytokinins on carotenes, with differential effects on xanthophylls in the green alga *Chlorella pyrenoidosa*, Chick. *Acta Soc. Bot. Pol.* 66:41–46.

Czerpak R., Krotke A., and A. Mical, 1999. Comparison of stimulatory effect of auxins and cytokinins on protein, saccharides and chlorophylls content of *Chlorella pyrenoidosa*, *Poloniae Arch. Hygrobiol.* 46: 71-82.

Czerpak R., Szamrej I. K. 2000. Metabolic activity of 11-deoxycorticosterone and prednisolone in the alga *Chlorella vulgaris* Beijerinck. *Acata Societatis Botanicorum Polonae* 69(1):25-19.

Czerpak R., Szamrej I. K. 2003. The effect of β-estradiol and corticosteroids on chlorophylls and carotenoids content in *Wolffia arrhiza* (L.) Wimm. (Lemnaceae) growing in municipal Bialystok tap water. *Polish J. of Environ. Studies* 12(6):677-684.

Czygan J. C., 1962. Blütenbildung bei *Lemna minor* nach Zusatz von Oestrogen [Flower formation in *Lemna minor* after the addition of edtrogen] *Naturwissenschaften* 49:285-286.

d'Ascenzo G., di Corcia A., Gentili A., Mancini R., Mastropasqua R., Nazzari M., and R. Samperi, 2003. Fate of natural estrogen conjugates in municipal sewage transport and treatment facilities. *Sci. Total Environ.* 302:199–209.

Dalrymple O. K., Halfhide T., Udom I., Gilles B., Wolan J., Zhang Q., and S. Ergas, 2013. Wastewater use in algae production for generation of renewable resources: a review and preliminary result, *Aqua. Biosyst.* 9(1):1-11.

Davies P. J., 2004. Regulatory factors in hormone action. Level, location and signal transduction. pp. 16-35 In: *Plant Hormones,* Kluwer Academic Publishers, 716pp. doi: 10.1186/2046-9063-9-2.

Della Greca M., Pinto G., Pistillo P., Pollio A., Previtera L., and F. Temussi, 2008. Biotransformation of ethinylestradiol by microalgae, *Chemosphere* 70:2047–53.

Department of Energy of the United States (DOE), 2010. *Energy Efficiency and Renewable Energy*, Golden Field Office, US.

De Vleesschauwer D., Van Buyten E., Satoh K., Balidion J., Mauleon R., Choi Il-R., Vera-Cruz C., Kikuchi S., and M. Höfte, 2012. Brassinosteroids antagonize gibberellin and salicylate mediated root immunity in rice, *Plant Physiol.* 158 (4):1833–1846. http://dx.doi.org/10.1104/pp.112.193672.

Du H., Ahmed F., Lin B., Li Z., Huang Y., Sun G., Ding H., Wang Ch., Meng Ch., and Zh. Gao 2017. The Effects of Plant Growth Regulators on Cell Growth, Protein, Carotenoid, PUFAs and Lipid Production of *Chlorella pyrenoidosa* ZF Strain, *Energies*, 10:1696. doi:10.3390/en10111696.

EC, SEC 1272, 1276, 2010: *Planning for a sustainable future: a Federal Sustainable Development Strategy for Canada*, 2010. Environment Canada, ISBN: 978-1-100-16796-1.

Esteban S., Gorga M., Petrovic M., Gonzalez-Alonso S., Barcelo D., and Y. Valcarcel, 2014. Analysis and occurrence of endocrine-disrupting compounds and estrogenic activity in the surface waters of central Spain, *Sci. Total Environ.* 466–467:939–951. doi:10.1016/j.scitotenv.2013.07.101.

Fan J., Yan C., Andre C., Shanklin J., Schwender J., and C. Xu, 2012. Oil accumulation is controlled by carbon precursor supply for fatty acid synthesis in *Chlamydomonas reinhardtii*, *Plant Cell Physiol.* 53(8):1380-1390.

Fernández-Sevilla J. M., Acién F. G., and E. Molina, 2010. Biotechnological production of lutein and its applications, *Appl. Microbiol. Biotechnol.* 86:27–40.

Finsterbusch A., Lindemann P., Grimm R., Eckerskorn C., and M. Luckner, 1999. δ5-3β-hydroxysteroid dehydrogenase from *Digitalis lanata* Ehrh. - a multifunctional enzyme in steroid metabolism, *Planta* 209:478-486.

Graham L. E., and L. W. Wilcox, 2000. *Algae*. Prentice Hall, Upper Saddle River, NJ.

Grossmann, 2010. Auxin herbicides: current status of mechanism andmode of action, *Pest Manag Sci.,* 66:113–120.

Guo Z., Liu Y., Guo H., Yan S. and J. Mu, 2013. Microalgae cultivation using an aquaculture wastewater as growth medium for biomass and biofuel production, *J. Environ. Sci.* (China) 25(Suppl 1):85–88.

Hagen G., Guilfoyle T. J., and W. M. Gray, 2004. Auxin signal transduction. pp. 282-303. In Davies 2004. *Plant Hormones*, Kluwer Academic Publishers, 716pp. doi:10.1186/2046-9063-9-2.

Haiyan R., Shulan J., ud din Ahmad N., Dao W., and C. Chengwu, 2007. Degradation characteristics and metabolic pathway of 17α-ethynylestradiol by Sphingobacterium sp. JCR5, *Chemosphere* 66: 340–346.

Hala D., Petersen L. H., Martinovic D., and D. B. Huggett, 2012. Constraints-based stoichiometric analysis of hypoxic stress on steroidogenesis in fathead minnows, *Pimephales promelas*, *The J. of Experim. Biol.* 215:1753-1765.

Han S., Jin W., Chen Y., et al., 2016. Enhancement of lipid production of *Chlorella Pyrenoidosa* cultivated in municipal wastewater by magnetic treatment, *Appl. Biochem. Biotechnol.* 180(6):1043-1055. doi:10.1007/s12010-016-2151-3.

Harrison M. A., 2012. Cross-talk between phytohormone signaling pathways under both optimal and stressful environmental conditions. In: Khan N. A., Nazar R., Iqbal N., Anjum N. A. (eds.) Phytohormones and abiotic stress tolerance in plants, *Springer,Berlin*: 49–76.

Hartmann S., Lacorn M., Steinhart H. 1998. Natural occurrence of steroid hormones in food, *Food Chemis.* 62:7-20.

Hartung W., 2010. The evolution of abscisic acid (ABA) and ABA function in lower plants, fungi and lichen, *Funct. Plant Biol.* 37:806–812.

Hedden P., and V. Sponsel, 2015. "A Century of Gibberellin Research," *J. of Plant Growth Regul,* 34 (4):740–60. doi:10.1007/s00344-015-9546-1

Hendrix J. W., and S. M. Guttman, 1968. Annulment of sterol-induced sexual reproduction by estradiol in *Pythium periplocum*, *Sci.* 161:1252.

Hirsch R., Hartung W., and H. Gimmler, 1989. Abscisic acid content of algae under stress, *Botan. Acta* 102(4):326–334. doi:10.1111/j.1438-8677.1989.tb00113.x.

Hunt R. W., Chinnasamy S., Bhatnagar A., and K. C. Das, 2010. Effect of biochemical stimulants on biomass productivity and metabolite content of the microalga, *Chlorella sorokiniana*, *Appl. Biochem. Biotechnol.* 162(8):2400–2414.

Iwamoto K. and Y. Shiraiwa, 2012.Characterization of intracellular iodine accumulation by iodine-tolerant microalgae, *Procedia Envir. Sci.* 15:34–42.

Janeczko A., Filek W., Biesaga-Koscielniak J., Marcinska I., and Z. Janeczko, 2003. The influence of animal sex hormones on the induction of flowering in *Arabidopsis thaliana*: comparison with the effect of 24-epibrassinolide, *Plant Cell Tissue Organ Cult*. 72:147-151.

Janeczko A., and A. Skoczowski, 2005. Mammalian sex hormones in plants, *Folia Histochemica et cytobiologica* 43(2):71-79.

Janeczko A. 2012. The presence and activity of progesterone in the plant kingdom, *Steroids,* 77:169-173.

Janik J. R., and J. H. Adler, 1984. Estrogen receptors in *Gladiolus ovules*, *Plant Physiol.* 75:135.

Jeannot R., Sabik H., Sauvard E., Dagnac T., and K. Dohrendorf, 2002. Determination of endocrine-disrupting compounds in environmental samples using gas and liquid chromatography with mass spectrometry, *J.of Chromatog. A*, 974:143-159.

Jin E., Polle J. E. W., Lee H. K., Hyun S. M., and M. Chang, 2003. Xanthophylls in microalgae: From biosynthesis to biotechnological mass production and application, *J. of Microb. and Biotech.*, 13(2):165-174.

Jobling S., and R. Owen, 2003. Ethinyl oestradiol in the aquatic environment, *Emerg. Lessons Ecosyst.* 13:279-307.

Johnson A. C., Belfroid A., and A. Di Corsa, 2000. Estimating steroid oestrogen inputs to activated sludge treatment works and observations on their removal from the effluent, *Sci. Total Environ.,* 256 (2-3):163-173.

Johnson A. C., and R. J. Williams, 2004. A model to estimate influent and effluent concentrations of estradiol, estrone, and ethinylestradiol at sewage treatment works, *Environ Sci Technol* 38:3649–3658.

Julius M. L., Stepanek J., Tedrow O'N, Gamble C., and H. L. Schoenfuss, 2007. Estrogen-receptor independent effects of two ubiquitous environmental estrogens on *Melosira varians* Agardh, a common component of the aquatic primary production community, *Aqua. Tox.* 85:19–27.

Julius M. L. and E. C. Theriot. 2011. The diatoms: A primer, In book: The Diatoms: Applications for the Environmental and Earth Sciences Edition: Second Chapter: 2 Publisher: Cambridge University Press Editors: John P. Smol, Eugene F. Stoermer. doi: 10.1017/CBO9780511763175.003.

Kang C. D., An J. Y., Park T. H., and S. J. Sim, 2006. Astaxanthin biosynthesis from simultaneous N and P uptake by the green alga *Haematococcus pluvialis* in primary-treated wastewater, *Biochem. Eng. J.* 31:234–238.

Karnachuk R. A., Golovatskaya I. F., Efimova M. V. and V. A. Khripach, 2002. Action of epibrassinolide on morphogenesis and hormonal balance in Arabidopsis seedlings at green light, *Plant Physiol.* 49:1-5.

Katsumi M., 1985. Interaction of a brassinosteroid with IAA and GA3 in the elongation of Cucumber hypocotyl sections, *Plant Cell Physiol* 26:615-625.

Khasina M., Cahoon R. B., Alvarezc S., Beckeris R., Eyunc S., Jiac O., Nickersona K. W., and W. R. Riekhofa, 2017. Synthesis, secretion, and perception of abscisic acid regulates stress responses in *Chlorella sorokiniana*. http://dx.doi.org/10.1101/180547.

Khripach V., Zhabinski V., and A. De Groot, 2000. Twenty years of brassinosteroids: steroidal plant hormones warrant better crops for the XXI century, *Annals of Botany* 86:441-447. doi:10.1006/anbo.2000.1227.

Kim M. K., Park J. W., Park C. S., Kim S. J., Jeune K. H., Chang M. U., and J. Acreman, 2007. Enhanced production of *Scenedesmus* spp. (green microalgae) using a new medium containing fermented swine wastewater, *Biores. Technol.* 98:2220–2228.

Kim J., Lingaraju B. P., Rheaume R., Lee J. Y., and K. F. Siddiqui, 2010. Removal of ammonia from wastewater effluent by *Chlorella vulgaris, Tsinghua Sci. and Technol.,* 15(4):391-396.

Kiseleva A. A., Tarachovskaya E. R., and M. F. Shishova, 2012. Biosynthesis of phytohormones in algae, *Russ. J. Plant Physiol.* 59:595–610.

Kjær, J., Olsen, P., Bach, K., Barlebo, H. C., Ingerslev, F., Hansen, M., Sørensen, B. H., 2007. Leaching of estrogenic hormones from manure-treated structured soils, *Environ. Sci. Technol.,* 41:3911–3917.

Koh, Y. K. K., Lester J. N. and M. D. Scrimshaw, 2005. Fate and behavior of alkylphenols and their polyethox-ylates in anactivated sludge plant, *Bullet. of Envir. Contam. and Toxicol.* 75:1098–1106. doi: 10.1007/s00128-005-0862-1.

Kolpin D. W., Furlong E. T., Meyer M. T., Thurman E. M., Zaugg S. D., Barber L. B., … Al., E. 2002. Pharmaceuticals, hormones, and other organic wastewater contaminants in U.S. streams, 1999 - 2000. *Environ. Sci. Technol.,* 36:1202–1211.

Kong Q., Li L., Martinez B., Chen P., and R. Ruan, 2010. Culture of microalgae *Chlamydomonas reinhardtii* in wastewater for biomass feedstock production, *Appl. Biochem. Biotechnol.* 160(1):9-18. doi:10.1007/s12010-009-8670-4.

Kopcewicz J., and Z. Porazinski Z, 1974. Effect of growth regulators, steroids and estrogen fraction from sage plants on flowering of a long-day plant, *Salvia splendens*, grown under non-inductive light conditions, *Biol. Plant (Praha)* 16:132-135.

Kozlova T. A., Hardy B. P., Krishna P., and D. B. Levin, 2017. Effect of phytohormones on growth and accumulation of pigments and fatty acids in the microalgae *Scenedesmus quadricauda, Algal Res.* 27:325-334. https://doi.org/doi:10.1016/j.algal.2017.09.020.

Kozlova T. A., Hardy B. P. and D. B. Levin 2018 The combined influence of 24-epibrassinolide and 3-indoleacetic acid on growth and accumulation of pigments and fatty acids in the microalgae Scenedesmus quadricauda (CPCC-158), *Algal Res.* 35:22-32. https://doi.org/10.1016/j.algal.2018.08.009.

Kozlova T. A., Hardy B. P., and D. B. Levin, 2019. Effect of Fish Steroids 17β-estradiol and 17,20β-dihydroxy-4-pregnen-3-one on Growth, Accumulation of Pigments, and Fatty Acid Profiles in the Microalgae

Scenedesmus quadricauda (CPCC-158), *Renew. Ener.*, https://doi.org/10.1016/j.renene.2019.10.165.

Kuch H. M., and K. Ballschmiter, 2001. Determination of endocrine-disrupting phenolic compounds and estrogens in surface and drinking water by HRGC-(NCI)-MS in the Picogram per liter range. *Environ. Sci. & Technol.* 35:3201-3206.

Larsson D. G. J, Mayer I., Hyllner S. J., and L. Forlin, 2002. Seasonal variation of vitelline envelope proteins, vitellogenin, and sex steroids in male and female eelpout (*Zoarces viviparus*), *Gen. Comp. Endocrin.* 125:184-196.

Lai K. M., Scrimshaw M. D., and J. N. Lester. 2002a. Prediction of the bioaccumulation factors and body burden of natural and synthetic estrogens in aquatic organisms in the river systems, *Sci. Total Environ.* 289(1-3):159-68.

Lai K., Scrimshaw M., and J. Lester, 2002b. Biotransformation and bioconcentration of steroid estrogens by *Chlorella vulgaris*, *Appl. Environ. Microbiol.* 68:859–864.

Lemoine Y., and B. Schoefs, 2010. Secondary ketocarotenoid astaxanthin biosynthesis in algae: A multifunctional response to stress, *Photosynth. Res.* 106:155–177.

Li J., Biswas M. G., Chao A., Russell D. W., and J. Chory, 1997. Conservation of function between mammalian and plant steroid 5α-reductases, *Proc. Natl. Acad. Sci. USA* 94:3554-3559.

Li J., Zhu D., Niu J., and 5 others, and G. Wang, 2011. An economic assessment of astaxanthin production by large-scale cultivation of *Haematococcus pluvialis*, *Biotech. Advan.* 29(6):568-74. doi: 10.1016/j.biotechadv.2011.04.001.

Li P., Mao X., Li R., and J. Zhong, 2011. In situ biodiesel production from fast-growing and high oil content *Chlorella pyrenoidosa* in rice straw hydrolysate, *J. of Biomed. and Biotech.* p.8. doi:10.1155/2011/141207.

Lino M., Nomura N., Tamaki Y., Yamada Y., Yoneyama K., Takeuchi Y., Mori M., Asami T., Nakano T., and T. Yokota, 2007. Progesterone: its occurrence in plants and involvement in plant growth, *Phytochem.* 68 (12):1664–73.

Liu Z. H., Kanjo Y., and S. Mizutani, 2009. Removal mechanisms for endocrine disrupting compounds (EDCs) in wastewater treatment d physical means, biodegradation, and chemical advanced oxidation: a review. *Sci. of Total Environ.* 407:731-748.

Liu J., Qiu W., and Y. Song, 2016. Stimulatory effect of auxins on the growth and lipid productivity of *Chlorella pyrenoidosa* and *Scenedesmus quadricauda*, *Algal Res.* 18:273–280.

Liu W., Chen Q., He N., Sun K., Sun D., Wu X., and S. Duan, 2018. Removal and biodegradation of 17β-Estradiol and diethylstilbestrol by the freshwater microalgae *Raphidocelis subcapitata*, *Int. J. Environ. Res. Public Health* 15:452. doi:10.3390/ijerph15030452.

Leech D. M., Snyder M. T., and R. G. Wetzel, 2009. Natural organic matter and sunlight accelerate the degradation of 17b-estradiol in water, *Sci. of Total Environ.* 407:2087-2092.

Lohr M., Schwender J., and J. E. W. Polle, 2012. Isoprenoid biosynthesis in eukaryotic phototrophs: a spotlight on algae, *Plant Sci.* 185-186:9–22.

Lopez M. L., Peralta-Videa J. R., Benitez T., and J. L. Gardea-Torresdey, 2005. Enhancement of lead uptake by alfalfa (*Medicago sativa*) using EDTA and a plant growth promoter, *Chemosph.* 61:595–598.

Ludwig-Müller J., 2011. Auxin conjugates: their role for plant development and in the evolution of land plants, *J. Exp. Bot.* 62(6):1757–1773. doi:10.1093/jxb/erq41.

Luo Y., Guo W., Ngo H. H., Nghiem L. D., Hai F. I., Zhang J., Liang S. and X. C. Wang, 2014. A review on the occurrence of micropollutants in the aquatic environment and their fate and removal during wastewater treatment, *Sci. of Total Environ.* 473:619-641.

Machida S. and I. Suzuki, 2018. Characterization of cyanobacterial cells synthesizing 10-methyl stearic acid, *Photosyn. Res.* 139(1):173-183.

Malca J., and F. Freire, 2006. Renewability and life-cycle energy efficiency of bioethanol and bio-ethyl tertiary butyl ether (bioETBE): Assessing the implications of allocation, *Energy* 31:3362–3380.

Malchi T., Maor Y., Tadmor G., Shenker M., and B. Chefetz, 2014. Irrigation of root vegetables with treated wastewater: evaluating uptake of pharmaceuticals and the associated human health risks, *Environ. Sci. Technol.* 48:9325–9333.

Mandava N. B., 1988. Plant growth-promoting brassinosteroids, *Ann. Rev. Plant Physiol. Plant Mol. Biol.* 39:23-52. http://dx.doi.org/10.1146/annurev.pp.39.060188.000323.

McGinn P. J., Dickinson K. E., Bhatti S., Frigon J. C., Guiot S. R., and S. J. B. O'Leary, 2011. Integration of microalgae cultivation with industrial waste remediation for biofuel and bioenergy production: opportunities and limitations, *Photosynth. Resear.* 109:231-247.

McGinn P. J., Dickinson K. E., Park K. C., Whitney C. G., MacQuarrie S. P., Black F. J., Frigon G. S., Guiot S. R., and S. G. B. O'Leary, 2012. Assessment of the bioenergy and bioremediation potentials of the microalga *Scenedesmus sp*. AMDD cultivated in municipal wastewater effluent in batch and continuous mode, *Algal Res.* 1(2):155-165. doi:10.1016/j.algal.2012.05.00.

McKendry P., 2002. Energy production from biomass (part 2): conversion technologies, *Biores. Technol.* 83:47–54.

McMurtry M. R., Snaders D. C., Cure J. D., and R. G. Hodson, 1997. Effects of biofilter/culture tank volume ratio on productivity of s recirculating fish/vegetable co-culture system, *J. Appl. Aquacult.* 7(4):33-51.

Milanesi L., Monje P., and R. Boland, 2001. Presence of estrogens and estrogen receptor-like proteins in *Solanum glaucophyllum*, *Biochem. Biophys. Res. Commun.* 289:1175-1179.

Mota V. C., Martins C. I. M., Eding E. H., Canário A. V. M., and J. A. J. Verreth, 2014. Steroids accumulate in the rearing water of commercial recirculating aquaculture systems, *Aquacult. Engineer.* 62:9–16.

Moya J. L., Ros R., and I. Picazo, 1995. Heavy metal-hormone interactions in rice plants: Effects on growth, net photosynthesis, and carbohydrate distribution, *J. Plant Growth Regul.* 14:61-67.

Müller B., 2011. Generic signal-specific responses: cytokinin and context-dependent cellular responses, *J. Exp. Bot.* 62:3273–3288.

Nagahama Y. and M. Yamashita, 2008. Regulation of oocyte maturation in fish, *Develop. Growth Differ.* 50:195-219.

Navarri P., 2015. Integration Software – Optimizing heat recovery and reducing waste heat generation in industrial processes, peer-review report CETC number: 2015-036, *Natural Resources Canada*.

Nemhauser J. L., T. C. Mockler, J. Chory, 2004. Interdependency of brassinosteroid and auxin signaling in Arabidopsis, *PLoS Biol.* 2: E258.

Nendza M., and A. Wenzel, 2006. Discriminating toxicant classes by mode of action, *Environ. Sci. Pollut. Res.* 13(3):192-203.

Nakamura A, Higuchi K, Goda H, Fujiwara M, Sawa S, Koshiba T, Shimada Y, and S. Yoshida, 2003. Brassinolide induces IAA5, IAA19, and DR5, a synthetic auxin response element in Arabidopsis, implying a cross talk point of brassinosteroid and auxin signaling, *Plant Physiol.* 133:1843–1851.

Nobel A., Kisiala A., Galer D., and R. J. N. Clysdale, 2014. Emery, *Euglena gracilis* (Euglenophyceae) produces abscisic acid and cytokinins and responds to their exogenous application singly and in combination with other growth regulators, *Eur. J. Phycol.* 49(2): 244–254.

Norvill Z. N., Toledocervantes A., Blanco S., Shilton A., Guieysse B., and R. Muñoz, 2017. Photodegradation and sorption govern tetracycline removal during wastewater treatment in algal ponds, *Bioresour. Technol.* 232:35–43.

Olaizola M., 2000. Commercial production of astaxanthin from *Haematococcus pluvialis* using 25,000-liter outdoor photobioreactors, *J. of Appl. Phycol.* 12(3-5):499-506.

Osborne D. J., and M. T. McManus, 2005. Hormones, signals and target cells in plant development, *Cambridge University Press.* p. 158. ISBN 978-0-521-33076-3.

Owen R., and S. Jobling, 2013. Ethyinlestradiol in the aquatic environment. In: *Late lessons from early warnings: science, precaution, innovation.* Copenhagen: Europ. Environ. Agen. (EEA):279–307.

Ozioko F. U., Chiejina N. V., and J. C. Ogbonna, 2015. Effect of some phytohormones on growth characteristics of *Chlorella sorokiniana*

IAM-C212 under photoautotrophic conditions, *Afric. J. Biotech.* 14(30):2367-2376. doi:10.5897/AJB2015. 14641.

Park J. B. K., Craggs R. J., and A. N. Shilton, 2011. Wastewater treatment high rate algal ponds for biofuel production, *Bioresour. Technol.* 102(1):35–42.

Park W., Yoo G., Moon M., Kim C. W., Choi Y. E., and J. W. Yang, 2013. Phytohormone supplementation significantly increases growth of *Chlamydomonas reinhardtii* cultivated for biodiesel broduction, *Appl. Biochem. Biotechnol.* 171:1128–1142. doi:10.1007/s12010-013-0386-9.

Parry A. D., Babiano M. J., and R. Horgan, 1990. The role of ciscarotenoids in abscisic acid biosynthesis, *Planta* 182:118–128.

Pastori G. M., and C. H. Foyer, 2002. Common components, networks, and pathways of cross-tolerance to stress. The central role of "redox" and abscisic acid-mediated controls, *Plant Physiol.* 129(2):460-468.

Pazirandeh A., Xue Y., Prestegaard T., Jondal M., and S. Okret, 2002. Effects of altered glucocorticoid sensitivity in the T cell lineage on thymocyte and T cell homeostasis, *FASEB Journal.* 16(7):727–9. doi:10.1096/fj.01-0891.

Peleg Z., and E. Blumwald, 2011. Hormone balance and abiotic stress tolerance in crop plants, *Curr. Opin. Plant Biol.* 14:290–295. http://dx.doi.org/10.1016/j.pbi.2011.02.001.

Perez-Garcia O., Escalante F. M. E., de-Bashan L. E., and Y. Bashan, 2011. Heterotrophic cultures of microalgae: metabolism and potential products, *Water Res.* 45(1):11-36.

Perez-Lopez P., Gonzalez-Garcia S., Jeffryes C., Agathos S. N., Mchugh E.,et al., 2014. Life cycle assessment of the production of the red antioxidant carotenoid astaxanthin by microalgae: from lab to pilot scale, *J. of Clean. Produc.* 64:332-344. ff10.1016/j.jclepro.2013.07.011ff.

Perron M. C., and P. Juneau, 2011. Effect of endocrine disruptors on Photosystem II energy fluxes of green algae and cyanobacteria, *Environ Res.* 111(4):520-529. doi: 10.1016/j.envres.2011.02.013.

Perrot-Rechenmann C., 2010. Cellular responses to auxin: division versus expansion, *Cold Spring, Harb. Perspect. Biol.* 2(5):1446. doi:10.1101/cshperspect.a001446.

Petrus L., and A. M. Noordermeer, 2006. Biomass to biofuels, a chemical perspective, *Green Chem.* 8(10):861-67. doi 10.1039/B605036K.

Pflugmacher S., Spangenberg M., and C. E. W. Steinberg, 1999. Dissolved organic matter (DOM) and effects on the aquatic macrophyte *Ceratophyllum demersum* in relation to photosynthesis, pigment pattern and activity of detoxication enzymes, *J Appl BotAngew Bot.* 73:184–190.

Piotrowska A., Czerpak R., Pietryczuk A., Olesiewicz A., and M. Wędołowska, 2008. The effect of indomethacin on the growth and metabolism of green alga *Chlorella vulgaris* Beijerinck, *Plant Grow. Regul.* 55:125–36.

Piotrowska-Niczyporuk A. and A. Bajguz, 2014. The effect of natural and synthetic auxins on the growth, metabolite content and antioxidant response of green alga *Chlorella vulgaris* (Trebouxiophyceae), *Plant Growth Regul.,* 73:57–66. DOI 10.1007/s10725-013-9867-7.

Pojana G., Gomiero A., Jonkers N. and A. Marcomini, 2007. Natural and synthetic endocrine disrupting compounds (EDCs) in water, sediment, and biota of a coastal lagoon, *Environ. Internat.* 33(7):929-936.

Pollio A., Pinto G., Della Greca M., Di Maio A., Fiorentino A., and L. Previtera, 1994. Progesterone bioconversion by microal-gal cultures, *Phytochem.* 37:1269–1272.

Pollard H. B., and E. Rojas, 1988. Ca^{2+}-activated synexin forms highly selective, voltage-gated Ca^{2+} channels in phosphatidylserine bilayer membranes, *Proc. Natl. Acad. Sci. USA* 85:2974-2978.

Rakocy J. E., Shultz R. C., Bailey D. S. and E. S. Thoman, 2004. Aquaponics production of tilapia and basil: comparing a batch and staggered cropping system, *Acta Hortic.* 648:63-69. doi: 10.17660/ActaHortic.2004.648.8.

Rakocy J. E., Losordo T. M., and M. P. Masser, 2006. Recirculating aquaculture tankmproduction systems: integrating fish and plant culture, southern region, *Aquacul. Center Pub.* 454:1–16.

Rawat I., Kumar R. R., Mutanda T., and F. Bux, 2013. Biodiesel from microalgae: A critical evaluation from laboratory to large scale production, *Appl. Energy* 103:444-467 doi: 10.1016/j.apenergy. 2012.10.004.

Reid J. B., Symons G. M. and J. J. Ross, 2004. Regulation of gibberellin and brassinosteroid biosynthesis by genetic, environmental and hormonal factors. In *Plant Hormones: Biosyn. Signal Transduc. Action.* 3 (P. J. Davies, ed.). London: Kluwer Academic Publishers: 179– 203.

Revollo J. R., and J. A. Cidlowski, 2009. Mechanisms generating diversity in glucocorticoid receptor signaling, *Annals of the New York Acad. of Sci.* 1179:167–78. doi:10.1111/j.1749-6632.2009.04986.

Robert-Seilaniantz A., Grant M. R., and J. D. G. Jones, 2011. Hormone crosstalk in plant disease and defense: more than just jasmonate-salicylate antagonism, *Annu. Rev. Phytopathol.* 49: 2621-2627.

Rober-Kleber N., Albrechtova J. T., Fleig S., Huck N., Michalke W., Wagner E., Speth V., Neuhaus G., and C. Fischer-Iglesias, 2003. Plasma membrane H^+-ATPase is involved in auxin-mediated cell elongation during wheat embryo development, *Plant Physiol.* 131:1302–1312.

Rock C. D., and X. Sun, 2005. Crosstalk between ABA and auxin signaling pathways in roots of *Arabidopsis thaliana* (L.), *Heynh. Planta* 222:98-106.

Romanenko E. A., Kosakovskaya I. V., and P. A. Romanenko, 2015. Phytohormones of microalgae: Biological Role and Involvement in the Regulation of Physiological Processes. Pt I. Auxins, Abscisic Acid, Ethylene. *Int. J. Algae* 17:275–289.

Rosati F., Danza G., Guarna A., Cini N., Racchi M. L., and M. Serio, 2003. New evidence of similarity between human and plant steroid metabolism: 5α-reductase activity in *Solanum malacoxylon*, *Endocrinol.* 144:220-229.

Safafar H., van Wagenen J., Møller P., and Ch. Jacobsen, 2015. Carotenoids, phenolic compounds and tocopherols contribute to the antioxidative properties of some microalgae species grown on industrial wastewater, *Mar. Drugs* 13(12):7339-7356. doi:10.3390/ md13127069.

Salama El-S., Kabra A. N., Ji M. K., Kim J. R., Min B., and B. H. Jeon, 2014. Enhancement of microalgae growth and fatty acid content under the influence of phytohormones, *Bioresour. Technol.* 172:97–103.

Sala C, Sala F (1985) Effect of brassinosteroid on cell division and enlargement in cultured carrot (*Daucus carota* L.) cells, *Plant Cell Rep* 4:144-147.

Sasaki Y., Asamizu E., 11 others and S. Tabata, 2001. Monitoring of methyl jasmonate-responsive genes in Arabidopsis by cDNA macroarray: self-activation of jasmonic acid biosynthesis and crosstalk with other phytohormone signaling pathways, *DNA Res.* 8:153–161.

Sasse J. M., 1999. Physiological actions of brassinosteroids. In: Sakurai A, Yokota T., Clouse S. D. Eds. *Brassinosteroids: steroidal plant hormones* Tokyo, Springer: pp 137–161.

Saygideger S., and O. Okkay, 2008. Effect of 2,4-dichlorophenoxyacetic acid on growth, protein and chlorophyll-a content of *Chlorella vulgaris* and *Spirulina platensis* cells, *J. Environ. Biol.* 29:175–178.

Savidov N., 2004. *Evaluation and development of aquaponics production and product market capabilities in Alberta ids initiatives fund final report*, Project #679056201, Crop Diversification Centre South Brooks, AB.

Schroeder J. I., Kwak J. M., and G. J. Allen, 2001. Guard cell abscisic acid signaling and engineering drought hardiness in plants, *Nature* 410:327-330.

Silva C. P., Otero M., and V. Esteves, 2012. Processes for the elimination of estrogenic steroid hormones from water: a review, *Environ. Pollut.* 165:38-58. doi:10.1016/j.envpol.2012.02.002.

Shi J, Podola B, and M. Melkonian, 2007. Removal of nitrogen and phosphorus from wastewater using microalgae immobilized on twin layers: An experimental study, *J. of Appl. Phycol.* 19(5):417-423.

Shi W., Wang L., Rousseau D. P., and P. N. Lens, 2010. Removal of estrone, 17-ethinylestradiol, and 17ß-estradiol in algae and duckweed-based wastewater treatment systems, *Environ. Sci. Pollut. Res.* 17:824–833.

Shore L. C. and M. Shemesh, 2003. Naturally produced steroid hormones and their release into the environment, *Pure Appl. Chem.* 75(11–12):1859–1871.

Sorensen P. W., and A. P. Scott, 1994. The evolution of hormonal sex pheromones in teleost fish: Poor correlation between the pattern of steroid release by goldfish and olfactory sensitivity suggests that these cues evolved as a result of chemical spying rather than signal specialization, *Acta Physiol. Scand.* 152:191–205.

Stacey N., Sorensen P. Reproductive pheromones. In: Sloman K. A., Wilson R. W., Balshine S., editors. *Behavior and Physiology of Fish.* Elsevier Press; Amsterdam: 2005. pp. 359–412.

Stepanova A. N., Robertson-Hoyt J., Yun J., Benavente L. M., Xie D. Y., Dolezal K., Schlereth A., Jurgens G., Alonso J. M. 2008. TAA1-mediated auxin biosynthesis is essential for hormone crosstalk and plant development. *Cell* 133:177–191.

Stirk W. A., and J. Van Staden, 1996. Comparison of cytokinin-and auxin-like activity in some commercially used seaweed extracts, *J. Appl. Phycol.* 8:503–508.

Stirk W. A., Ördög V., Novák O., Rolčík J., Strnad M., Bálint P., and J. van Staden, 2013. Auxin and cytokinin relationships in 24 microalgal strains, *J. Phycol.* 49:459–467.

Swarup R., Parry G., Graham N., Allen T., and M. Bennett, 2002. Auxin cross-talk: integration of signaling pathways to control plant development, *Plant Molec. Biol.* 49:411-426.

Sui Q., Cao X., Lu S., Zhao W., Qiu Z., and G. Yu, 2015. Occurrence, sources and fate of pharmaceuticals and personal care products in the groundwater: a review. *Emerg. Contamin.*1:14–24.

Sun Q., Lv M., Hu A., Yang X., and C. P. Yu, 2014. Seasonal variation in the occurrence and removal of pharmaceuticals and personal care products in a wastewater treatment plant in Xiamen, *China. J. Hazard. Mater.* 277:69–75.

Takeda F., Adler P. R., and D. M. Glenn, 1997. Strawberry production linked to aquaculture wastewater treatment, *Acta Hort*. 439:673-677.

Talarek-Karwel M., Bajguz A., Piotrowska-Niczyporuk A., and I. Rajewska, 2018. The effect of 24-epibrassinolide on the green alga *Acutodesmus obliquus* (Chlorophyceae), *Plant Physiol. Biochem.* 124:175–183. http://dx.doi.org/10.1016/j.plaphy.2018.01.016.

Tarakhovskaya E. R., Maslov Y. I., and M. F. Shishova, 2007. Phytohormones in algae, *Russ. J. Plant Physiol.* 54:163–70.

Tarakhovskaya E. R., Kang E. J., Kim K. Y., and D. J. Barbary, 2013. Influence of phytohormones on morphology and chlorophyll a fluorescence parameters in embryos of F*ucus vesiculosus* L. (Phaeophyceae), *Russ. J. of Plant Physiol.* 60(2):176-183.

Tate J., Gutierrez-Wing M. T., Rusch K., and M. Benton, 2013. The effects of plant growth substances and mixed cultures on growth and metabolite production of green algae *Chlorella* sp.: a review, *J. Plant Growth Regul.* 32:417–28.

Tatkowska E., and J. Buczek, 1980. Effect of phytohormones on the growth of *Scenedesmus quadricauda* (Turp.), *Breb. Acta Soc. Bot. Pol.* 49:211–20.

Tatsuki M., Nakajima N., Fujii H., Shimada T., Nakank M., Hayashi K. I., Hayama H., Yoshioka H., and Y. Nakamura, 2013. Increased levels of IAA are required for system 2 ethylene synthesis causing fruit softening in peach (Prunus persica L. Batsch), *J. Exp. Bot*. 64:1049-1059.

Taylor T. C., Backlund A., Bjorhall K., Spreitzer J., and I. Andersson, 2001. First crystal structure of RuBisCo from a green algae *Chlamydomonas reinhardtii*, *The J. of Biol. Chem.* 276(51):48159-48164. doi:10.1074/jbc.M107765.

Teal W. D., Paponov I. A., and K. Palme, 2006. Auxin in action: signalling, transport and the control of plant growth and development, *Nat. Rev. Mol. Cell Biol.* 7:847-859.

Thaler J. S., Humphrey P. T., and N. K. Whiteman, 2012. Evolution of jasmonate and salicylate signal crosstalk, *Trends Plant Sci*. 17(5):260-270. http://dx.doi.org/10.1016/j.tplants. 2012.02.010.

Tsavkelova E. A., Klimova S. Y., Cherdyntseva T. A., and A. I. Netrusov, 2006. Microbial producers of plant growth stimulators and their practical use: a review, *Appl. Biochem. Microbiol.* 42:117–126.

Tuteja N., 2007. Abscisic acid and abiotic stress signaling, *Plant Signal Behav.* 2(3):135-138.

Upadhyay P., and C. Maier, 2016. Effects of 17β-estradiol on growth primary metabo-lism,phenylpropanoid-flavonoid pathways and pathogen resistance in *Arabidopssis thaliana*, *Am. J. Plant Sci.*1693–1710. https://doi.org/10.4236/ajps.2016.713160.

Van der Does D., Leon-Reyes A., Koornneef A., Van Verk M. C., Rodenburg N., Pauwels L., Goossens A., Körbes A. P., Memelink J., Ritsema T., Van Wees S. C. M., and C. M. J. Pieterse, 2013. Salicylic acid suppresses jasmonic acid signaling downstream of SCFCOI1-JAZ by targeting GCC promoter motifs via transcription factor ORA59, *Plant Cell* 112. http://dx.doi.org/10.1105/tpc.112.108548.

Vance B. D. 1987. Phytohormone effects on cell division in *Chlorella pyrenoidosa* Chick (TX-7-11-05) (Chlorellaceae), *J. Plant Growth Regul* 5:169–173.

Verhage A., van Wees S. C. M., and C. M. J. Pieterse, 2010. Plant immunity: It's the hormones talking, but what do they say? *Plant Physiol.* 154:536–40.

Vermeirssen E. L. M., and A. P. Scott, 1996. Excretion of free and conjugated steroids in rainbow trout (*Oncorhynchus mykiss*): evidence for branchial excretion of the maturation-inducing steroid, 17,20-dihydroxy-4-pregnen-3-one, *Gen. Comp. Endocrinol.* 101:180–194.

Vert G., Walcher L. C., Chory J., and J. L. Nemhauser, 2008. Integration of auxin and brassinosteroid pathways by auxin response factor 2, *PNAS,* 105(28):9829–9834.

Vethaak A. D., Lahr J., Schrap S. M., Belfroid A. C., Rijs G. B. J., Gerritsen A., et al. An integrated assessment of estrogenic contamination and biological effects in the aquatic environment of The Netherlands, *Chemosphere* 59:511–24.

Vidhyasekaran P., 2015. Plant hormone signaling systems in plant innate immunity, *Springer Sci. Busi. Media.* 458.

Wang Y., Duanmu D., and M. H. Spalding, 2011. Carbon dioxide concentrating mechanism in *Chlamydomonas reinhardtii*: inorganic carbon transport and CO_2 recapture, *Photosynth. Resear.* 109:115-122.

Wang P., Wong Y. S., and F. Y. Tam, 2017. Green microalgae in removal and biotransformation of estradiol and ethinylestradiol, *J. Appl. Phycol.* 29:263–273.

Williams C., and G. M. Stancel, 1996. Estrogens and progestins, p. 1411–1426. In J. G. Hardman, L. E. Limbird, P. B. Molinoff, and R. W. Ruddon (ed.), *The pharmacological basis of therapeutics,* 9th ed. The McGraw-Hill Book Co., New York, N.Y.

Wu G., Gao Z., Du H., 8 others and Ch. Meng, 2017. The effects of abscisic acid, salicylic acid and jasmonic acid on lipid accumulation in two freshwater *Chlorella* strains, *J. Gen. Appl. Microbiol.* doi 10.2323/jgam.2017.06.001.

Xu W., Jia L., Shi W., Liang J., Zhou F., Li Q., and J. Zhang, 2013. Abscisic acid accumulation modulates auxin transport in the root tip to enhance proton secretion for maintaining root growth under moderate water stress, *New Phytol.* 197(1):139-150.

Xu J., and G. Shen, 2011. Growing Duckweed in swine wastewater for nutrient recovery and biomass production, *Biores. Technol.* 102:848-853.

Yang P., Diener D. R., Yang C., Kohno T., Pazour G. J., Dienes J. M., Agrin N. S., King S. M., Sale W. S., Kamiya R., Rosenbaum J. L., and G. B. Witman, 2006. Radial spoke proteins of *Chlamydomonas flagella*, *J. of Cell Sci.* 119:1165-1174. doi: 10.1242/jcs.02811.

Yang D. L., Yang Y., and Z. He, 2013. Roles of plant hormones and their interplay in rice immunity, *Mol Plant* 6:675-685.

Yi A. X., Leung P. T. Y., and K. M. Y. Leung, 2014. Photosynthetic and molecular responses of the marine diatom *Thalassiosira pseudonana* to triphenyltin exposure, *Aquat. Toxicol.* 154:48–57.

Ylstra B., Touraev A., Brinkmann A. O., Heberle-Bors E., and A. Tunen, 1995. Steroid hormones stimulate germination and tube growth of in vitro matured tobacco pollen, *Plant Physiol.* 107:639-643.

Zhang X., Gao Y., Li Q., Li G., Guo Q., and C. Yan, 2011. Estrogenic compounds and estrogenicity in surface water, sediments, and organisms from Yundang lagoon in Xiamen, China, *Arch. Environ. Contam. Toxicol.* 61:93–100.

Zhang J., Zeng G., Chen Y., Yu M., Huang H., Fan C., Zhu Y., Li H., Liu Z., Chen M., and M. Jiang, 2013. Impact of *Phanerochaete chrysosporium* inocula-tion on indigenous bacterial communities during agricultural waste composting, *Appl. Microbiol Biotechnol.* 97(7): 3159–3169.

Zeevaart J. A. D. and R. A. Creelman, 1988. Metabolism and physiology of abscisic acid, *Annu. Rev. Plant Physiol. Plant Mol. Biol.* 39:439–473

Zhong-han Y., Yin T., Zong-xun C., and T. H. Tsao, 1994. The changes of steroidal sex hormone - testosterone contents in reproductive organs of *Lilium davidii* Duch, *Acta Bot. Sin.* 36:215-220.

Zhou W., Li Y., Min M., Hu B., Chen P., and R. Ruan, 2011. Local bioprospecting for high-lipid producing microalgal strains to be grown on concentrated municipal wastewater for biofuel production, *Biores. Technol.* 102(13):6909–6919.

Zhou G. J., Peng F. Q., Yang B. and G. G. Ying, 2013. Cellular responses and bioremoval of nonylphenol and octylphenol in the freshwater green microalga Scene*desmus obliquus*, *Ecotoxicol. Environ. Saf.* 87:10–6.

Zweig R. D., Morton J. D. and M. M. Stewart, 1999. *Source water quality for aquaculture: A guide for Assessment*. Washington D.C., The World Bank.

Table 1 References

(1) Arnon S., Dahan O., Elhanany S., Cohen K., Pankratov I., Gross A., Ronen Z., Baram S. and L. S. Shore, 2008. Transport of testosterone and estrogen from dairy-farm waste lagoons to groundwater, *Environ. Sci. Technol.* 42:5521–5526.

(2) Aris AZ., Shamsuddin AS., and SM Praveena, 2014. Occurrence of 17α-ethynylestradiol (EE2) in the environment and effect on exposed biota: a review, *Environ. Internat*, 69:104–119.

(3) Barry T. P., Riebe J. D., Parrish J. J. and J. A. Malison, 1997. Effects of 17α, 20β-Dihydroxy-4-pregnen-3-one on cortisol production by Rainbow Trout interrenal tissue *in vitro*, *Gener.and Comparat. Endocrin.* 107(2):172-81.

(4) Besse J. P. and J. Garric, 2009. Progestagens for human use, exposure and hazard assessment for the aquatic environment, *Environ. Pollut.* 157(12):3485-3494.

(5) Blanchfield P. J., Kidd K. A., Docker M. F., Palace V. P., Park B. J. and L. D. Postma, 2015. Recovery of a wild fish population from whole-lake additions of a synthetic estrogen, *Environ. Sci. Technol.* 49:3136–3144.

(6) Bolong N., Ismail A., Salim M. R. and T. Matsuura, 2009. A review of the effects of emerging contaminants in wastewater and options for their removal, *Desalination* 239: 229–246.

(7) Boyle T. P., 1980. Effects of the aquatic herbicide 2,4-D DMA on the ecology of experimental ponds, *Environ. Pollut,* 21:35–49.

(8) Caldwell D. J., Mastrocco F., Anderson P. D., Länge R. and J. P. Sumpter, 2012. Predicted-no-effect concentrations for the steroid estrogens estrone, 17β-estradiol, estriol, and 17α- ethinylestradiol, *Environ. Toxicol. Chem.* 31:1396–1406.

(9) Casey F. X., Larsen G. L., Hakk H., and J. Šimunek, 2003. Fate and transport of 17β-estradiol in soil-water systems, *Environ. Sci. Technol.* 37:2400–2409.

(10) Chen C. Y., Wen T. Y., Wang G. S., Cheng H. W., Lin Y. H. and G. W. Lien, 2007. Determining estrogenic steroids in Taipei waters and

removal in drinking water treatment using high-flow solid phase extraction and liquid chromatography/tandem mass spectrometry, *Sci Total Environ.* 378:352–65.
(11) Cobb A., 1992. Herbicides and plant physiology, *Auxin-type herbicides*, Chapman Hall: 82–106.
(12) Dalrymple O. K., Halfhide T., Udom I., Gilles B., Wolan J., Zhang Q., and S. Ergas, 2013. Wastewater use in algae production for generation of renewable resources: a review and preliminary result, *Aqua. Biosyst.* 9(1): 1-11.
(13) Duncan L. A., Tyner J. S., Buchanan J. R., Hawkins S. A. and J. Lee, 2015. Fate and transport of 17β-estradiol beneath animal waste holding ponds, *J. Environ. Qual.* 44:982–988.
(14) Esteban S., Gorga M., Petrovic M., Gonzalez-Alonso S., Barcelo D. and Y. Valcarcel, 2014. Analysis and occurrence of endocrine-disrupting compounds and estrogenic activity in the surface waters of central Spain, *Sci. Total Environ.* 466–467:939–951.
(15) Frank R. and L. Logan, 1988. Pesticide and industrial chemical residues at the mouth of the Grand, Saugeen and Thames rivers, Ontario, Canada, 1981-85, *Arch. Environ. Contam. Toxicol.* 17:741.
(16) Franks C. G., 2006. Phytoremediation of pharmaceuticals with *Salix exigua. M.Sc. Thesis* Lethbridge, Alta. University of Lethbridge, Faculty of Arts and Science, Canada.
(17) Furlong E. T., Kinney C. A., Ferrer I., Werner S., Cahill J. and G. Ratterman, 2004, August. Pharmaceuticals and personal care products in solids: Analysis and field results for sediment, soil and biosolid samples. In *Abstracts of Papers of The American Chemical Society* (Vol. 228, pPp. U620-U620). 1155 16th St, NWw, Washington, Dc 20036 USAsa: Amer Chemical Soc.
(18) Gee R. H., Rocket L. S. and P. C. Rumsby, 2015. Considerations of endocrine disruptors in drinking water. In: Darbre, P. D. (Ed.), *Endocr. Disrup. and Hum. Health.* Academic Press, London, N.Y. http:// dx.doi.org/10.1016/B978-0-12-801139-3.00018-1.
(19) Grossmann, 2010. Auxin herbicides: current status of mechanism andmode of action, *Pest Manag Sci.*, 66:113–120.

(20) Hala D., Petersen L. H., Martinovic D. and D. B. Huggett, 2012. Constraints-based stoichiometric analysis of hypoxic stress on steroidogenesis in fathead minnows, *Pimephales promelas*, *The J. of Experim. Biol.* 215:1753-1765.

(21) Hartung W, 2010. The evolution of abscisic acid (ABA) and ABA function in lower plants, fungi and lichen, *Funct. Plant Biol.* 37:806–812.

(22) Hirsch R., Hartung W. and H. Gimmler, 1989. Abscisic acid content of algae under stress, *Botan. Acta* 102(4):326–334. doi:10.1111/j.1438-8677.1989.tb00113.x.

(23) Huggett D. B., Ericson J. F., Cook J. C. and R. T. Williams, 2004. Plasma concentrations of human pharmaceuticals as predictors of pharmacological responses in fish. In *Pharmaceuticals in the Environment* (pp. 373-386). Springer Berlin Heidelberg.

(24) Islam F., Wang J., Ahsan M., Show F., all 10 authors and W. Zhou, 2017. Potential impact of the herbicide 2,4-dichlorophenoxyacetic acid on human and ecosystems, *Environ. Internat.* 111. doi: 10.1016/ j.envint.2017.10.020.

(25) Jenkins R. L., Wilson E. M., Angus R. A., Howell W. M., Kirk M., Moore R. and A. Brown, 2004. Production of androgens by microbial transformation of progesterone in vitro: A model for androgen production in rivers receiving paper mill effluent. *Environ. Health Perspect.* 112(15):1508–1511.

(26) Jobling S. and R. Owen, 2003. Ethinyl oestradiol in the aquatic environment, *Emerg. Lessons Ecosyst.* 13:279-307.

(27) Jürgens M. D., Holthaus K. I., Johnson A. C., Smith J. J., Hetheridge M. and R. J. Williams, 2002.The potential for estradiol and ethinylestradiol degradation in English rivers, *Environ. Toxicol. Chem.* 21:480–488.

(28) Khanal S. K., Xie B., Thompson M. L., Sung S., Ong S. K. and J. Van Leeuwen, 2006. Fate, transport and biodegradation of natural estrogens in the environment and engineered systems, *Environ. Sci. Technol.* http://doi.org/10.1021/es0607739.

(29) Khripach V., Zhabinski V., and A. De Groot, 2000. Twenty years of brassinosteroids: steroidal plant hormones warrant better crops for the XXI century, *Annals of Botany* 86:441-447. doi:10.1006/anbo. 2000.1227.

(30) Kidd K. A., Blanchfield P. J., Mills K. H., Palace V. P., Evans R. E., Lazorchak J. M. and R. W. Flick, 2007. Collapse of a fish population after exposure to a synthetic estrogen, *Proc. Nat. Acad. Sci.* 104:8897–8901.

(31) Kobayashi M., Sorensen P. W. and N. E. Stacey, 2002. Hormonal and pheromonal control of spawning behavior in the goldfish, *Fish Physiol. Biochem.* 26(1):71-84.

(32) Kolodziej E. P., Gray J. L. and D. L. Sedlak, 2003. Quantification of steroid hormones with pheromonal properties in municipal wastewater effluent, *Environ. Toxicol. Chem.* 22(11):2622-2629.

(33) Kolodziej E. P. and D. L. Sedlak, 2007. Rangeland grazing as a source of steroid hormones to surface waters, *Environ.Sci. Technol.* 41(10): 3514-3520.

(34) Kolpin D. W., Furlong E. T., Meyer M. T., Thurman E. M., Zaugg S. D., Barber L. B. and H. T. Buxton, 2002. Pharmaceuticals, hormones, and other organic wastewater contaminants in US streams, 1999–2000: A national reconnaissance, *Environ. Sci. Techn.* 36(6):1202-1211.

(35) Kosma C. I., Lambropoulou D. A. and T. A. Albanis, 2015. Comprehensive study of the antidiabetic drug metformin and its transformation product guanylurea in Greek wastewaters, *Water Res.* 70:436–448.

(36) Kostich M., Flick R. and J. Martinson, 2013. Comparing predicted estrogen concentrations with measurements in US waters, *Environ. Pollut.* 178:271–277.

(37) Kuch H. M. and K. Ballschmiter, 2001. Determination of endocrine-disrupting phenolic compounds and estrogens in surface and drinking water by HRGC-(NCI)-MS in the Picogram per liter range, *Environ. Sci. & Technol.* 35:3201-3206.

(38) Kuster M., de Alda M. J. L., Hernando M. D., Petrovic M., Martín-Alonso J. and D. Barceló, 2008. Analysis and occurrence of pharmaceuticals, estrogens, progestogens and polar pesticides in sewage treatment plant effluents, river water and drinking water in the Llobregat river basin (Barcelona, Spain), *J. of Hydrol.* 358(1):112-123.

(39) Labadie P. and E. M. Hill, 2007. Analysis of estrogens in river sediments by liquid chromatography-electrospray ionisation mass spectrometry. Comparison of tandem mass spectrometry and time-of-flight mass spectrometry, *J Chromatogr A.* 1141(2):174-81.

(40) Laurenson J. P., Bloom R. A., Page S. and N. Sadrieh, 2014. Ethinyl estradiol and other human pharmaceutical estrogens in the aquatic environment: a review of recent risk assessment data, *AAPS J.* 16:299–310.

(41) Liu S., Chen H., Xu X. R., Liu S. S., Sun K. F., Zhao J. L. and G. G. Ying, 2015a. Steroids in marine aquaculture farms surrounding Hailing Island, South China: Occurrence, bioconcentration, and human dietary exposure, *Sci. Total Environ.* 502:400–407.

(42) Liu S. S., Ying G. G., Liu Y. S., Yang Y. Y., He L. Y., Chen J. and J. L. Zhao, 2015b. Occurrence and removal of progestagens in two representative swine farms: Effectiveness of lagoon and digester treatment, *Water Res.* 77:146–154.

(43) Liu S., Ying G. G., Zhao J. L., Chen F., Yang B., Zhou L. J. and H. J. Lai, 2011. Trace analysis of 28 steroids in surface water, wastewater and sludge samples by rapid resolution liquid chromatography–electrospray ionization tandem mass spectrometry. *Journal of Chromatography A*, 1218(10):1367-1378.

(44) López de Alda M. J. L., Gil A., Paz E. and D. Barceló, 2002. Occurrence and analysis of estrogens and progestogens in river sediments by liquid chromatography-electrospray-mass spectrometry, *Analyst* 127(10):1299-1304.

(45) Lopez B., Ollivier P., Togola A., Baran N. and J. P. Ghestem, 2015. Screening of French groundwater for regulated and emerging contaminants, *Sci. Total Environ.* 518–519: 562–573.

(46) Lopez M. L., Peralta-Videa J. R., Benitez T. and J. L. Gardea-Torresdey, 2005. Enhancement of lead uptake by alfalfa (*Medicago sativa*) using EDTA and a plant growth promoter, *Chemosph.* 61:595–598.

(47) Martins C. I. M., Ochola D., Ende S. S. W., Eding E. H., and J. A. J. Verreth, 2009. Is growth retardation present in Nile tilapia *Oreochromis niloticus* cultured in low water exchange recirculating aquaculture systems? *Aquacult.* 298:43–50.

(48) Mota V. C., Martins C. I. M., Eding E. H., Canário A. V. M., and J. A. J. Verreth, 2014. Steroids accumulate in the rearing water of commercial recirculating aquaculture systems, *Aquacult. Engineer.* 62:9–16.

(49) Moya J. L., Ros R., and I. Picazo, 1995. Heavy metal-hormone interactions in rice plants: Effects on growth, net photosynthesis, and carbohydrate distribution, *J. Plant Growth Regul.* 14:61-67.

(50) Mullison W. R., 1987. Environmental fate of phenoxy herbicides. Fate of pesticides in the environment, *California Agricultural Experiment Station Publication* 3320:121–131.

(51) Munira S., Farenhorst A., Sapkota K., 5 other authors, and C. Sheedy, 2018. Auxin herbicides and pesticide mixtures in groundwater of a Canadian Prairie province, *J. of Environ. Qual.* 47(6). doi: 10.2134/ jeq2018.05.0202.

(52) Nobel A., Kisiala A., Galer D., and R. J. N. Clysdale, 2014. Emery, *Euglena gracilis* (Euglenophyceae) produces abscisic acid and cytokinins and responds to their exogenous application singly and in combination with other growth regulators, *Eur. J. Phycol.* 49(2):244–254.

(53) Pauwels B., Noppe H., De Brabander H. and W. Verstraete, 2008. Comparison of steroid hormone concentrations in domestic and hospital wastewater treatment plants, *J. of Environ. Engin.* 134(11):933-936.

(54) Pojana G., Gomiero A., Jonkers N. and A. Marcomini, 2007. Natural and synthetic endocrine disrupting compounds (EDCs) in water,

sediment, and biota of a coastal lagoon, *Environ. Internat.* 33(7):929-936.

(55) Rodriguez-Navas C., Björklund E., Halling-Sørensen B. and M. Hansen, 2013. Biogas final digestive byproduct applied to croplands as fertilizer contains high levels of steroid hormones, *Environ. Pollut.* 180:368–371.

(56) Sarikaya R. and M. Yılmaz. 2003. Investigation of acute toxicity and the effect of 2,4-D (2,4-dichlorophenoxyacetic acid) herbicide on the behavior of the common carp (*Cyprinus carpio* L., 1758; Pisces, Cyprinidae), *Chemosph.* 52:195-201.

(57) Savidov N., 2004. Evaluation and development of aquaponics production and product market capabilities in Alberta ids initiatives fund final report, Project #679056201, Crop Diversification Centre South Brooks, AB.

(58) Shore L. C., and M. Shemesh, 2003. Naturally produced steroid hormones and their release into the environment, *Pure Appl. Chem.* 75(11–12):1859–1871.

(59) Silva C. P., Otero M., and V. Esteves, 2012. Processes for the elimination of estrogenic steroid hormones from water: a review, *Environ. Pollut.* 165:38-58. doi:10.1016/j.envpol.2012.02.002.

(60) Stacey N. and P. Sorensen, 2005. Reproductive pheromones. In: Sloman K. A., Wilson R. W., Balshine S., editors. Behavior and Physiology of Fish. Elsevier Press; Amsterdam: 359–412.

(61) Stearma G., Kim and M. J. M. Wells Leaching and Runoff of Simazine, 2,4-D,and Bromide from Nursery Plots, *J. Soil Water Conserv.* 52:137-144.

(62) Sumpter J. P. and S. Jobling, 2013. The occurrence, causes, and consequences of estrogens in the aquatic environment, *Environ. Toxicol. Chem.* 32:249–251.

(63) Vethaak A. D., Lahr J., Schrap S. M., Belfroid A. C., Rijs G. B. J., Gerritsen A., et al. 2005. An integrated assessment of estrogenic contamination and biological effects in the aquatic environment of The Netherlands, *Chemosph.* 59:511–24.

(64) Vulliet E., Wiest L., Baudot R. and M. F. Grenier-Loustalot, 2008. Multi-residue analysis of steroids at sub-ng/L levels in surface and ground-waters using liquid chromatography coupled to tandem mass spectrometry, *J. of Chromatog. A,* 1210(1):84-91.

(65) Wafa T., Ikbal C., and M. Hammami, 2007. Environmental fate and effects of 2,4-dichlorophenoxyacetic herbicide, In: *Herbicides: Properties, Crop Protection...* ISBN: 978-1-61209-301-7 Editor: Karl D Piotrowski, pp 245-262 ©2011 Nova Science Publishers, Inc.

(66) Wenzel A., Müller J. and T. Ternes, 2003. Study on endocrine disrupters in drinking water. *Final Report* ENV.D.1/ETU/2000/0083. Schmallenberg and Wiesbaden, Germany.

(67) Yu C. P., Deeb R. A. and K. H. Chu, 2013. Microbial degradation of steroidal estrogens. *Chemosph.* 91(9):1225–1235.

(68) Zhang X., Gao Y., Li Q., Li G., Guo Q. and C. Yan, 2011. Estrogenic compounds and estrogenicity in surface water, sediments, and organisms from Yundang lagoon in Xiamen, *China. Arch. Environ. Contam. Toxicol.* 61:93–100.

(69) Zhou W., Li Y., Min M., Hu B., Chen P., and R. Ruan, 2011. Local bioprospecting for high-lipid producing microalgal strains to be grown on concentrated municipal wastewater for biofuel production, *Biores. Technol.* 102(13):6909–6919.

(70) Zuo Y., Zhang K. and S. Zhou, 2013. Determination of estrogenic steroids and microbial and photochemical degradation of 17α-ethinylestradiol (EE2) in lake surface water, a case study, *Environ. Sci. Proc. Impacts* 15:1529–1535.

In: A Closer Look at Hormones
Editor: Alexander V. Knyazev
ISBN: 978-1-53617-254-6

Chapter 8

ROLES OF BRASSINOSTEROIDS IN PLANT GROWTH AND DEVELOPMENT UNDER CHANGING ENVIRONMENT

Husna Siddiqui, Fareen Sami and Shamsul Hayat [*]
Plant Physiology Section, Department of Botany,
Faculty of Life Sciences, Aligarh Muslim University,
Aligarh, U.P., India

ABSTRACT

Brassinosteroids (BRs), sharing structural similarity with animal steroid hormones are polyhydroxylated derivatives of sterol that participate in various physiological processes of plants such as cell elongation and differentiation, male fertility, senescence, and leaf development. Photosynthesis and abiotic stress response are the two major areas of research regarding effects of exogenously applied BRs on plants. Different modes of BR application have been studied including seed soaking, foliar spray and root dipping. BRs regulate seed

[*] Corresponding Author's Email: hayat_68@yahoo.co.in.

germination, primary photochemistry, photosynthesis, different biochemical enzymes, elemental composition, carbohydrate metabolism and redox status of plants. The BR-generated effects are evident in presence/absence of stress.

Keywords: brassinosteroids, photosynthesis, antioxidant system, abiotic stress, yield

Introduction

Brassinosteroids (BRs) are steroidal plant hormone and found throughout the plant kingdom. BRs occur in all plant parts for instance roots, shoot, pollen, flower buds, fruits, seeds and leaves. Plants require BRs for normal growth and development. They are involved in a wide array of metabolic processes undergoing inside the plants such as root and shoot growth, flower and fruit formation. (Hayat and Ahmad 2003; Sasse 2003). BRs are known to possess specialized role against abiotic stress in plants. Moreover, the phytohormone is also known to alter the response of plants towards biotic stress (Krishna, 2003).

Plants respond to changes in environment that occur in the form of stress. Stress could be of different types for example, salinity, drought, temperature, heavy metal, and pathogen or pest attack. Abiotic stress alters the morpho-physiology of plants. Abiotic stress causes oxidative stress and induces denaturation of efficient and structural proteins. Various environmental stresses trigger parallel cascade of signals and reactions inside the cells, for instance stress proteins and antioxidants production, and accumulation of compatible solutes. Alteration in phytohormone level is one of the biochemical adaptations in plants towards changing environmental conditions that defines the aptitude of an organism to survive and reproduce it (Fujita et al., 2006; Smirnoff 1995).

Reactive oxygen species (ROS) acts as signaling molecules in different cellular processes that affects growth and development of plants. ROS mediates the interaction of plants with biotic and abiotic stresses. These are readily generated during the process of photosynthesis and respiration, however, their level is firmly regulated by the defensive mechanisms present inside the cell. Imbalance in the ROS production and protective mechanism may lead to oxidative stress and cell death. Occurrence of stress disrupts the cellular homeostasis and enhances ROS production (Gapper and Dolan 2006).

Application of BRs to the plants growing in presence of stress, altered enzymatic and non-enzymatic antioxidants such as superoxide dismutase (SOD), catalase (CAT), glutathione peroxidase (GPOX) and ascorbate peroxidase (APX), proline, glutathione, vitamins and carotenoids (Li et al., 1998; Nunez et al., 2003; Ozdemir et al., 2004; Thomashow 1999).

When BR-treated seedlings of maize were exposed to water stress the rise in the SOD, CAT and APX was observed moreover, content of non-enzymatic antioxidants also increased (Li et al., 1998). In contrast, there was a reduction in the POX and ascorbic acid oxidase activity in Sorghum plants treated with BRs under osmotic stress (Thomashow, 1999).

Under saline conditions, the BR-treated rice seedlings presented a rise in the glutathione reductase (GR), SOD, CAT and APX activity (Nunez et al., 2003). A comparative study on salt-sensitive and resistant variety of Oryza sativa treated with EBL was conducted by (Ozdemir et al., 2004). Where, EBL strengthen the tolerance of salt-sensitive cultivars towards salt stress. The enhanced activity of APX under saline conditions salt-sensitive rice cultivar suggests that EBL induces the resistance in these plants.

Role of BRs in Presence of Abiotic Stress

Osmotic stress occurs due to drought, salinity, high and low temperature. To combat with dehydration and low temperature, plants have evolved various physiological and developmental mechanisms. Accumulation of osmo-protectants is one such mechanism. Plants produce high levels of abscisic acid (ABA) in presence of drought and high salinity. ABA when supplied exogenously triggers a wide array of genes that respond to dehydration and cold stress. Induction of genes by dehydration and low temperature has been reported many times hence, it is plausible, that there is an existence of both ABA-independent and dependent signal transduction cascades (Upreti and Murti, 2004).

BRs regulate root nodulation and endogenous content of ABA and cytokinin trans- zeatin riboside (ZR) in *Phaseolus vulgaris* (Vardhini and Rao, 2003). Under non-stressed conditions, BRs promote root nodulation, ZR content and mitigate the stress-triggered decline in root nodulation. Though, BR did not alter the ABA contents in the nodules of control or stressed plants. Water stress mediated debility in root nodulation is related to elevated ABA content and reduction in cytokinin content. BR-induced root nodulation in presence of stress could be an effect generated by influencing cytokinin level in the nodulated roots.

Application of BRs enhanced the seedling biomass of Sorghum varieties (both susceptible and resistant to water stress) under osmotic stress (Thomashow, 1999). Similarly, drought stressed lead to a marked reduction in weight of taproot in sugar-beet. However, application of BR completely neutralized the toxic effects and restored the values for plant biomass, protein and proline content that got reduced due to drought stress (Thomashow, 1999). Contrary, the increase in biomass was directly proportional to the elevation in leaf acid invertase activity, which enhanced the supply of assimilates to the plants (Schilling et al., 1991). Similar to EBL, homo-brassinolide (HBL) also promoted the growth of drought-tolerant and susceptible varieties of wheat growing under stressful environment (Sairam, 1994). It raised the relative water content (RWC), chlorophyll (chl) content, photosynthesis along with nitrate reductase (NR) activity under both normal and stressful conditions and this effect was more pronounced in the drought- tolerant wheat variety. The promotive effect of BR on these parameters resulted in enhanced growth and development which ultimately resulted in higher yield of plants (Sairam, 1994).

Salinity Stress

Saline soil possesses 4 dS m^{-1} or higher electrical conductivity. Salinity increases toxic Na^+ accumulation that disintegrates chl molecule and reduces its content (Yang et al., 2011). Furthermore, salinity also elevates the chl degrading enzymes synthesis (Reddy and Vora, 1986). Reduction in chl content upon exposure to saline conditions has been reported in *Helianthus annuus, Triticum aestivum, Cicer arietinum, Brassica juncea, Ricinus communis* (Ashraf and Sultana 2000; Arfan et al., 2007; Ali et al., 2007, 2008a; Perveen et al., 2010; Pinheiro et al., 2008). In presence of salinity, EBL improves seed germination and seedling growth of *Eucalyptus camaldulensis*. In rice seedlings, BRs proved beneficial in removing the salinity-induced inhibition of seed germination and seedling growth. The reduction NR activity salt-stressed leaves occur due to salinity-mediated inhibition of nitrate transport to aerial parts, which hinders the NO_3^- uptake and xylem loading (Ali et al., 2008a). Chickpea seeds soaked in HBL elevates the nitrogen fixing capability of plants in presence of NaCl (Ali et al., 2007). At harvest, the HBL-treated plants showed better seed yield as compared to the ones exposed to sole NaCl stress. Root and foliar application of HBL enhanced the growth, nucleic acid content, ethylene and seed yield of mustard plants grown from seeds soaked in NaCl (Anuradaha and Rao, 2003; Hayat et al., 2007a,b).

Under harsh environmental conditions such as salinity, drought and osmotic stress, BRs control the level of a cereal lectin naming, wheat germ agglutinin (WGA). Its synthesis and accumulation is controlled by ABA. Since, BRs did not affect ABA content however it enhanced the accumulation of lectin under salt stress in wheat roots it could be inferred that BRs regulate lectin accumulation via ABA-independent pathway. BRs lowered down the salinity-induced WGA accumulation by 50% in roots of wheat. It is plausible that BRs might be involved in the hormonal regulation of WGA and ABA level. BRs protect wheat seedlings by lessening the NaCl-induced ABA and WGA accumulation in roots (Shakirova and Bezrukova 1998, 2002).

BRs reinstates the loss of pigment under changing environment (Anuradha and Rao, 2003; Sharma et al., 2013). BRs eliminated the NaCl toxicity in *Lactuca sativa* (Ekinci et al., 2012). Similarly, BR treatment reversed the salinity generated toxicity on photosynthesis, stomatal conductance, transpiration rate and water use efficiency in mustard and wheat (Siddiqui et al., 2018a; Alyemeni et al., 2013; Shahbaz et al., 2008). Dubey et al., (2005) and Siddiqui et al., (2018b, c) proposed the BR-mediated enhancement in photosynthesis might depend on stomatal or non-stomatal factors.

Temperature Stress

BR activity is evaluated using lamina-inclination test. Rice seedlings treated with BRs demonstrated cell elongation under low temperature conditions (15°C). However, indole-3-acetic acid (IAA) had minimal effect on cell elongation, but the duo of BRs and IAA stimulated it synergistically. The rate of germination and growth of paddy growing at low temperature was enhanced in presence of BRs (Fujii and Saka, 2001; Hotta et al., 1998). Similarly, role of BRs in improving seed germination and seedling growth of *Zea mays* and *Cucumis sativus* under chilling conditions has been established (He et al., 1991; Khripach et al., 1999).

Both high and low temperatures generate stress symptoms in leaves of banana. Though, treatment with BR analogue significantly reduces the severity of thermal stress (Gonza' lez-Olmedo et al., 2005). EBL slightly improve tolerance of brome grass towards freezing stress however significantly promote cell viability after exposure to high temperature regime. EBL triggers the accumulation of ABA-induced heat-stable proteins. The increase in heat-shock proteins (hsps), hsp90 transcript levels by EBL whereas its reduction by ABA suggests that EBL provide stress tolerance to plant cells (Wilen et al., 1995). Moreover, accumulation of hsps such as hsp100 and hsp70, and other low-molecular-weight hsps also increases in EBL treated seedlings with respect to untreated ones. The increase in accumulation of hsps upon EBL supplementation does not coincides with hsp mRNA levels during the recovery phase. Thus, it could be inferred that EBL restricts the loss of translational components during persistent heat stress and promote the expression of certain translational components throughout recovery. It associates with enhanced hsp synthesis under extreme hot conditions, a further swift recommencement of protein synthesis inside the cell after heat stress, and a better survival rate (Dhaubhadel et al., 2002, 1999). Expression of mitochondrial hsps was induced by EBL in tomato leaves. Plants treated with BR yielded better photosynthetic efficiency. At 35°C (heat-stress inductive temperature) pollen germination and pollen tube growth in tomato was enhanced in presence of BR, suggesting a potential role of BR in plant growth and development. BR application also increase the fruit yield under heat-stress conditions (Singh and Shono, 2005).

Change in the protein level and leaf ultrastructure was examined in BR treated and non-treated plants. Where, BR maintained the protein level at 43°C quite similar to those at 23°C. however, there was a marked reduction of 2.5-fold at 43°C in non-treated plants with respect to the control temperature (Kulaeva et al., 1991).

In presence of severe heat conditions, small hsps amass to form well-organized cytoplasmic complexes that are known as heat shock granules (HSGs). HSGs act as shielding sites for housekeeping mRNPs, released subsequent to stress removal (Kulaeva et al., 1991). The size and number of HSGs cluster is known to be higher in BR-treated leaves as compared to the untreated ones (Kulaeva et al., 1991). In a study by Kagale et al., (2007) it was showed that despite of the role of BRs in conferring heat tolerance, it is not requisite for expression of hsp during heat stress in BR-deficient mutants in Arabidopsis (det2-1 and dwf4). Suggesting that BR deficiency itself act as a cellular stress and triggers the activation of the hsp genes.

BR promotes epicotyl elongation of *Vigna radiata* under normal conditions and mitigates the low temperature generated effects and allows the plants to recover from stress-injury. The proteins that get down-regulated at low temperature are re-upregulated by BR treatment. These proteins are known to be responsible for cell growth, wall formation, ATP production, the stress response, and methionine assimilation (Huang et al., 2006).

Heavy Metal Stress

Plants have developed specialized mechanisms to uptake and accumulate heavy metals from surrounding environment. Heavy metal (HM) stress is regarded as one of the most toxicity generating abiotic stresses that damages plant growth and development of plant. Due to rapidly increasing industrialization and urbanization these HM have become a serious threat to the environment and regarded as major environment pollutant. Heavy metals cause toxicity symptoms in plants such as enzyme denaturation and inactivation, disruption in transport of essential nutrients and ions, and loss of membrane integrity. Moreover, metals also induce toxic effects via free radical formation (Smirnoff, 1995). HM toxicity stimulates a range of adaptive responses in plants. EBL has been found to cut the absorption of HM in various plants for example tomato, barley, sugar beet, and radish (Khripach et al., 1996). EBL blocks the accumulation of HMs like cadmium (Cd), lead (Pb), copper (Cu) and zinc (Zn) in *Chlorella vulgaris* cells (Bajguz, 2000). EBL mitigated the toxicity generated by Zn, manganese (Mn), cobalt (Co), and nickel (Ni) in *Brassica juncea* by increasing growth and reducing accumulation of HMs (Sharma and Bhardwaj, 2007). BR plays an important role in generating response against HMs by activating enzymes like SOD, CAT, APX and GR (Bajguz, 2010). Moreover, BR reduces the HMs stress on growth, chlorophyll and protein content and alongside increase the phytochelatin synthesis (Bajguz, 2011). BRs promote seed germination and seedling growth of maize growing under Zn and Ni stress besides it also reduced HM-induced electrolyte leakage from cells (Lukatkin et al., 2013).

A specialized mechanism for HM detoxification is the chelation of the metal ion by a ligand such as organic acids, phytochelatins (PCs), metallothioneins (MT) and amino acids (Anjum et al., 2015; Manara 2012;

Rauser and Curvetto 1980). In *Chlorella vulgaris,* BRs induce the synthesis of PCs as a response towards Pb stress (Bajguz 2002).

BRs reduce the impact of HMs stress on growth and prevent chlorophyll, sugar and protein loss and stimulate PC synthesis cells of *Chlorella vulgaris* and the effect seems to be concentration-dependent (Bajguz, 2000, 2002). BRs reduce Cd and Cu accumulation in the winter rape and Indian mustard seedlings, respectively (Janeczko et al., 2005, Sharma et al., 2007). BR diminishes the Cd-induced damage to reaction centers and oxygen evolving complex and maintains the efficiency of photosynthetic electron transport (Janeczko et al., 2005). The effect of BR on metal content depends on the developmental stage of plant at which the treatment was given. BRs improve the performance of mustard, chickpea, mung bean and mustard growing in presence of different HMs (Hayat et al., 2007c; Hasan et al., 2008; Alam et al., 2007; Ali et al., 2008b).

Biosorption is a novel mechanism for metal removal. Its mechanism usually relies on physico-chemical interactions between metal ions and the functional groups like carboxylic acid, hydroxyl, amine and phosphoryl present on the cell surface, for example electrostatic interactions, ion exchange and metal ion chelation. Biosorption is highly sensitive to pH and the presence of other ions in the medium as they precipitate down the HMs as insoluble salts, however, metabolic inhibitors or light/ dark cycles have no effect on this mechanism (Vı´lchez et al., 1997). Water pH directly affects HM toxicity in algae, such as *Chlorella* sp. It has been established that HM toxicity decreases with the decreasing pH. And in turn, acidity or alkalinity of the medium restrains the HM toxicity. Lower pH may increase the bioavailability of metal ions increases at lower pH which results in enhanced toxicity (Bajguz, 2000).

CONCLUSION

BRs boost-up the immune system of plants in a concentration dependent manner and stage of plant development. BRs provide tolerance towards changing environment like soil salinity, fluctuation in temperature regime and HM pollution. It enhances anti-oxidative defense system of plants to counter free radicals generated during stress. BRs activate and suppress various enzymatic reactions, and induce the synthesis of protein and different chemicals used in defense mechanism. Application of BRs to plants confer resistance against harsh environmental conditions hence, its stress mitigating property could be exploited in different horticultural crops to improve their quality and yield.

References

Alam, M. M., Hayat, S., Ali, B., & Ahmad, A. (2007). Effect of 28-homobrassinolide treatment on nickel toxicity in *Brassica juncea*. *Photosynthetica*, *45*(1), 139-142.

Ali, B., Hasan, S. A., Hayat, S., Hayat, Q., Yadav, S., Fariduddin, Q., & Ahmad, A. (2008b). A role for brassinosteroids in the amelioration of aluminium stress through antioxidant system in mung bean (*Vigna radiata* L. Wilczek). *Environmental and Experimental Botany*, 62(2), 153-159.

Ali, B., Hayat, S., & Ahmad, A. (2007). 28-Homobrassinolide ameliorates the saline stress in chickpea (*Cicer arietinum* L.). *Environmental and experimental botany*, *59*(2), 217-223.

Ali, B., Hayat, S., Fariduddin, Q., & Ahmad, A. (2008b). 24-Epibrassinolide protects against the stress generated by salinity and nickel in *Brassica juncea*. *Chemosphere*, *72*(9), 1387-1392.

Alyemeni, M. N., Hayat, S., Wijaya, L., & Anaji, A. (2013). Foliar application of 28-homobrassinolide mitigates salinity stress by increasing the efficiency of photosynthesis in *Brassica juncea*. *Acta Botanica Brasilica*, *27*(3), 502-505.

Anjum, N. A., Hasanuzzaman, M., Hossain, M. A., Thangavel, P., Roychoudhury, A., Gill, S. S. & Duarte, A. C. (2015). Jacks of metal/metalloid chelation trade in plants—an overview. *Frontiers in plant science*, *6*, 192.

Anuradha, S., & Rao, S. S. R. (2003). Application of brassinosteroids to rice seeds (*Oryza sativa* L.) reduced the impact of salt stress on growth, prevented photosynthetic pigment loss and increased nitrate reductase activity. *Plant Growth Regulation*, *40*(1), 29-32.

Arfan, M., Athar, H. R., & Ashraf, M. (2007). Does exogenous application of salicylic acid through the rooting medium modulate growth and photosynthetic capacity in two differently adapted spring wheat cultivars under salt stress? *Journal of Plant Physiology*, *164*(6), 685-694.

Ashraf, M., & Sultana, R. (2000). Combination effect of NaCl salinity and nitrogen form on mineral composition of sunflower plants. *Biologia plantarum*, *43*(4), 615-619.

Bajguz, A. (2000). Blockade of heavy metals accumulation in *Chlorella vulgaris* cells by 24-epibrassinolide. *Plant Physiology and Biochemistry*, 38(10), 797-801.

Bajguz, A. (2002). Brassinosteroids and lead as stimulators of phytochelatins synthesis in *Chlorella vulgaris*. *Journal of plant physiology*, *159*(3), 321-324.

Bajguz, A. (2010). An enhancing effect of exogenous brassinolide on the growth and antioxidant activity in *Chlorella vulgaris* cultures under heavy metals stress. *Environmental and Experimental Botany*, *68*(2), 175-179.

Bajguz, A. (2011). Suppression of *Chlorella vulgaris* growth by cadmium, lead, and copper stress and its restoration by endogenous brassinolide. *Archives of environmental contamination and toxicology*, *60*(3), 406-416.

Dhaubhadel, S., Browning, K. S., Gallie, D. R., & Krishna, P. (2002). Brassinosteroid functions to protect the translational machinery and heat-shock protein synthesis following thermal stress. *The Plant Journal*, *29*(6), 681-691.

Dhaubhadel, S., Chaudhary, S., Dobinson, K. F., & Krishna, P. (1999). Treatment with 24-epibrassinolide, a brassinosteroid, increases the basic thermotolerance of *Brassica napus* and tomato seedlings. *Plant molecular biology*, *40*(2), 333-342.

Dubey, R. S. (2005). Photosynthesis in plants under stressful conditions. *Handbook of photosynthesis*, 717-737.

Ekinci, M., Yildirim, E., Dursun, A., & Turan, M. (2012). Mitigation of salt stress in lettuce (*Lactuca sativa* L. var. Crispa) by seed and foliar 24-epibrassinolide treatments. *Hort Science*, *47*(5), 631-636.

Fujii, S., & Saka, H. (2001). The promotive effect of brassinolide on lamina joint-cell elongation, germination and seedling growth under low-temperature stress in rice (*Oryza sativa* L.). *Plant production science*, *4*(3), 210-214.

Fujita, M., Fujita, Y., Noutoshi, Y., Takahashi, F., Narusaka, Y., Yamaguchi-Shinozaki, K., & Shinozaki, K. (2006). Crosstalk between abiotic and biotic stress responses: a current view from the points of convergence in the stress signaling networks. *Current opinion in plant biology*, *9*(4), 436-442.

Gapper, C., & Dolan, L. (2006). Control of plant development by reactive oxygen species. *Plant physiology*, *141*(2), 341-345.

González-Olmedo, J. L., Córdova, A., Aragón, C. E., Pina, D., Rivas, M., & Rodríguez, R. (2005). Effect of an analogue of brassinosteroid on FHIA-18 plantlets exposed to thermal stress. *Info Musa*, *14*(1), 18-20.

Hasan, S. A., Hayat, S., Ali, B., & Ahmad, A. (2008). 28-Homobrassinolide protects chickpea (*Cicer arietinum*) from cadmium toxicity by stimulating antioxidants. *Environmental pollution*, *151*(1), 60-66.

Hayat, S., Ahmad, A., & Ahmad, A. (Eds.). (2003). *Brassinosteroids: bioactivity and crop productivity*. Springer Science & Business Media.

Hayat, S., Ali, B., Hasan, S. A., & Ahmad, A. (2007c). Brassinosteroid enhanced the level of antioxidants under cadmium stress in *Brassica juncea*. *Environmental and Experimental Botany*, *60*(1), 33-41.

Hayat, S., Ali, B., Hasan, S. A., & Ahmad, A. (2007a). Effect of 28-homobrassinolide on salinity-induced changes in *Brassica juncea*. *Turkish Journal of Biology*, *31*(3), 141-146.

Hayat, S., Ali, B., Hasan, S. A., Hayat, Q., & Ahmad, A. (2007b). Effect of 28-Homobrassinolide on salinity induced changes in growth, ethylene and seed yield of mustard. *Indian Journal of Plant Physiology*, *12*(2), 207.

He, R. Y., Wang, G. J., & Wang, X. S. (1991). Effects of brassinolide on growth and chilling resistance of maize seedlings. In *ACS Symposium series American Chemical Society*.

Hotta, Y., Tanaka, T., Bingshan, L., Takeuchi, Y., Konnai, M. (1998). Improvement of cold resistance in rice seedlings by 5-aminolevulinic acid. *Journal of pesticide science*, *23*(1), 29-33.

Huang, B., Chu, C. H., Chen, S. L., Juan, H. F., & Chen, Y. M. (2006). A proteomics study of the mung bean epicotyl regulated by brassinosteroids under conditions of chilling stress. *Cellular & molecular biology letters*, *11*(2), 264.

Janeczko, A., Koscielniak, J., Pilipowicz, M., Szarek-Lukaszewska, G., & Skoczowski, A. (2005). Protection of winter rape photosystem 2 by 24-epibrassinolide under cadmium stress. *Photosynthetica*, *43*(2), 293-298.

Kagale, S., Divi, U. K., Krochko, J. E., Keller, W. A., & Krishna, P. (2007). Brassinosteroid confers tolerance in *Arabidopsis thaliana* and *Brassica napus* to a range of abiotic stresses. *Planta*, *225*(2), 353-364.

Khripach, V. A., Zhabinskii, V. N., & de Groot, A. E. (1998). *Brassinosteroids: a new class of plant hormones*. Academic Press.

Krishna, P. (2003). Brassinosteroid-mediated stress responses. *Journal of Plant Growth Regulation*, *22*(4), 289-297.

Kulaeva, O. N., Burkhanova, E. A., Fedina, A. B., Khokhlova, V. A., Bokebayeva, G. A., Vorbrodt, H. M., & Adam, G. (1991). *Effect of brassinosteroids on protein synthesis and plant-cell ultrastructure under stress conditions*.

Li, L., Staden, J. V., & Jäger, A. K. (1998). Effects of plant growth regulators on the antioxidant system in seedlings of two maize cultivars subjected to water stress. *Plant Growth Regulation*, *25*(2), 81-87.

Lukatkin, A. S., Kashtanova, N. N., & Duchovskis, P. (2013). Changes in maize seedlings growth and membrane permeability under the effect of epibrassinolide and heavy metals. *Russian agricultural sciences*, *39*(4), 307-310.

Manara, A. (2012). Plant responses to heavy metal toxicity. In *Plants and heavy metals* (pp. 27-53). Springer, Dordrecht.

Nunez, M., Mazzafera, P., Mazorra, L. M., Siqueira, W. J., & Zullo, M. A. T. (2003). Influence of a brassinosteroid analogue on antioxidant enzymes in rice grown in culture medium with NaCl. *Biologia Plantarum*, *47*(1), 67-70.

Özdemir, F., Bor, M., Demiral, T., & Türkan, İ. (2004). Effects of 24-epibrassinolide on seed germination, seedling growth, lipid peroxidation, proline content and antioxidative system of rice (*Oryza sativa* L.) under salinity stress. *Plant Growth Regulation*, *42*(3), 203-211.

Perveen, S., Shahbaz, M., & Ashraf, M. (2010). Regulation in gas exchange and quantum yield of photosystem II (PSII) in salt-stressed and non-stressed wheat plants raised from seed treated with triacontanol. *Pak. J. Bot*, *42*(5), 3073-3081.

Pinheiro, H. A., Silva, J. V., Endres, L., Ferreira, V. M., de Albuquerque Câmara, C., Cabral, F. F., ... & dos Santos Filho, B. G. (2008). Leaf gas exchange, chloroplastic pigments and dry matter accumulation in castor bean (*Ricinus communis* L) seedlings subjected to salt stress conditions. *Industrial Crops and Products*, *27*(3), 385-392.

Rauser, W. E., & Curvetto, N. R. (1980). Metallothionein occurs in roots of Agrostis tolerant to excess copper. *Nature*, *287*(5782), 563.

Reddy, M. P. (1986). Changes in pigment composition. Hill reaction activity and saccharides metabolism in bajra (*Penisetum typhoides*) leaves under NaCl salinity. *Photosynthetica*, *20*, 50-55.

Sairam, R. K. (1994). Effects of homobrassinolide application on plant metabolism and grain yield under irrigated and moisture-stress conditions of two wheat varieties. *Plant Growth Regulation*, *14*(2), 173-181.

Sasse, J. M. (2003). Physiological actions of brassinosteroids: an update. *Journal of plant growth regulation*, *22*(4), 276-288.

Schilling, G., Schiller, C., Otto S. (1991) Influence of brassinosteroids on organ relations and enzyme activities of sugar-beet plants, In: H. G. Cutler, T. Yokota, G. Adam (Eds.), *Brassinosteroids: Chemistry, Bioactivity and Applications, American Chemical Society*, Washington, DC, pp. 208–219.

Shahbaz, M., & Ashraf, M. (2008). Does exogenous application of 24-epibrassinolide ameliorate salt induced growth inhibition in wheat (*Triticum aestivum* L.)?. *Plant Growth Regulation*, *55*(1), 51-64.

Shakirova, F. M., & Bezrukova, M. V. (1998). Effect of 24-epibrassinolide and salinity on the levels of ABA and lectin. *Russian journal of plant physiology*, *45*(3), 388-391.

Shakirova, F. M., Bezrukova, M. V., Aval'baev, A. M., & Gimalov, F. R. (2002). Stimulation of wheat germ agglutinin gene expression in root seedlings by 24-epibrassinolide. *Russian journal of plant physiology*, *49*(2), 225-228.

Sharma, I., Ching, E., Saini, S., Bhardwaj, R., & Pati, P. K. (2013). Exogenous application of brassinosteroid offers tolerance to salinity by altering stress responses in rice variety Pusa Basmati-1. *Plant Physiology and Biochemistry*, *69*, 17-26.

Sharma, P., & Bhardwaj, R. (2007). Effect of 24-epibrassinolide on seed germination, seedling growth and heavy metal uptake in *Brassica juncea* L. *General and Applied Plant Physiology*, *33*(1-2), 59-73.

Sharma, P., & Bhardwaj, R. (2007). Effects of 24-epibrassinolide on growth and metal uptake in *Brassica juncea* L. under copper metal stress. *Acta Physiologiae Plantarum*, *29*(3), 259-263.

Siddiqui, H., Ahmed, K. B. M., & Hayat, S. (2018b). Comparative effect of 28-homobrassinolide and 24-epibrassinolide on the performance of different components influencing the photosynthetic machinery in Brassica juncea L. *Plant physiology and biochemistry*, *129*, 198-212.

Siddiqui, H., Hayat, S., & Bajguz, A. (2018c). Regulation of photosynthesis by brassinosteroids in plants. *Acta physiologiae plantarum*, 40(3), 59.

Siddiqui, H., Yusuf, M., Faraz, A., Faizan, M., Sami, F., & Hayat, S. (2018a). 24-Epibrassinolide supplemented with silicon enhances the photosynthetic efficiency of *Brassica juncea* under salt stress. *South African journal of botany*, *118*, 120-128.

Singh, I., & Shono, M. (2005). Physiological and molecular effects of 24-epibrassinolide, a brassinosteroid on thermotolerance of tomato. *Plant Growth Regulation*, *47*(2-3), 111.

Smirnoff, N. (1995). *Environment and plant metabolism: flexibility and acclimation*. BIOS Scientific publishers.

Thomashow, M. F. (1999). Plant cold acclimation: freezing tolerance genes and regulatory mechanisms. *Annual review of plant biology*, *50*(1), 571-599.

Upreti, K. K., & Murti, G. S. R. (2004). Effects of brassmosteroids on growth, nodulation, phytohormone content and nitrogenase activity in French bean under water stress. *Biologia Plantarum*, *48*(3), 407-411.

Vardhini, B. V., & Rao, S. S. R. (2003). Amelioration of osmotic stress by brassinosteroids on seed germination and seedling growth of three varieties of sorghum. *Plant Growth Regulation*, *41*(1), 25-31.

Vilchez, C., Garbayo, I., Lobato, M. V., & Vega, J. (1997). Microalgae-mediated chemicals production and wastes removal. *Enzyme and Microbial Technology*, *20*(8), 562-572.

Wilen, R. W., Sacco, M., Gusta, L. V., & Krishna, P. (1995). Effects of 24-epibrassinolide on freezing and thermotolerance of bromegrass (*Bromus inermis*) cell cultures. *Physiologia Plantarum*, *95*(2), 195-202.

Yang, J. Y., Zheng, W., Tian, Y., Wu, Y., & Zhou, D. W. (2011). Effects of various mixed salt-alkaline stresses on growth, photosynthesis, and photosynthetic pigment concentrations of *Medicago ruthenica* seedlings. *Photosynthetica*, *49*(2), 275-284.

About the Editor

Alexander Knyazev
Professor, Dean of the Faculty of Chemistry
Lobachevsky State University of Nizhni Novgorod (UNN)
Nizhni Novogorod, Russia

Alexander Knyazev is a Russian crystal chemist and physical chemist, Professor of the Lobachevsky University, Dean of the Faculty of Chemistry, Head of the Department of Medicinal Chemistry, Head of the postgraduate research school "New materials based on inorganic compounds." He graduated from the Faculty of Chemistry, Lobachevsky University (1998) and PhD (2000), Doctor of Chemistry (2009), academic rank of professor (2011). Author of more than 230 articles in international academic journals, three chapters in books and three textbooks, his scientific school includes 9 PhD. Scientific interests are the synthesis and crystal chemistry of inorganic compounds and biologically active substances, as well as a comprehensive thermodynamic study of these objects.

INDEX

B

C

D

E

F

G

H

I

J

K

L

M

N

O

P

Q

R

S

T

U

V

W

X

Y

Z